Discrete Techniques of Parameter Estimation

The Equation Error Formulation

CONTROL THEORY

A Series of Monographs and Textbooks

Editor

JERRY M. MENDEL

McDonnell Douglas Astronautics Company
Huntington Beach, California

Associate Editors

Michael Athans
Massachusetts Institute of Technology
Cambridge, Massachusetts

David G. Luenberger
Stanford University
Stanford, California

Volume 1: Discrete Techniques of Parameter Estimation: The Equation Error Formulation, JERRY M. MENDEL

OTHER VOLUMES IN PREPARATION

Discrete Techniques of Parameter Estimation

The Equation Error Formulation

JERRY M. MENDEL
McDonnell Douglas Astronautics Company
Huntington Beach, California

MARCEL DEKKER, INC. New York

MARCEL DEKKER, INC.

270 Madison Avenue, New York, New York 10016

LIBRARY OF CONGRESS CATALOG CARD NUMBER 72-76062

ISBN: 0-8247-1455-5

Current printing (last digit):
10 9 8 7 6 5 4 3 2

PRINTED IN THE UNITED STATES OF AMERICA

TO

Letty, Jonathan, Aileen

and

Professor John G. Truxal

Preface

This book brings together four estimation techniques and presents them from a unified point of view. The techniques are: generalized least-squares (also known as weighted least-squares), which includes least-squares estimation as an important but special case; unbiased minimum-variance (also known as best linear unbiased estimation); deterministic-gradient; and stochastic-gradient (a special case of stochastic approximation). The unification is possible when one uses the Equation Error Method as the basis for his problem formulation.

Classical batch processing estimation algorithms, which are more familiar to statisticians and econometricians interested in modeling, and, sequential estimation algorithms, which are more familiar to control engineers interested in optimal estimation theory and parameter identification, are both presented. The book brings together some concepts which are more familiar to statisticians and others which are more familiar to control engineers. In this manner, it is hoped that a rather well rounded and complete coverage of the four estimation techniques is achieved.

The book is written assuming the reader is familiar with probability, random variables, and random (stochastic) processes, linear differential equations, some vector and matrix notions, and some notions from the theory of linear systems, such as state space. Some background material is included for completeness in the appendixes.

There are at least two ways to present the results that are contained in this book. One approach is to solve one- and two-dimensional (i. e., one and two unknown parameters) problems in great detail, and to then state generalizations for higher dimensional problems. A second approach is to derive general results in a formal manner and to then illustrate them for one- and two-dimensional examples. We have adopted the second approach. Often, in the first approach algebraic manipulations that are only suitable for the one- and two-dimensional situations must be used to obtain the desired results. Usually, these same manipulations cannot be used for higher dimensional situations for any one of a number of reasons. They may become too cumbersome, or, they may just not work. The second approach is a general one. It makes use of unifying notation, through the concepts of vectors and matrices, and unifying mathematics, through the vector calculus.

The price paid for generalized results is formalism. Most of the book's important results are presented as formal theorems or corollaries. In this way, they can be easily pulled out from the main body of the text.

Often one is interested in estimating parameters for very difficult situations. For example, the author has spent many years studying the problem of estimating time-varying aerodynamic stability derivatives for high-performance aerospace vehicles. This problem is complicated not only by the often rapid time variations in the stability derivatives, but also by the random disturbances that act upon the vehicles, such as wind, and by noise in the measurements of signals which are used as the basis for estimating the stability derivatives. One can always try ad hoc estimation procedures for such difficult problems, and perhaps due to insight, luck, or a combination of both, he will get reasonably good estimates. However, change a few conditions for the same system (e. g., decrease signal to noise ratios) and these ad hoc procedures may no longer work. It is felt that a solid analytical understanding of an estimation technique is absolutely necessary, or else a user will not know where to begin to modify an estimation algorithm when he attempts to apply it to difficult and new situations. How, for example, do you remove a bias in an estimate if you have not determined whether or not a bias will occur and how much bias will be present?

Our approach is to study in depth the problem of estimating constant parameters when measurements are either perfect or noisy. A vast literature exists for this problem, probably because it is only for the estimation of constant parameters that definitive theoretical results can be proved. The first five chapters of this book approach the constant parameter estimation problem from four different estimation techniques. The last chapter examines how these techniques can be extended to the problem of estimating time-varying parameters. For a more detailed outline of the book, see Section 1.4.

Applications such as curve fitting, estimating the coefficients in a superposition summation, estimating aerodynamic coefficients for a high performance adaptive aerospace control system, and estimating coefficients in a finite-difference equation are carried through the entire book.

Problems are included at the end of each chapter. These problems supplement the textual material in four ways: they require the reader (1) to gain familiarity with important concepts and techniques by solving numerical examples, (2) to extend concepts from simple situations to more difficult situations, (3) to learn new concepts, and (4) to fill in missing details.

Certain important topics have been omitted so as not to take us too far afield from the main stream of ideas. For example, there are no discussions on determining confidence limits on parameter estimates and estimation errors, since such discussions require considerably more background than is required for the rest of the book. In addition, the very general theory of stochastic approximation is not presented, since it really is much too general for our purposes. Furthermore, it is not our intention to survey the field of parameter estimation (identification). Many excellent surveys exist and the interested reader is referred to them[†, ‡, ¶, §].

The field of parameter estimation is very dynamic. Its many aspects are being explored by researchers throughout the world and, no doubt, different solutions to certain problems examined in this book will be obtained in the not too distant future, or have already been obtained. For example, after the book was written, the author came across some interesting results on the problem of unbiasedness for least-squares estimation, obtained by Åström.[#] These results give new conditions under which it may be possible, for example, to obtain unbiased least-squares estimates of coefficients of stochastic finite-difference equations. Although new answers may be obtained for how to apply the book's four estimation techniques to specific applications, it is felt that the essence of these techniques will not change; therefore, it is the essence of these techniques that is emphasized throughout the book.

---

† P. Eykhoff, Process parameter and state estimation, *Automatica*, 4, 205-233 (1968).

‡ A. V. Balakrishnan and V. Peterka, Identification in automatic control systems, *Automatica*, 5, 817-829 (1969).

¶ G. A. Bekey, System identification - An introduction and a survey, *Simulation*, 151-166 (October 1970).

§ K. J. Åström and P. Eykhoff, Systems identification, a survey, *Automatica*, 7, 123-162 (March 1971).

\# K. J. Åström, *Lectures on the Identification Problem - The Least Squares Method*, Report No. 6806, Lund Institute of Technology, Division of Automatic Control, Sweden, September 1968.

This book is for the most part an outgrowth of a professional development course given by the author in the Fall of 1970 at McDonnell Douglas Astronautics Company, Huntington Beach, in affiliation with the University of California at Irvine, and in the Winter of 1970-1971 at McDonnell Douglas Astronautics Company, Santa Monica, in affiliation with UCLA. Some of the material in the book, especially in Chapters 4, 5, and 6, is based on the author's research activities during the past five years in the Guidance and Flight Mechanics Department at McDonnell Douglas Astronautics Company.

The author wishes to thank his many co-workers and students who encouraged him to write this book. Special thanks go to Dr. S. Sholar, who worked out solutions to many of the exercises and reviewed portions of the manuscript, and to Professor G. Bekey of USC who also reviewed portions of the manuscript.

The author wishes to express his sincere thanks to his wife and children for their patience and understanding during the preparation of this book. Finally, the author wishes to acknowledge motivation for his interest in the area of parameter estimation, and automatic control in general, to his teacher Professor John G. Truxal, of the Polytechnic Institute of Brooklyn.

Los Angeles, California Jerry M. Mendel

Contents

Discrete Techniques of Parameter Estimation

The Equation Error Formulation

1

Equation Error Formulation of Parameter Estimation Problems

1.1 INTRODUCTION

Man creates models of his natural and man-made environments to understand and explain them better and as prelude to action. He models the economy in order to learn about and control inflation, cost of living, balance of payments, etc. He models the stock market to get the best return on an investment. He models his natural environment to make this Earth a better place to live. He models the solar system for many reasons, some of which are: (1) to understand the interactions between the Sun and the celestial bodies that revolve about it; (2) to understand tidal phenomena so that he can predict the occurrence of tidal waves (tsunamis) and other tidal conditions; and (3) to send spacecrafts to the moon and planets. He models his bodily processes (e. g., homeostatic mechanisms), organs (e. g., heart and brain), and his entire self for countless reasons, some of which are: (1) to understand the diffusion and administration of drugs in the blood stream; (2) to understand brain waves, so that, for example, epileptic patients can be forewarned of oncoming seizures; and (3) to optimize the comfort of passengers in an aircraft. He models his machines and other man-made dynamic processes to explain their functions and to optimize their performance characteristics before they are built (e. g., heart-lung machine, turbine, automobile, space station), synthesized (e. g., atomic fission, atomic fusion, photosynthesis), and so on.

Modeling encompasses four problems: representation, measurement, estimation, and validation. The *representation problem* deals with how something should be modeled. We shall be interested only in mathematical models. Within this class of models we need to know whether the model should be static or dynamic, linear or nonlinear, deterministic or random, continuous or discretized, fixed or varying, lumped or distributed (continuous-time lumped parameter systems can be described by ordinary differential equations, whereas continuous-time distributed parameter systems can be described by partial differential equations), in the time domain or in the frequency domain (a time series is a time-domain representation, whereas a transfer function is a frequency-domain representation), etc.

In order to verify a model, physical quantities must be measured. We distinguish between two types of physical quantities, *signals* and *parameters*. Parameters express a relation between signals. Some examples will help clarify the distinction between signals and parameters.

Example 1. Newton's Law states that $F(t) = MA(t)$, where $F(t)$ is force, $A(t)$ is acceleration, and M is mass. Force and acceleration are signals, both of which are often easily measured, and mass is a parameter. ▲

Example 2. Ohm's Law states that $E(t) = RI(t)$, where $E(t)$ is voltage, $I(t)$ is current, and R is resistance. Voltage and current are signals, whereas resistance is a parameter. ▲

Example 3. Heat flows from body 1 to body 2 at a rate $q(t)$ proportional to the temperature difference between body 1 and body 2; that is to say,

Parameter estimation is often referred to as parameter identification. We shall use the terms parameter estimation and parameter identification interchangeably.

1.2 EQUATION ERROR IDENTIFICATION SYSTEMS

1.2.1 Identification Representation

Without some a priori mathematical representation, the parameter estimation problem is meaningless. We shall be interested in linear, discrete representations.

A *linear, discrete representation* is a mathematical representation in which one or more signals can be expressed as linear combinations of other signals, at discrete instants of time (or some other independent variable). A block diagram for one such representation is depicted in Fig. 1.2-1.† We shall refer to the signals

---

† A block diagram is a pictorial way of representing the interrelations between the various quantities present in a system as well as the input or inputs applied to the system from outside. The following block diagram elements are used in this book.

A. Transfer element. A transfer element represents the functional relationship (g_{12}) between a single input signal (x_1) and a single output signal (x_2), in which the input signal, indicated by the arrow, is the independent variable [i. e., $x_2 = g_{12}(x_1)$].

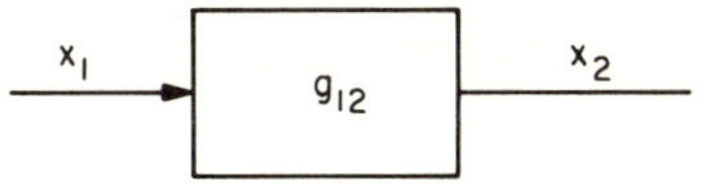

B. Summing point. A summing point indicates the algebraic addition of two or more signals to produce one output signal ($x_3 = x_1 - x_2$).

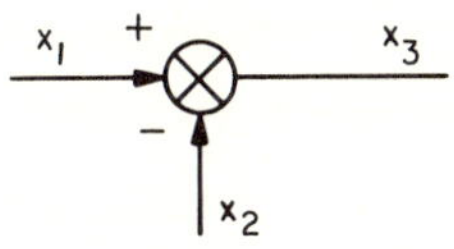

(Cont.)

$y(k)$ and $\underline{x}(k)$ as the *actual output* and *actual input vector*, respectively, for the *Identification Representation* block. The symbol k is shorthand for time t_k.

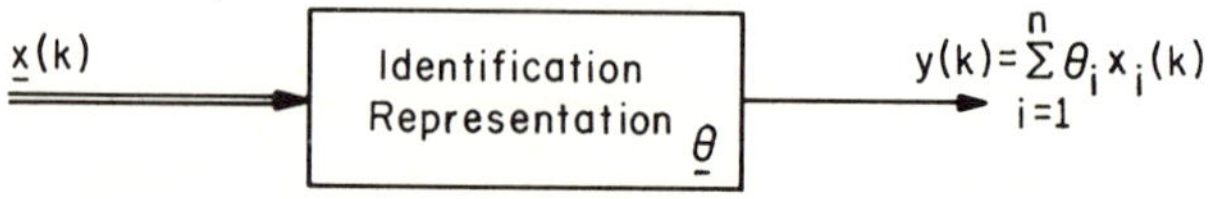

Fig. 1.2-1. Linear, discrete representation.

Although the linearity constraint is quite restrictive, there are, as we shall see in Section 1.3, many important applications which fit within its theoretical framework.

We shall be interested in the identification of time-varying parameters $\theta_i(k)$, as well as constant parameters θ_i. In the former situation, $\underline{x}(k)$ and $y(k)$ are related by the expression

$$y(k) = \sum_{i=1}^{n} x_i(k)\theta_i(k) \tag{1.2-1}$$

whereas, in the latter situation

$$y(k) = \sum_{i=1}^{n} x_i(k)\theta_i \quad . \tag{1.2-2}$$

Letting

$$\underline{x}(k) = (x_1(k), x_2(k), \ldots, x_n(k))' \quad , \tag{1.2-3}$$

$$\underline{\theta}(k) = (\theta_1(k), \theta_2(k), \ldots, \theta_n(k))' \quad , \tag{1.2-4}$$

---

† (Continued from preceding page.)

C. Branch point. A branch point indicates that a signal is distributed to two or more points.

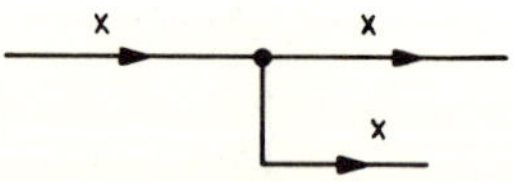

and

$$\underline{\theta} = (\theta_1, \theta_2, \ldots, \theta_n)' \quad , \tag{1.2-5}$$

we can write Eqs. (1.2-1) and (1.2-2) more compactly, as

$$y(k) = \underline{x}'(k)\underline{\theta}(k) \tag{1.2-6}$$

and

$$y(k) = \underline{x}'(k)\underline{\theta} \tag{1.2-7}$$

where $(\cdot)'$ denotes the transpose of $(\cdot)$.

In order to identify $\underline{\theta}(k)$ (or $\underline{\theta}$), the signals $y(k)$ and $\underline{x}(k)$ must be available either from direct measurements or signal estimation. *We shall assume that* $y(k)$ *and* $\underline{x}(k)$ *are known a priori.* How they are obtained will not concern us, since such questions belong more properly to the modeling problems of measurement and signal estimation, which fall outside the scope of this book.

Regardless of how $y(k)$ and $\underline{x}(k)$ are obtained, two situations are possible: (1) they can be obtained without errors, in which case we shall say that they can be obtained perfectly; and (2) they can only be obtained to within certain random errors, in which case we shall say that they can be obtained noisily, and the noise shall be called measurement noise.†

1.2.2 Scalar Stochastic Equation Error

Let us direct our attention at the noisy measurement situation first. We denote the *measured value of* $\underline{x}(k)$ as $\underline{r}(k)$, where (see Fig. 1.2-2a)

$$\underline{r}(k) = \underline{x}(k) + \underline{n}(k) \quad . \tag{1.2-8}$$

The symbol $\underline{n}(k)$ is an $n \times 1$ discrete vector random sequence. We denote the *measured value of* $y(k)$ as $z(k)$, where

$$z(k) = y(k) + v(k) \tag{1.2-9}$$

† Strictly speaking, we should distinguish between random measurement errors and random estimation errors.

and v(k) is a scalar discrete random sequence. The statistical properties of $\underline{n}(k)$ and v(k) are defined in later chapters, as they are needed.

Combining Eqs. (1.2-6) and (1.2-9) we find

$$z(k) = \underline{x}'(k)\underline{\theta}(k) + v(k) \quad . \tag{1.2-10}$$

We shall call this equation the *scalar stochastic measurement equation* (a vector counterpart is described in Section 1.2.4). It is the starting point for the estimation techniques that are discussed in Chapters 2, 3, and 5.

To estimate $\underline{\theta}(k)$, we approximate z(k) by $\hat{z}(k)$, where (see Fig. 1.2-2b)

$$\hat{z}(k) = \underline{r}'(k)\hat{\underline{\theta}}(k) \quad . \tag{1.2-11}$$

$\hat{z}(k)$ is a structural model of z(k), in which $\hat{\underline{\theta}}(k)$ denotes the estimated value of $\underline{\theta}(k)$. $\underline{r}'(k)$, rather than $\underline{x}'(k)$, premultiplies $\hat{\underline{\theta}}(k)$ because in the noisy measurement situation only $\underline{r}(k)$ is known to the analyst.

The *scalar stochastic equation error* between z(k) and $\hat{z}(k)$ denoted $\tilde{z}(k)$, is

$$\tilde{z}(k) = z(k) - \hat{z}(k) \quad . \tag{1.2-12}$$

It is measurable and is used along with $\underline{r}(k)$ to obtain $\hat{\underline{\theta}}(k)$ (see Fig. 1.2-2c) from an identification algorithm.†

By an algorithm is meant *(1-13)* "a list of instructions specifying a sequence of operations which will give the answer to any problem of a given type." As Trakhtenbrot points out, "This is not a precise mathematical definition of the term, but it gives the sense of such a definition. It reflects the concept of algorithm which arose naturally and has been used in mathematics since ancient times."

---

† In the literature on signal estimation, $\tilde{z}(k)$ is referred to as the measurement residual *(1-1)*. In the parameter estimation literature, $\tilde{z}(k)$ is almost always referred to as equation error.

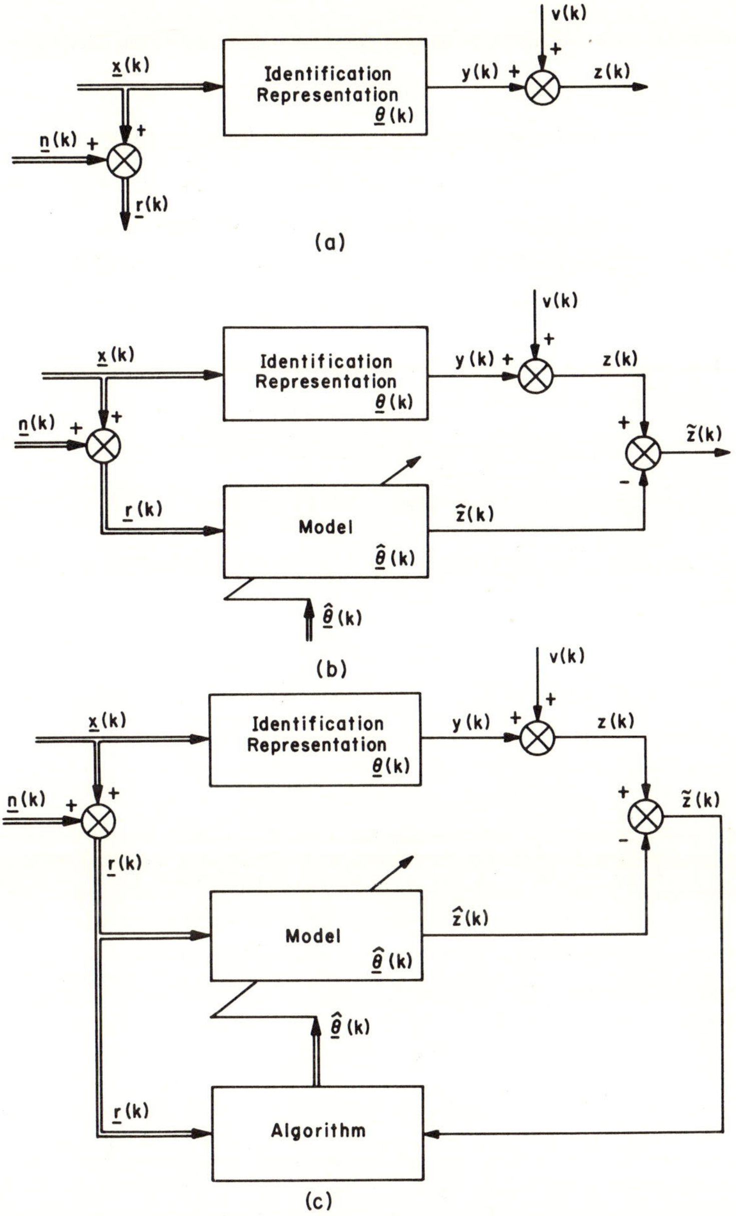

Fig. 1.2-2. Scalar stochastic equation-error identification system (double lines denote transmission of vector quantities, single lines denote transmission of scalar quantities). (a) Identification Representation, plus (b) Model, plus (c) Algorithm.

Different identification algorithms are discussed in Chapters 2 through 6, and are the main topic of this book.

1.2.3 Scalar Deterministic Equation Error

In the special but important case when $\underline{x}(k)$ and $y(k)$ are obtained perfectly, the scalar stochastic measurement equation, Eq. (1.2-10), reduces to the following *scalar deterministic measurement equation*:

$$y(k) = \underline{x}'(k)\underline{\theta}(k) \quad . \tag{1.2-13}$$

This equation is the starting point for the estimation technique that is described in Chapter 4.

In this case, $y(k)$ is approximated by $\hat{y}(k)$, where

$$\hat{y}(k) = \underline{x}'(k)\hat{\underline{\theta}}(k) \tag{1.2-14}$$

and $\hat{\underline{\theta}}(k)$ is obtained from an algorithm that uses $\underline{x}(k)$ and the following *scalar deterministic equation error* between $y(k)$ and $\hat{y}(k)$:

$$\tilde{y}(k) = y(k) - \hat{y}(k) \quad . \tag{1.2-15}$$

Clearly, when $\underline{x}(k)$ and $y(k)$ are obtained perfectly, the identification system in Fig. 1.2-2c simplifies to the structure depicted in Fig. 1.2-3.

1.2.4 Vector Equation Error

Here we generalize the scalar equation-error formulations in the preceding sections to vector equation-error formulations. Now the output of the "identification representation" block in Fig. 1.2-1 is the $m \times 1$ vector $\underline{y}(k)$, where

$$\underline{y}(k) = X(k)\underline{\theta}(k) \tag{1.2-16}$$

and

$$X(k) = \begin{pmatrix} x_{11}(k) & x_{12}(k) & \cdots & x_{1n}(k) \\ x_{21}(k) & x_{22}(k) & \cdots & x_{2n}(k) \\ \vdots & \vdots & \ddots & \vdots \\ x_{m1}(k) & x_{m2}(k) & \cdots & x_{mn}(k) \end{pmatrix} \quad . \tag{1.2-17}$$

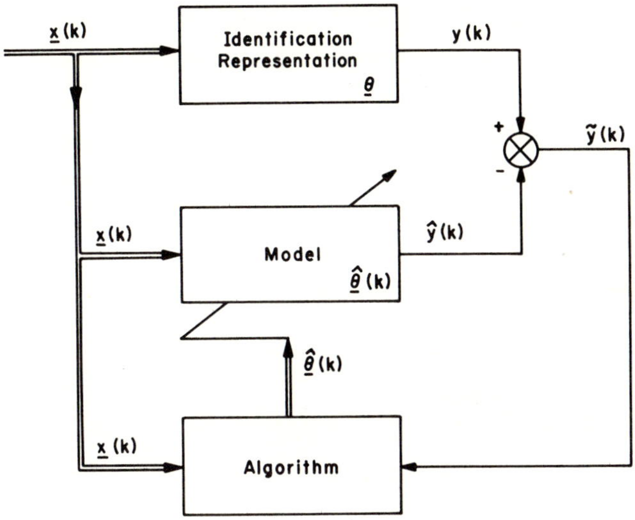

Fig. 1.2-3. Scalar deterministic equation-error identification system.

In the stochastic case, when $\underline{y}(k)$ and $X(k)$ cannot be obtained perfectly, their measured values are

$$\underline{z}(k) = \underline{y}(k) + \underline{v}(k) \tag{1.2-18}$$

and

$$\Xi(k) = X(k) + N(k) \quad . \tag{1.2-19}$$

The *vector stochastic measurement equation* is obtained by combining Eqs. (1.2-18) and (1.2-16), as

$$\underline{z}(k) = X(k)\underline{\theta}(k) + \underline{v}(k) \quad . \tag{1.2-20}$$

This equation is used frequently in Chapters 2 and 3.

To estimate $\underline{\theta}(k)$, we approximate $\underline{z}(k)$ by $\hat{\underline{z}}(k)$, where

$$\hat{\underline{z}}(k) = \Xi(k)\hat{\underline{\theta}}(k) \quad . \tag{1.2-21}$$

The vector stochastic equation error between $\underline{z}(k)$ and $\hat{\underline{z}}(k)$ is $\tilde{\underline{z}}(k)$, where

$$\tilde{\underline{z}}(k) = \underline{z}(k) - \hat{\underline{z}}(k) \quad . \tag{1.2-22}$$

$\tilde{\underline{z}}(k)$, which is measurable, is used along with $\Xi(k)$ to obtain $\hat{\underline{\theta}}(k)$ from an identification algorithm. The vector stochastic equation-error formulation that we have just described is summarized in Fig. 1.2-4.

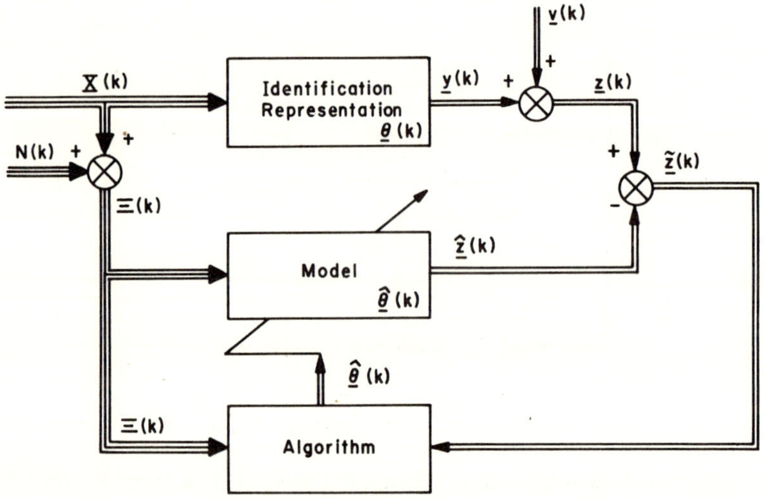

Fig. 1.2-4. Vector stochastic equation-error identification system (triple lines denote transmission of matrix quantities, double lines denote transmission of vector quantities).

We leave the formulation of the vector deterministic equation-error identification system to the reader.

1.3 APPLICATIONS

Here we shall demonstrate how some specific applications can be cast into an equation-error formulation.

1.3.1 Curve Fitting by Arbitrary Functions

The general problem of finding equations of approximating curves which fit given sets of data is called *curve fitting* (*1-2*).

Let us assume that a signal $y(\zeta)$ is measured for q values of ζ, ζ_1, ζ_2, ..., and ζ_q, and that $z(\zeta_1)$, $z(\zeta_2)$, ..., and $z(\zeta_q)$ denote the measured values of $y(\zeta_i)$ $(i = 1, 2, \ldots, q)$. The measured values of $y(\zeta)$ are corrupted by some unknown additive measurement noise $v(\zeta)$. The $y(\zeta)$ could be *time-series data*, such as an electroencephlogram (EEG) voltage record, in which case it is more common to use t for the independent variable than ζ; or $y(\zeta)$ could be *cross-section data*, such as 1971 entertainment expenditure, in dollars, for a sample of 1000 suburban households. *Pooled data*, i. e., combinations of time-series and cross-section data, are also possible.

In general, no single function can describe all of the data exactly because of the measurement noise. The question of approximating the data by arbitrary functions (e. g., polynomials, trigonometric functions, Walsh functions, spline functions, etc.) naturally arises. Indeed, this is the basic idea of the *curve-fitting problem*. It can be formally stated as follows: Find coefficients $\hat{\theta}_1$, $\hat{\theta}_2$, ..., and $\hat{\theta}_N$ so that the quantities $\hat{z}(\zeta)$ computed from

$$\hat{z}(\zeta) = \sum_{j=1}^{N} \hat{\theta}_j \phi_j(\zeta) \qquad (1.3\text{-}1)$$

best approximates the measured data $z(\zeta_i)$, $i = 1, 2, \ldots, q$.

There are two different interpretations which we can associate with $z(\zeta)$ in Eq. (1.3-1). In the first interpretation, which is the classical curve-fitting or data-fitting interpretation, we do not assume any structure for the mechanism that generates the data. The data are available, and we wish to express them in a more compact form for purposes of interpolation or extrapolation.

In the second interpretation, we assume that the data have been generated by a mechanism whose structure is $\theta_1\phi_1(\zeta) + \cdots + \theta_N\phi_N(\zeta) \triangleq z(\zeta)$. In this case, the right-hand side of Eq. (1.3-1) is structurally similar to $z(\zeta)$.

Regardless of which interpretation we prefer, the classical approach for obtaining $\hat{\underline{\theta}}$ is to choose it by minimizing some measure of the error $\tilde{z}(\zeta)$ between $z(\zeta)$ and $\hat{z}(\zeta)$, where

$$\tilde{z}(\zeta) = z(\zeta) - \hat{z}(\zeta) \quad . \qquad (1.3\text{-}2)$$

Comparing Eqs. (1.3-1) and (1.2-11), it is clear that by setting

$$\hat{z}(\zeta_k) = \hat{z}(k) \quad , \tag{1.3-3}$$

$$\phi_j(\zeta_k) = x_j(k) \qquad j = 1, 2, \ldots, N \quad , \tag{1.3-4}$$

and

$$\hat{\underline{\theta}} = \hat{\underline{\theta}}(k) \tag{1.3-5}$$

we are able to view the curve-fitting problem as the scalar stochastic equation-error identification system in Fig. 1.2-2c. For this application,

$$\underline{r}(k) = \underline{x}(k) \tag{1.3-6}$$

since the elements of $\underline{x}(k)$ are computed functions of the independent variable ζ, and ζ (e. g., time) is usually assumed to be known perfectly in the curve-fitting problem.

1.3.2 Superposition Summation

Given two discrete sequences $\{w(j),\ j = 0, 1, \ldots, k\}$ and $\{f_j,\ j = 0, 1, \ldots, k\}$, we form the summation

$$g(k) = \sum_{i=0}^{k} f_i w(k - i) \quad . \tag{1.3-7}$$

This summation defines a function $g(k)$ known as the *convolution* of $\{w(j),\ j = 0, 1, \ldots, k\}$ and $\{f_j,\ j = 0, 1, \ldots, k\}$ for all $k \geq 0$. It is known as a convolution or *superposition summation (1-3)*.

The superposition summation occurs in the theory of linear discrete-time systems and is one of the most important results in that theory.

It is well known, for example, that a deterministic linear time-invariant system is completely characterized by its impulse response $h(t)$. If $w(k)$ and $g(k)$ denote the input and output signals of such a system, both sampled at t_k, then

$$g(k) = \sum_{i=-\infty}^{\infty} h(i)w(k - i) \quad . \tag{1.3-8}$$

If the system is causal (which means it does not respond to an input before the input is applied) and the input starts at $t_0 = 0$, then

$$g(k) = \sum_{i=0}^{k} h(i)w(k - i) \quad . \tag{1.3-9}$$

Additionally, if the impulse response decays to zero in a finite duration

$$h(t) \simeq 0 \qquad \text{for} \quad t > N\Delta t \tag{1.3-10}$$

where $\Delta t = t_{k+1} - t_k$, then for $k \geq N + 1$

$$g(k) = \sum_{i=1}^{N} h(i)w(k - i) \quad .^{\dagger} \tag{1.3-11}$$

This equation can also be used as the basis for estimating h(1), h(2), ..., and h(N) from measured values of g(k), w(k - 1), w(k - 2), ..., and w(k - N). It is put into the form of Eq. (1.2-2) by setting

$$g(k) = y(k) \quad , \tag{1.3-12}$$

$$h(i) = \theta_i \qquad \text{for} \quad i = 1, 2, \ldots, N \quad , \tag{1.3-13}$$

and

$$w(k - i) = x_i(k) \qquad \text{for} \quad i = 1, 2, \ldots, N \quad . \tag{1.3-14}$$

If, for example, g(k) and w(k - i) can all be measured perfectly, then we have the situation that is described in Section 1.2.3: g(k) is approximated by $\hat{g}(k)$, where

$$\hat{g}(k) = \sum_{i=1}^{N} \hat{h}(i)w(k - i) \tag{1.3-15}$$

for $k \geq N + 1$, and the identification of h(1), h(2), ..., and h(N) falls within the framework of the scalar deterministic equation-error identification system depicted in Fig. 1.2-3. On the other hand, if g(k) and w(k - i) cannot be measured perfectly we have the situation that is described in Section 1.2.2.

---

† The impulse response has the property that h(0) = 0.

1.3.3 Finite-Difference Equations

We shall now describe two distinctly different approaches for identifying the 2n coefficients, a_1, a_2, ..., a_n, b_1^o, b_2^o, ..., and b_n^o in the following nth-order finite-difference equation:

$$y(k+n) + a_1 y(k+n-1) + \cdots + a_n y(k)$$
$$= b_n^o\, m(k+n-1) + b_{n-1}^o\, m(k+n-2) + \cdots + b_1^o\, m(k)$$
$$+ d_n^o\, \omega(k+n-1) + d_{n-1}^o\, \omega(k+n-2) + \cdots + d_1^o\, \omega(k) \quad .^\dagger \tag{1.3-16}$$

In this equation m represents a signal, such as a test signal or feedback control signal, that can be measured and can be chosen by a designer according to established principles; ω, on the other hand, represents a signal, such as a random disturbance, that cannot be measured and subsequently cannot be freely chosen by a designer. For the sake of brevity, we shall often refer to m as a controllable signal and ω as an uncontrollable signal. In all of our work, we shall assume that d_1^o, d_2^o, ..., and d_n^o are known ahead of time.

Some special cases of Eq. (1.3-16) are

$$y(k+n) + a_1 y(k+n-1) + \cdots + a_n y(k) = d_1^o\, \omega(k) \tag{1.3-17}$$

and

$$y(k+n) + a_1 y(k+n-1) + \cdots + a_n y(k)$$
$$= b_n^o\, m(k+n-1) + b_{n-1}^o\, m(k+n-2) + \cdots + b_1^o\, m(k) \quad . \tag{1.3-18}$$

Clearly, many other cases are possible and can be obtained from Eq. (1.3-16) by judiciously choosing the b^o and d^o coefficients.

---

† The reader may be wondering why, in this equation, the a parameters are not superscripted, whereas the b and d parameters are superscripted. This finite-difference equation and its associated notation are fully discussed in Appendix A.

Formulation Based on Structure of Equation

Observe, in Eq. (1.3-16), that the unknown a and b^o parameters appear linearly, and that y(k + n) can be viewed as a linear transformation on the signals y(k + n - 1), ..., y(k), m(k + n - 1), ..., and m(k); that is to say,

$$y(k + n) = -a_1 y(k + n - 1) - \cdots - a_n y(k) + b_n^o\, m(k + n - 1) + b_{n-1}^o\, m(k + n - 2) + \cdots + b_1^o\, m(k) + \omega^* \tag{1.3-19}$$

where

$$\omega^* = d_n^o\, \omega(k + n - 1) + \cdots + d_1^o\, \omega(k) \quad . \tag{1.3-20}$$

Generally, y(k + n), y(k + n - 1), ..., y(k), m(k + n - 1), ..., and m(k) can only be obtained noisily; i. e., only measured values (denoted with a subscript m)

$$y_m(k + n) = y(k + n) + n_y(k + n) \quad , \tag{1.3-21}$$

$$y_m(k + n - j) = y(k + n - j) + n_y(k + n - j) \quad , \tag{1.3-22}$$

and

$$m_m(k + n - j) = m(k + n - j) + n_m(k + n - j) \quad , \tag{1.3-23}$$

where j = 1, 2, ..., n, are available.

The *scalar stochastic measurement equation* from which the a and b^o parameters can be estimated is obtained from Eqs. (1.3-21) and (1.3-19), and is

$$y_m(k + n) = -a_1 y(k + n - 1) - \cdots - a_n y(k) + b_n^o\, m(k + n - 1) + \cdots + b_1^o\, m(k) + \omega^* + n_y(k + n) \quad . \tag{1.3-24}$$

By defining

$$\underline{\theta} = (-a_1, -a_2, \ldots, -a_n, b_n^o, b_{n-1}^o, \ldots, b_1^o)' \quad , \tag{1.3-25}$$

$$z(k) = y_m(k + n) \quad , \tag{1.3-26}$$

$$\underline{x}(k) = [y(k + n - 1), y(k + n - 2), \ldots, y(k), m(k + n - 1), \ldots, m(k)]' \quad , \tag{1.3-27}$$

and

$$v(k) = \omega^* + n_y(k + n) \tag{1.3-28}$$

we are able to rewrite Eq. (1.3-24) as

$$z(k) = \underline{x}'(k)\underline{\theta} + v(k) \quad , \tag{1.3-29}$$

which is exactly the same as Eq. (1.2-10) when $\underline{\theta}(k)$ in that equation is constant. Following the arguments prior to Eq. (1.2-11), we conclude that the model of $y_m(k + n)$ is $\hat{y}_m(k + n)$, where

$$\hat{y}_m(k + n) = -\hat{a}_1(k + n)y_m(k + n - 1) - \cdots - \hat{a}_n(k + n)y_m(k) + \hat{b}_n^o(k + n)m_m(k + n - 1) + \cdots + \hat{b}_1^o(k + n)m_m(k) \quad . \tag{1.3-30}$$

Clearly, the vector $\underline{r}(k)$ which appears in Eq. (1.2-11) is

$$\underline{r}(k) = [y_m(k + n - 1), \ldots, y_m(k), m_m(k + n - 1), \ldots, m_m(k)]' \quad . \tag{1.3-31}$$

At this point, let us pause to make two important observations. First, the arguments of $\hat{a}_i$ and $\hat{b}_j^o$ are k + n and not k; that is to say,

$$\hat{\underline{\theta}} = \hat{\underline{\theta}}(k + n) = [-\hat{a}_1(k + n), \ldots, -\hat{a}_n(k + n), \hat{b}_n^o(k + n), \ldots, \hat{b}_1^o(k + n)]' \quad . \tag{1.3-32}$$

The argument of $\hat{\underline{\theta}}$ *is chosen to agree with the argument of the most recent datum that is used to estimate* $\underline{\theta}$. In this application, the most recent datum is y(k + n); hence, $\hat{\underline{\theta}} = \hat{\underline{\theta}}(k + n)$. In the two preceding applications, the most recent datum occurred at t_k and, therefore, in those applications $\hat{\underline{\theta}} = \hat{\underline{\theta}}(k)$.

Our second observation is made to dispel any confusion that might exist in the reader's mind about apparent inconsistencies in the arguments of the functions on the two sides of Eqs. (1.3-26)-(1.3-28) and (1.3-31). Why, for example, have we chosen to write Eq. (1.3-26) as $z(k) = y_m(k + n)$ instead of $z(k + n) = y_m(k + n)$? The symbols on the left-hand sides of these four equations are *generic symbols* that are associated with the generic scalar stochastic equation-error identification system in Fig. 1.2-2c. The symbols on the right-hand sides of these equations are the ones that are associated with this special application, and are the ones that would be used to implement any identification algorithm. As long as we understand that the generic symbols are only being used as a shorthand notation for the actual symbols, no confusion can occur.

Solutions of Finite-Difference Equations

The second approach to the identification of the 2n a and b^o parameters in Eq. (1.3-16) is based on the solution to that equation. Before presenting details of this second approach, let us investigate solutions of Eq. (1.3-16).

In Appendix A, we show that Eq. (1.3-16) can also be written in so-called *state space form*, as

$$\underline{x}(k + 1) = \Phi\underline{x}(k) + \underline{\psi}m(k) + \underline{\gamma}\omega(k) \tag{1.3-33}$$

where

$$\underline{x}(k) = [x_1(k), x_2(k), \ldots, x_n(k)]' \quad , \tag{1.3-34}$$

$$\Phi = \left(\begin{array}{c|ccc} 0 & & & \\ 0 & & I_{n-1} & \\ \vdots & & & \\ \hline -a_n & -a_{n-1} & \cdots & -a_1 \end{array}\right) \tag{1.3-35}$$

[I_{n-1} is the $(n - 1) \times (n - 1)$ identity matrix],

$$\underline{\psi} = (b_1 \;\; b_2 \;\; \cdots \;\; b_n)' \quad , \tag{1.3-36}$$

and

$$\underline{\gamma} = (d_1 \quad d_2 \quad \cdots \quad d_n)' \quad . \tag{1.3-37}$$

Relationships between the components of $\underline{x}(k)$ and $y(k)$, $m(k)$ and $\omega(k)$, the components of $\underline{\psi}$ and the b^o parameters, and the components of $\underline{\gamma}$ and the d^o parameters are given in Eqs. (A-3) through (A-6).

Our approach for obtaining $y(k)$ is to obtain the solution $\underline{x}(k)$, of Eq. (1.3-33), and to then obtain $y(k)$ from the fact [see first equation in Eq. (A-3)] that

$$y(k) = \underline{h}'\underline{x}(k) \tag{1.3-38}$$

where

$$\underline{h} = (1 \quad 0 \quad 0 \quad \cdots \quad 0)' \quad . \tag{1.3-39}$$

Equation (1.3-33) is a special case of the following linear, time-varying discrete-time state equation:

$$\underline{x}(k + 1) = \Phi(k + 1, k)\underline{x}(k) + \Psi(k + 1, k)\underline{m}(k) + \Gamma(k + 1, k)\underline{\omega}(k) \tag{1.3-40}$$

in which

$\{\underline{x}(1), \underline{x}(2), \ldots\} \equiv n \times 1$ state vector sequence

$\{\underline{m}(0), \underline{m}(1), \ldots\} \equiv p \times 1$ test vector sequence

$\{\underline{\omega}(0), \underline{\omega}(1), \ldots\} \equiv r \times 1$ disturbance vector sequence

$\Phi(k + 1, k) \equiv n \times n$ state transition matrix†

$\Psi(k + 1, k) \equiv n \times p$ test transition matrix

$\Gamma(k + 1, k) \equiv n \times r$ disturbance transition matrix.

In later chapters, we will need the solution to Eq. (1.3-40); thus, we shall obtain it first and then specialize it to the time-invariant (constant parameter) Eq. (1.3-33).

---

† In the time-varying situation, Φ, Ψ, and Γ are functions of t_k and t_{k+1}. In the constant-parameter situation, they are functions only of $t_{k+1} - t_k$.

Since the last two terms on the right-hand side of Eq. (1.3-40) are forcing functions, we shall simplify Eq. (1.3-40) by writing it as

$$\underline{x}(k+1) = \Phi(k+1, k)\underline{x}(k) + \underline{f}(k+1, k) \tag{1.3-41}$$

where

$$\underline{f}(k+1, k) = \Psi(k+1, k)\underline{m}(k) + \Gamma(k+1, k)\underline{\omega}(k) \quad . \tag{1.3-42}$$

Theorem 1-1. *Two solutions of Eq. (1.3-41) are*

$$\underline{x}(k) = \Phi(k, 0)\underline{x}(0) + \sum_{i=1}^{k} \Phi(k, i)\underline{f}(i, i-1) \tag{1.3-43}$$

or

$$\underline{x}(k) = \Phi(k, j)\underline{x}(j) + \sum_{i=j+1}^{k} \Phi(k, i)\underline{f}(i, i-1) \tag{1.3-44}$$

where

$$\Phi(k, i) = \Phi(k, k-1)\Phi(k-1, k-2) \cdots \Phi(i+1, i) \tag{1.3-45}$$

and $k = 0, 1, \ldots .$

Proof. Both parts of this proof make use of some of the following well-known properties of the state transition matrix (*1-3*):

(i) $\Phi(t, t) = I_n$;

(ii) $\Phi(t_2, t_1) = \Phi(t_2, \tau)\Phi(\tau, t_1)$ for all t_1, t_2, and $\tau \geq t_0$;

(iii) $\Phi^{-1}(t_1, t_2) = \Phi(t_2, t_1)$ for all $t_1, t_2 \geq t_0$.

(A) Equation (1.3-43). For k = 0, 1, and 2, we find, from Eq. (1.3-41) and property (ii), that

$$\underline{x}(1) = \Phi(1, 0)\underline{x}(0) + \underline{f}(1, 0) \tag{1.3-46}$$

$$\begin{aligned}\underline{x}(2) &= \Phi(2, 1)\underline{x}(1) + \underline{f}(2, 1) \\ &= \Phi(2, 1)\Phi(1, 0)\underline{x}(0) + \Phi(2, 1)\underline{f}(1, 0) + \underline{f}(2, 1) \\ &= \Phi(2, 0)\underline{x}(0) + \Phi(2, 1)\underline{f}(1, 0) + \underline{f}(2, 1)\end{aligned} \tag{1.3-47}$$

and

$$\begin{aligned} \underline{x}(3) &= \Phi(3, 2)\underline{x}(2) + \underline{f}(3, 2) \\ &= \Phi(3, 2)\Phi(2, 0)\underline{x}(0) + \Phi(3, 2)\Phi(2, 1)\underline{f}(1, 0) \\ &\quad + \Phi(3, 2)\underline{f}(2, 1) + \underline{f}(3, 2) \\ &= \Phi(3, 0)\underline{x}(0) + \Phi(3, 1)\underline{f}(1, 0) + \Phi(3, 2)\underline{f}(2, 1) \\ &\quad + \underline{f}(3, 2) \quad . \end{aligned} \tag{1.3-48}$$

Observe that $\underline{x}(3)$ can also be written as

$$\underline{x}(3) = \Phi(3, 0)\underline{x}(0) + \sum_{i=1}^{3} \Phi(3, i)\underline{f}(i, i - 1) \tag{1.3-49}$$

which suggests that

$$\underline{x}(k) = \Phi(k, 0)\underline{x}(0) + \sum_{i=1}^{k} \Phi(k, i)\underline{f}(i, i - 1) \tag{1.3-50}$$

for $k = 0, 1, \ldots$. That Eq. (1.3-50) is true for all k is easily proved by means of *mathematical induction*. We have shown that it is true for $k = 1$ (2 and 3). Assuming it is true for k, let us demonstrate it is also true for $k + 1$. From Eqs. (1.3-41) and (1.3-50) and properties (i) and (ii), we find that

$$\begin{aligned} \underline{x}(k + 1) &= \Phi(k + 1, k)\left[\Phi(k, 0)\underline{x}(0) + \sum_{i=1}^{k} \Phi(k, i)\underline{f}(i, i - 1)\right] \\ &\quad + \underline{f}(k + 1, k) \\ &= \Phi(k + 1, 0)\underline{x}(0) + \sum_{i=1}^{k} \Phi(k + 1, i)\underline{f}(i, i - 1) \\ &\quad + \Phi(k + 1, k + 1)\underline{f}(k + 1, k) \quad . \end{aligned} \tag{1.3-51}$$

Combining the last two terms on the right-hand side of Eq. (1.3-51), we find

$$\underline{x}(k + 1) = \Phi(k + 1, 0)\underline{x}(0) + \sum_{i=1}^{k+1} \Phi(k + 1, i)\underline{f}(i, i - 1) \tag{1.3-52}$$

which is exactly what we obtain from Eq. (1.3-50) when we replace k by k + 1. This completes the proof for $\underline{x}(k)$ in Eq. (1.3-43).

(B) Equation (1.3-44). From Eq. (1.3-43) we find for k = j, that

$$\underline{x}(j) = \Phi(j, 0)\underline{x}(0) + \sum_{i=1}^{j} \Phi(j, i)\underline{f}(i, i-1) \quad ; \qquad (1.3\text{-}53)$$

hence,

$$\underline{x}(0) = \Phi^{-1}(j, 0)\underline{x}(j) - \Phi^{-1}(j, 0) \sum_{i=1}^{j} \Phi(j, i)\underline{f}(i, i-1) \qquad (1.3\text{-}54)$$

which can be simplified by means of properties (ii) and (iii) to

$$\underline{x}(0) = \Phi(0, j)\underline{x}(j) - \sum_{i=1}^{j} \Phi(0, i)\underline{f}(i, i-1) \quad . \qquad (1.3\text{-}55)$$

We leave it to the reader to show that Eq. (1.3-44) is a direct consequence of substituting Eq. (1.3-55) into Eq. (1.3-43) (Problem 1-3). ▲

For time-invariant systems, $\Phi(k+1, k) = \Phi$; hence,

$$\Phi(k, i) = \Phi^{k-i} \qquad (1.3\text{-}56)$$

and we obtain the following corollary to Theorem 1-1.

Corollary 1-1. *Given, the linear time-invariant system*

$$\underline{x}(k+1) = \Phi\underline{x}(k) + \underline{f}(k+1, k) \quad . \qquad (1.3\text{-}57)$$

Two solutions for x(k) *are*

$$\underline{x}(k) = \Phi^{k}\underline{x}(0) + \sum_{i=1}^{k} \Phi^{k-i}\underline{f}(i, i-1) \qquad (1.3\text{-}58)$$

or

$$\underline{x}(k) = \Phi^{k-j}\underline{x}(j) + \sum_{i=j+1}^{k} \Phi^{k-i}\underline{f}(i, i-1) \qquad (1.3\text{-}59)$$

where k = 0, 1, ... ▲

We say that the unforced system $\underline{x}(k+1) = \Phi\underline{x}(k)$ is asymptotically stable[†] if $\underline{x}(k) \to \underline{0}$ as $k \to \infty$. Clearly now, xince $\underline{x}(k) = \Phi^k\underline{x}(0)$, this must mean that $\Phi^k \to 0$ as $k \to \infty$. A state transition matrix having this property will be called a *stable matrix*.

Theorem 1-2. *Given, the linear time-invariant discrete-time system*

$$\underline{x}(k+1) = \Phi\underline{x}(k) + \underline{f}(k+1, k) \quad . \tag{1.3-60}$$

If Φ *is a stable matrix, then*

$$\underline{x}(k) = \sum_{i=0}^{\infty} \Phi^i\underline{f}(k-i, k-i-1) \tag{1.3-61}$$

for $k = 1, 2, \ldots$.

Proof. To begin, we shall demonstrate that

$$\underline{x}(0) = \Phi^L\underline{x}(-L) + \sum_{i=-L}^{-1} \Phi^{-(i+1)}\underline{f}(i+1, i) \quad . \tag{1.3-62}$$

This is easily proved by setting $k = -L$ in Eq. (1.3-60), in which case that equation becomes

$$\underline{x}(1-L) = \Phi\underline{x}(-L) + \underline{f}(1-L, -L) \quad ; \tag{1.3-63}$$

thus, we have for $L = 1, 2,$ and 3 that

$$\underline{x}(0) = \underline{x}(-1) + \underline{f}(0, -1) \quad , \tag{1.3-64}$$

$$\underline{x}(-1) = \Phi\underline{x}(-2) + \underline{f}(-1, -2) \quad , \tag{1.3-65}$$

and

$$\underline{x}(-2) = \Phi\underline{x}(-3) + \underline{f}(-2, -3) \quad . \tag{1.3-66}$$

Substituting Eq. (1.3-66) into Eq. (1.3-65), and the result into Eq. (1.3-64), it is easily shown that

$$\underline{x}(0) = \Phi^3\underline{x}(-3) + \Phi^2\underline{f}(-2, -3) + \Phi\underline{f}(-1, -2) + \underline{f}(0, -1) \tag{1.3-67}$$

which suggests that

---

† For more precise definitions of different types of stability, see Section 4.4.2.

$$\underline{x}(0) = \Phi^L \underline{x}(-L) + \Phi^{L-1}\underline{f}(1 - L, -L) + \Phi^{L-2}\underline{f}(2 - L, 1 - L)$$

$$+ \cdots + \Phi\underline{f}(-1, -2) + \underline{f}(0, -1)$$

$$= \Phi^L \underline{x}(-L) + \sum_{i=-L}^{-1} \Phi^{-(i+1)}\underline{f}(i + 1, i) \quad . \qquad (1.3\text{-}68)$$

We leave a general proof of this result to the reader (Problem 1-4).

Next, let $L = \infty$ in Eq. (1.3-68) and use the property for a stable matrix, that $\Phi^\infty \rightarrow 0$, to obtain

$$\underline{x}(0) = \sum_{i=-\infty}^{-1} \Phi^{-(i+1)}\underline{f}(i + 1, i) \quad . \qquad (1.3\text{-}69)$$

Substituting this expression for $\underline{x}(0)$ into Eq. (1.3-58), we find

$$\underline{x}(k) = \sum_{i=-\infty}^{-1} \Phi^{k-i-1}\underline{f}(i + 1, i) + \sum_{j=1}^{k} \Phi^{k-j}\underline{f}(j, j - 1) \quad . \qquad (1.3\text{-}70)$$

Letting $j = i + 1$ in the first summation, we obtain

$$\underline{x}(k) = \sum_{j=-\infty}^{0} \Phi^{k-j}\underline{f}(j, j - 1) + \sum_{j=1}^{k} \Phi^{k-j}\underline{f}(j, j - 1)$$

$$= \sum_{j=-\infty}^{k} \Phi^{k-j}\underline{f}(j, j - 1) \quad . \qquad (1.3\text{-}71)$$

Finally, let $i = k - j$ to obtain the desired result in Eq. (1.3-61).

▲

Formulation Based on Solution of Equation

Let us now direct our attention at the identification of the 2n a and b^o parameters in Eq. (1.3-16) from the solution of its equivalent state-space formulation given in Eqs. (1.3-33) through (1.3-39). The state equation's parameters are $a_1, a_2, \ldots, a_n$ and $b_1, b_2, \ldots,$ and b_n. Our approach to the identification of the a and b^o parameters is to first identify the a and b parameters and to then determine the b^o parameters from the equation

$$\hat{\underline{b}}^o = \hat{T}\hat{\underline{b}} \qquad (1.3\text{-}72)$$

which is obtained from Eq. (A-4) of Appendix A and the fact that the matrix T is a function of a_1, a_2, ..., and a_{n-1}.

Equation (1.3-33) is equivalent to the finite-difference equation (1.3-16), and, its solution is given in Eq. (1.3-61). y(k), the solution to the original difference equation, is related to $\underline{x}(k)$ by the first equation in Eq. (A-3) and is given, therefore, by the expression

$$y(k) = \sum_{i=0}^{\infty} \underline{h}'\Phi^i[\underline{\psi}m(k - i - 1) + \underline{\gamma}\omega(k - i - 1)] \qquad (1.3\text{-}73)$$

which can also be written as

$$y(k) = \sum_{i=0}^{2n-1} \underline{h}'\Phi^i\underline{\psi}m(k - i - 1) + v^*(k) \quad , \qquad (1.3\text{-}74)$$

where

$$v^*(k) = \sum_{i=2n}^{\infty} \underline{h}'\Phi^i\underline{\psi}m(k - i - 1) + \sum_{i=0}^{\infty} \underline{h}'\Phi^i\underline{\gamma}\omega(k - i - 1) \quad . \qquad (1.3\text{-}75)$$

Defining

$$\underline{\theta} = (\underline{h}'\underline{\psi}, \underline{h}'\Phi\underline{\psi}, \underline{h}'\Phi^2\underline{\psi}, \ldots, \underline{h}'\Phi^{2n-1}\underline{\psi})' \qquad (1.3\text{-}76)$$

and

$$\underline{m}(k) = [m(k - 1), m(k - 2), \ldots, m(k - 2n)]' \quad , \qquad (1.3\text{-}77)$$

Eq. (1.3-74) can be written as

$$y(k) = \underline{m}'(k)\underline{\theta} + v^*(k) \quad . \qquad (1.3\text{-}78)$$

Denoting the measured values of y(k) and $\underline{m}(k)$, $y_m(k)$ and $\underline{m}_m(k)$, respectively, where

$$y_m(k) = y(k) + n_y(k) \qquad (1.3\text{-}79)$$

and

$$\underline{m}_m(k) = \underline{m}(k) + \underline{n}_m(k) \quad , \qquad (1.3\text{-}80)$$

we obtain the following *scalar stochastic measurement equation* from which the 2n θ parameters can be estimated:

$$y_m(k) = \underline{m}'(k)\underline{\theta} + v(k) \tag{1.3-81}$$

where

$$v(k) = v^*(k) + n_y(k) \quad . \tag{1.3-82}$$

Two important observations are in order. First, we note from Eq. (1.3-76) that the elements of $\underline{\theta}$ appear to be rather complicated functions of $\underline{h}$, Φ, and $\underline{\psi}$. After the 2n θ parameters have been identified, we must extract the 2n a and b parameters from the 2n relationships in Eq. (1.3-76). It certainly is not at all obvious that this can be accomplished.

We assert that Eq. (1.3-76) can also be written as (*1-4*)

$$\underline{\theta} = \Lambda(\underline{\theta})\underline{p} \tag{1.3-83}$$

where $\underline{p}$ denotes a 2n parameter vector that is related to the 2n a and b parameters. Clearly, if Eq. (1.3-83) is true, then

$$\hat{\underline{p}} = \Lambda^{-1}(\hat{\underline{\theta}})\hat{\underline{\theta}} \quad . \tag{1.3-84}$$

The parameter transformation matrix, $\Lambda(\hat{\underline{\theta}})$, is computed in Appendix C for two systems, an abstract digital system (i. e., a system that is totally discrete) and a discrete-time representation of a continuous-time system.

Our second observation has to do with the statistical independence of $\underline{m}(k)$ and $v^*(k)$. The $\underline{m}(k)$ contains $m(k - 1)$, ..., and $m(k - 2n)$, whereas $v^*(k)$ contains $m(k - 2n - 1)$, $m(k - 2n - 2)$, ..., $\omega(k - 1)$, $\omega(k - 2)$, ..., etc. This observation is used in Chapter 5 to establish the present formulation as a Class-1 stochastic identification problem.

Let us summarize the steps which must be taken to identify the a and b^o parameters: (1) identify the 2n θ parameters, (2) determine the estimated a and b parameters from Eq. (1.3-84), and (3) determine the estimated b^o parameters from Eq. (1.3-72).

1.3.4 Differential Equations

We shall now consider two distinctly different approaches for identifying the 2n coefficients $a_1, a_2, \ldots, a_n, b_1, b_2, \ldots$, and b_n in the following nth-order differential equation:

$$\begin{aligned} y^{(n)}(t) + a_1 y^{(n-1)}(t) + \cdots + a_{n-1}\dot{y}(t) + a_n y(t) \\ = b_1 m^{(n-1)}(t) + \cdots + b_{n-1}\dot{m}(t) + b_n m(t) \\ + d_1 \omega^{(n-1)}(t) + \cdots + d_{n-1}\dot{\omega}(t) + d_n \omega(t) \end{aligned} \qquad (1.3\text{-}85)$$

where m(t) represents a controllable and measurable signal, whereas $\omega(t)$ represents an uncontrollable and unmeasurable signal. As in the preceding section, we shall assume that the n d parameters, $d_1, d_2, \ldots$, and d_n, are known to us a priori.

Formulation Based on Structure of Equation

Paralleling our development of Eq. (1.3-19) in Section 1.3.3, we solve Eq. (1.3-85) for $y^{(n)}(t)$, to obtain

$$\begin{aligned} y^{(n)}(t) = -a_1 y^{(n-1)}(t) - \cdots - a_n y(t) \\ + b_1 m^{(n-1)}(t) + \cdots + b_n m(t) \\ + d_1 \omega^{(n-1)}(t) + \cdots + d_n \omega(t) \end{aligned} \quad . \qquad (1.3\text{-}86)$$

Let us assume, for illustrative purposes only, that $y(t), \dot{y}(t), \ldots, y^{(n)}(t), m(t), \dot{m}(t), \ldots$, and $m^{(n-1)}(t)$ are *measured perfectly* at discrete instants of time $t_1, t_2, \ldots, t_k, \ldots$. Symbols y(k), m(k), ..., etc., denote the measured values of y(t), m(t), ..., etc., at $t = t_k$; hence, at these times, we can rewrite Eq. (1.3-86) as

$$\begin{aligned} y^{(n)}(k) = -a_1 y^{(n-1)}(k) - \cdots - a_n y(k) \\ + b_1 m^{(n-1)}(k) + \cdots + b_n m(k) \\ + d_1 \omega^{(n-1)}(k) + \cdots + d_n \omega(k) \end{aligned} \quad . \qquad (1.3\text{-}87)$$

This equation can now be written in the form of Eq. (1.2-10) as

$$z(k) = \sum_{i=1}^{2n} \theta_i x_i(k) + v(k) \qquad (1.3\text{-}88)$$

where

$$\underline{\theta} = (-a_1, -a_2, \ldots, -a_n, b_1, b_2, \ldots, b_n)' \quad , \qquad (1.3\text{-}89)$$

$$\underline{x}(k) = [y^{(n-1)}(k), y^{(n-2)}(k), \ldots, y(k), m^{(n-1)}(k), m^{(n-2)}(k), \ldots, m(k)]' \quad , \qquad (1.3\text{-}90)$$

$$z(k) = y^{(n)}(k) \quad , \qquad (1.3\text{-}91)$$

and

$$v(k) = d_1\omega^{(n-1)}(k) + \cdots + d_n\omega(k) \quad . \qquad (1.3\text{-}92)$$

Equation (1.3-88) is a scalar stochastic measurement equation, which is the starting point for the estimation of the 2n a and b parameters in $\underline{\theta}$. Its stochastic nature is due solely to the driving noise terms in the original differential equation.

Although the goal has been achieved, the end result may not in itself be very useful because along the way the assumption had to be made that y(t) and its n time derivatives, and m(t) and its n - 1 time derivatives are all measurable. Often it is not practical to obtain higher-order derivatives of y(t) and m(t). One approach for circumventing this difficulty is to pass measured values of y(t) and m(t) through banks of so-called "state variable filters" (*1-5* - *1-7*) and to then work with their outputs, termed "auxiliary signals," instead of with $\underline{z}(k)$ and $\underline{r}(k)$ [or y(k) and $\underline{x}(k)$]. This technique is explored in Problems 1-1 and 1-2.

Formulation Based on Discretization

It is possible, by means of suitable transformations, to obtain a finite-difference equation that is approximately equivalent to the differential equation in Eq. (1.3-85). The finite-difference equation representation of the differential equation becomes more and more accurate the smaller the sampling interval becomes. Appendix B presents all of the theory necessary to obtain a finite-difference equation from Eq. (1.3-85), in the following three steps: (1) obtain a suitable continuous-time state space representation for Eq. (1.3-85); (2) obtain an approximate discrete-time state space representation of the representation determined in step 1; and (3) obtain a finite-difference equation that is equivalent to the representation determined in step 2. The main results obtained in Appendix B for each of these steps are summarized in Theorem B-2, Corollary B-1, and Theorem B-5, respectively. The approximately equivalent finite-difference equation for Eq. (1.3-85) is given in Eq. (B.3-23), as

$$y(k+n) + \alpha_1 y(k+n-1) + \cdots + \alpha_n y(k) = \beta_1 m(k+n-1) + \cdots + \beta_n m(k) + \gamma_1 \omega(k+n-1) + \cdots + \gamma_n \omega(k) \tag{1.3-93}$$

where

$$\alpha_{n-i} = \sum_{j=i}^{n-1} (-1)^{j-i} \binom{j}{j-i} a_{n-j} T^{n-j} + (-1)^{n-i} \binom{n}{n-i}, \tag{1.3-94}$$

$$\beta_{n-i} = \sum_{j=i}^{n-1} (-1)^{j-i} \binom{j}{j-i} b_{n-j} T^{n-j}, \tag{1.3-95}$$

$$\gamma_{n-i} = \sum_{j=i}^{n-1} (-1)^{j-i} \binom{j}{j-i} d_{n-j} T^{n-j}, \tag{1.3-96}$$

for $i = 0, 1, \ldots, n-1$, and

$$T = t_{k+1} - t_k . \tag{1.3-97}$$

The approach to identifying the original a and b parameters from Eq. (1.3-93) is: (1) identify the 2n α and β parameters using either one of the techniques described in Section 1.3.3; and (2) obtain estimate of the 2n a and b parameters from the 2n equations in Eqs. (1.3-94) and (1.3-95). Step 2 in this two-step procedure is purely algebraic.

1.3.5 Adaptive Control System

Basically, there are two types of control systems: open loop, and closed loop (feedback). An open-loop control system (*1-8*) is a system in which the output has no effect upon the input signal. A closed-loop control system is a system in which the output affects the input quantity in a manner which maintains desired system performance. Our attention will be directed to closed-loop control systems because they make use of the well-known feedback principle which is important and useful in the formulation of advanced control systems.

Space limitations prevent extensive elaboration on closed-loop control systems. (See *1-8 - 1-10* for more comprehensive coverage.) For our purposes, it is convenient to view a closed-loop control system as shown in Fig. 1.3-1. Examples of process- or plant-environment combinations are: a missile disturbed by gusts of wind, a spacecraft acted upon by the moon's gravity while orbiting around the moon, and a gun following the motions of targets while mounted on the deck of a ship. The control law may take the form of a shaping network, a decision surface, or a special-purpose digital computer. The output of the control law acts as input to a device or subsystem which is able to exert control over the process. In the missile, spacecraft, and fire-control examples given, the device might be a reaction-jet engine, an inertia wheel, and a hydraulic amplifier, respectively.

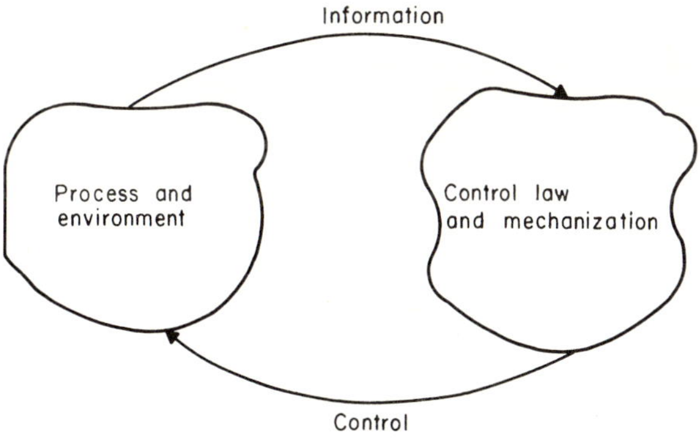

Fig. 1.3-1. Representation of a closed-loop control system.

If everything about the environment and process is known a priori, the design of the control law is straightforward and can be accomplished by means of proved techniques. On the other hand, if the environment or process is poorly defined, more advanced and sometimes less proved techniques must be used to design the law. In the latter situation, control specialists have devised adaptive control systems.

An adaptive control system is one which is provided with (1) a means of continuously monitoring its own performance relative to desired performance conditions, and (2) a means of modifying a part of its control law, by closed-loop action, so as to approach these conditions (*1-11*). The adaptive control law encompasses three major functions (*1-9*): identification, decision, and modification. Identification is the process by which the plant or environment is characterized, or by which the performance of the system is monitored. In the decision process, the performance measurements are used to decide how system performance relates to the desired performance conditions. If performance is inadequate, corrective adjustments are made according to an established strategy. Modification is the process of changing parameters - within the control law - toward their optimum settings. It is controlled by the identification and decision processes.

Conventional control systems are designed to meet certain performance specifications under known environment and process conditions; however, should these conditions change, performance will change as a result. In an adaptive control system, the system's actual performance is compared to the desired performance, and the difference is then used to drive actual performance toward desired performance. In a parameter adaptive control system, this is accomplished by adjusting feedback control parameters, such as control gains, as explicit functions of identified plant parameters.

Figure 1.3-2 depicts simplified third-order pitch-plane dynamics for a typical, high-performance, aerodynamically controlled aerospace vehicle. Cross-coupling and body-bending effects are neglected. Normal acceleration control is considered with feedback on normal acceleration and angle-of-attack rate. Stefani (*1-12*) shows that if the system gains are chosen as

$$K_{Ni} = \frac{C_2}{100\ M_\delta Z_\alpha} \quad , \tag{1.3-98}$$

$$K_{\dot{\alpha}} = \frac{C_1 - 100\left(\dfrac{Z_\alpha 1845}{\mu}\right) + M_\alpha}{100\ M_\delta} \quad , \tag{1.3-99}$$

and

$$K_{Na} = \frac{C_2 + 100\ M_\alpha}{100\ M_\delta Z_\alpha} \quad , \tag{1.3-100}$$

then[†]

$$\frac{N_a}{N_i}(s) = \frac{C_2}{s^3 + 100s^2 + C_1 s + C_2} \quad . \tag{1.3-101}$$

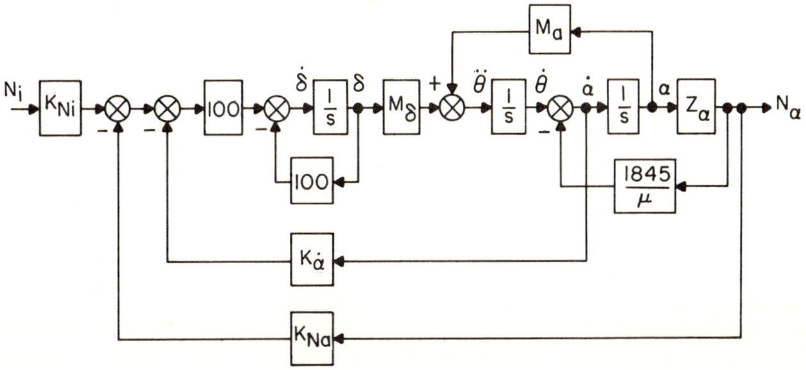

Fig. 1.3-2. Pitch-plane dynamics and nomenclature, N_i, input normal acceleration along the negative Z axis; K_{Ni}, gain on N_i; δ, control-surface deflection; M_δ, control-surface effectiveness; $\ddot{\theta}$, rigid-body acceleration; α, angle-of-attack; M_α, aerodynamic moment effectiveness; $K_{\dot{\alpha}}$, control gain on $\dot{\alpha}$; Z_α, normal acceleration force coefficient; μ, axial velocity; N_a, system-achieved normal acceleration along the negative Z axis; K_{Na}, control gain on N_a.

Stefani assumes $Z_\alpha 1845/\mu$ is relatively small, and chooses $C_1 = 1400$ and $C_2 = 14{,}000$. The closed-loop response resembles that of

---

[†] The s is the Laplace transform operator for d/dt.

a second-order system with a bandwidth of 2 cps and a damping ratio of 0.6 that responds to a step command of input acceleration with zero steady-state error.

If system response is to remain the same for all values of M_α, M_δ, and Z_α, it is necessary to estimate these parameters so that K_{Ni}, $K_{\dot{\alpha}}$, and K_{Na} can be adapted to keep C_1 and C_2 invariant at their designed values. In general, M_α, M_δ, and Z_α are dynamic parameters and all vary through a large range of values. Also, M_α may be positive (unstable vehicle) or negative (stable vehicle). For present purposes we shall assume that these parameters are frozen at specific values. From Fig. 1.3-2,

$$\ddot{\theta}(t) = M_\alpha \alpha(t) + M_\delta \delta(t) \tag{1.3-102}$$

and

$$N_a(t) = Z_\alpha \alpha(t) \quad . \tag{1.3-103}$$

Our attention is directed at the identification of M_α and M_δ in Eq. (1.3-102). We leave it as an exercise for the reader to explore the identification of Z_α in Eq. (1.3-103). Measured values of $\ddot{\theta}(t)$, $\alpha(t)$, and $\delta(t)$ are assumed available at $t = t_1, t_2, \ldots, t_k, \ldots$, etc., and are given by the expressions

$$\ddot{\theta}_m(k) = \ddot{\theta}(k) + v_{\ddot{\theta}}(k) \quad , \tag{1.3-104}$$

$$\alpha_m(k) = \alpha(k) + n_\alpha(k) \quad , \tag{1.3-105}$$

and

$$\delta_m(k) = \delta(k) + n_\delta(k) \quad . \tag{1.3-106}$$

The scalar stochastic measurement equation from which M_α and M_δ will be estimated is obtained by combining Eqs. (1.3-104) and (1.3-102), and is

$$\ddot{\theta}_m(k) = M_\alpha \alpha(k) + M_\delta \delta(k) + v_{\ddot{\theta}}(k) \quad . \tag{1.3-107}$$

If measurements of $\ddot{\theta}(t)$, $\alpha(t)$, and $\delta(t)$ can be made without errors, then M_α and M_δ are estimated from the scalar deterministic measurement equation

$$\ddot{\theta}(k) = M_\alpha \alpha(k) + M_\delta \delta(k) \quad . \tag{1.3-108}$$

In Chapter 6, we shall study the situation where M_α, M_δ, and Z_α are time-varying, in which case $M_\alpha(t)$ and $M_\delta(t)$ are estimated from

$$\ddot{\theta}(k) = M_\alpha(k)\alpha(k) + M_\delta(k)\delta(k) \tag{1.3-109}$$

if measurements are perfect, or

$$\ddot{\theta}_m(k) = M_\alpha(k)\alpha(k) + M_\delta(k)\delta(k) + v_{\ddot{\theta}}(k) \tag{1.3-110}$$

if measurements are noisy.

1.4 OUTLINE OF CONTENTS OF THE BOOK

The remaining chapters in this book are concerned with four parameter estimation techniques: generalized least squares (Chapter 2), unbiased minimum variance (Chapter 3), deterministic gradient (Chapter 4), and stochastic gradient (Chapter 5). Attention, for the most part, is directed at the identification of constant parameters; however, Chapter 6 examines the extensions of many of the constant parameter results to the case of time-varying parameters.

Generalized least-squares and unbiased minimum-variance estimations both fall within the framework of the vector stochastic equation-error identification system depicted in Fig. 1.2-4. Deterministic- and stochastic-gradient estimations both fall within the framework of scalar equation-error formulations.

Chapters 2 through 5 each begin with a careful statement of the parameter estimation problem central to that chapter. Examples are given to illustrate the different formulations. Some of the material in these sections may be redundant; however, it is our intent to make each of these chapters as independent of each other as possible; hence, some redundance is inevitable.

Chapter 2 begins with a construction of the classical concatenated measurement equation. This vector measurement equation is not only the basic starting point for least-squares estimation but is also the basic starting point for minimum-variance estimation.

Chapter 2 continues with a derivation of the classical batch processing generalized least-squares parameter estimation algorithm;

that is to say, the algorithm in which all measurements are processed simultaneously. Statistical properties of parameter estimation error, such as bias and covariance, are then examined and it is shown why it is not possible to use this estimation algorithm to obtain unbiased estimates for certain important applications, such as the identification of coefficients in finite-difference equations that are excited by random forcing functions.

Our attention is then directed to sequential versions of the generalized least-squares algorithm and various problems associated with these algorithms, such as a sequential startup technique, and two types of data processing techniques for statistically independent measurements or batches of measurements. The sequential algorithms are first-order, time-varying, vector finite-difference equations. As such, we study the question of convergence of the parameter identification-error system to zero in terms of the stability of the parameter identification-error finite-difference equation. Most of our attention is directed at stability in the mean so that we may learn under what conditions the sequential estimates will be asymptotically unbiased.

Chapter 2 concludes with discussions on computational aspects of both the batch processing and sequential processing algorithms, and applicability of generalized least-squares estimation.

Chapter 3 presents parallel sections for much of the material discussed in Chapter 2. The basic difference between the formulations adopted in Chapters 2 and 3 is that the covariance of the signal measurement noise is assumed to be known a priori in Chapter 3, whereas it is not assumed known in Chapter 2. One of the main results in Chapter 3 is Corollary 3-1 which states: *All results obtained in Chapter 2 for generalized least-squares estimates of* $\underline{\theta}$ *apply as well to minimum-variance estimates of* $\underline{\theta}$*, when the weighting matrix used to weight the measurements in Chapter 2 is replaced by the inverse of the measurement noise covariance matrix that is associated with the concatenated measurement noise vector in Chapter 3.*

Chapter 3 continues by relating sequential unbiased minimum-variance algorithms to a special case of discrete Kalman filtering, and unbiased minimum-variance estimation, in general, to maximum-likelihood estimation.

Modeling errors that appear either in the concatenated measurement equation or in the measurement noise covariance matrix are studied by means of sensitivity analysis. Chapter 3 then concludes with discussions on computational aspects and applicability of its algorithms.

Motivation for gradient parameter-estimation algorithms comes from gradient hill-climbing (optimization) algorithms. The situation when all signals are deterministic is examined in Chapter 4, whereas the stochastic situation is examined in Chapter 5. Convergence questions for both situations are discussed in great detail. The stability properties of the deterministic parameter identification-error system are examined in Chapter 4 by means of the Second Method of Lyapunov. Mean-square convergence and with probability one convergence of the stochastic parameter identification-error system are examined in Chapter 5. Background material for all of these analyses is given in Appendixes D through G.

Chapters 4 and 5 also describe ways to accelerate convergence by choosing, in some optimal manner, the weighting matrix that appears in the gradient estimation algorithms.

Chapter 5 distinguishes between three types of stochastic identification problems: (i) *Class 1*, in which $\underline{x}(k)$ (see Fig. 1.2-2c) is independent of $v(k)$ and $\hat{\underline{\theta}}(k)$; (ii) *Class 2* , in which $\underline{x}(k)$ is independent of $\hat{\underline{\theta}}(k)$, but is dependent upon $v(k)$; and (iii) *Class 3*, in which $\underline{x}(k)$ is dependent upon both $v(k)$ and $\hat{\underline{\theta}}(k)$. Class 1 and Class 2 Identification Problems are explored in detail; however, Class 3 problems are beyond the scope of this book, since their identification-error systems are nonlinear and are, therefore, fraught with all of the difficulties associated with nonlinear systems. The problem of identifying the coefficients in a finite-difference equation is pinpointed as a Class 2 identification problem; and a rather large section in Chapter 5 is devoted to how unbiased estimates of these coefficients can be obtained.

Chapter 6 begins by classifying a time-varying parameter as Type A, Type B, or Type C, depending upon how much is known about it a priori. Type A information means no structural information is known about $\underline{\theta}(k)$. Type B information means $\underline{\theta}(k)$ can be decomposed, as $\underline{\theta}(k) = L(k)\underline{\beta}(k)$, where $L(k)$ is a known information matrix and $\underline{\beta}(k)$ is a less rapidly varying vector than $\underline{\theta}(k)$. This decomposition may be possible from physical principles associated with the meaning of $\underline{\theta}(k)$, or it can be induced in an artificial manner. Type C information means $\underline{\theta}(k)$ is modeled by a dynamical system

$$\underline{\theta}(k + 1) = \Phi(k + 1, k)[\underline{\theta}(k) - \underline{\theta}_N(k)] + \underline{\theta}_N(k + 1) + \underline{\omega}(k) \qquad (1.4\text{-}1)$$

where $\Phi(k + 1, k)$ is known a priori, $\underline{\theta}_N(k)$ is a known nominal time history of $\underline{\theta}(k)$, and $\underline{\omega}(k)$ is a discrete vector white sequence that provides the model with different qualities of uncertainty.

Estimation of $\underline{\theta}(k)$ from the three types of information is explored in Chapter 6. Results get better and better as more and more information that is known about $\underline{\theta}(k)$ is used. Minimum-variance estimation of $\underline{\theta}(k)$ from Type-C information leads to a full-blown discrete Kalman filter.† The derivation of the Kalman filter is purely algebraic and does not need any concepts that are outside of this book.

In addition to many independent examples that appear throughout the book, the applications of curve fitting, superposition summations, finite-difference equations, and adaptive control system, that were described in Sections 1.3.1, 1.3.2, 1.3.3, and 1.3.5, respectively, appear in many of the chapters. For the reader who is interested in these applications, we present the following guide to their locations:

---

† A discrete Kalman filter is an algorithm that gives minimum-variance estimates of signals.

Curve fitting: Sections 2.2, 2.4.1 - 2.4.3, 2.5, and 4.2;

Superposition summations: Sections 2.2, 2.5, 4.2, and 4.7.1;

Finite-difference equations: Sections 2.5, 4.2, 5.1, 5.4.3, and 5.6;

Adaptive control system: Sections 2.4.4, 2.5, 4.2, 4.7.2, 5.1, 5.7, and 6.3.

PROBLEMS

1-1. Here we shall investigate the identification of the 2n coefficients in the nth-order differential equation

$$y^{(n)}(t) = -a_1 y^{(n-1)}(t) - \cdots - a_n y(t) + b_1 m^{(n-1)}(t) + \cdots + b_n m(t)$$

when only y(t) and m(t) can be measured. Our approach is: to generate *auxiliary signals* by passing y(t) and m(t) through banks of *state variable filters*; to identify 2n coefficients, $\alpha_1, \alpha_2, \ldots, \alpha_n, \beta_1, \beta_2, \ldots,$ and β_n that are associated with the auxiliary signals; and, to extract $a_1, a_2, \ldots, a_n, b_1, b_2, \ldots,$ and b_n from algebraic relationships that link the a and b parameters to the α and β parameters.

The T_i symbolizes a linear time-invariant filter, which has the form

$$T_i(s) = H(s)(s + c)^i$$

where H(s) is an arbitrary filter chosen by the designer. Actually, as Lion (*1-6*) points out, the filters may have a still more general form; e. g., the "free" zeros [those excluding H(s)] may be different for each value of i. (See Fig. P1-1.) Let

$$y_j = T_j y \qquad \text{for } j = 0, 1, \ldots, n$$

and

$$m_i = T_i m \qquad \text{for } i = 0, 1, \ldots, n - 1 \quad .$$

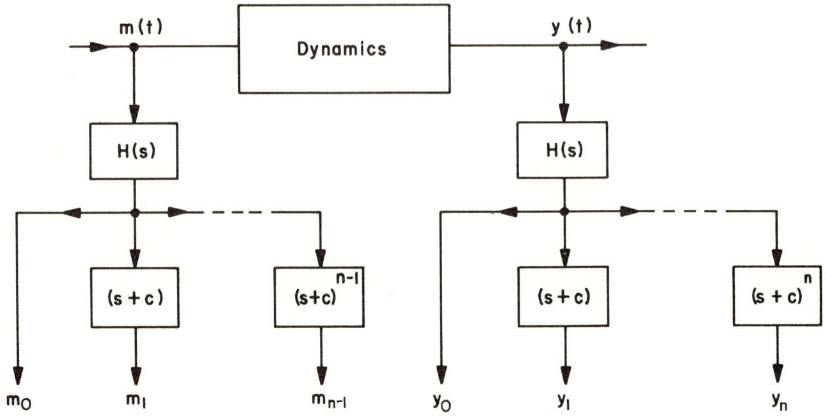

Fig. P1-1.

We define the following *auxiliary-measurement equation*:

$$y_n = \alpha_1 y_{n-1} + \cdots + \alpha_n y_0 + \beta_1 m_{n-1} + \cdots + \beta_n m_0 \quad .$$

(a) Explain how $\alpha_1, \ldots, \alpha_n, \beta_1, \ldots,$ and β_n can be identified.

(b) Let $\hat{\alpha}_1, \ldots, \hat{\alpha}_n, \hat{\beta}_1, \ldots,$ and $\hat{\beta}_n$ denote the final estimated values of $\alpha_1, \ldots, \alpha_n, \beta_1, \ldots,$ and β_n. If $c = 0$, show that $\hat{a}_i = \hat{\alpha}_i$ and $\hat{b}_i = \hat{\beta}_i$ for $i = 1, 2, \ldots, n$.

(c) Consider the second-order system

$$\ddot{y}(t) + a_1\dot{y}(t) + a_0 y(t) = b_1\dot{m}(t) + b_0 m(t) \quad .$$

If $H(s) = 1/(s + 1)^2$, show that

$$\hat{a}_1 = \hat{\alpha}_1 + 2 \quad , \quad \hat{a}_0 = \hat{\alpha}_0 + \hat{\alpha}_1 + 1 \quad ,$$

$$\hat{b}_1 = -\hat{\beta}_1 \quad , \quad \text{and} \quad \hat{b}_0 = -(\hat{\beta}_0 + \hat{\beta}_1) \quad .$$

1-2. Obtain the general relationships between the a and b parameters and the α and β parameters in Problem 1-1 for the following choices of H(s):

(a) $H(s) = 1$ (b) $H(s) = s^{-n}$ (c) $H(s) = (s + c)^{-n}$.

1-3. Show that Eq. (1.3-44) results when Eq. (1.3-55) is substituted into Eq. (1.3-43).

1-4. Demonstrate the truth of Eq. (1.3-68) by means of mathematical induction.

1-5. The choice one makes for the state variables of a system often is crucial to the success (possibility) or failure (impossibility) in identifying all of the parameters associated with a system's model. This is also related to the fact that different sets of parameters can be used to characterize the same system, and some sets are easier to identify than others.

Consider the system depicted in Fig. P1-5. The a, b, and c are the system's *basic parameters*.

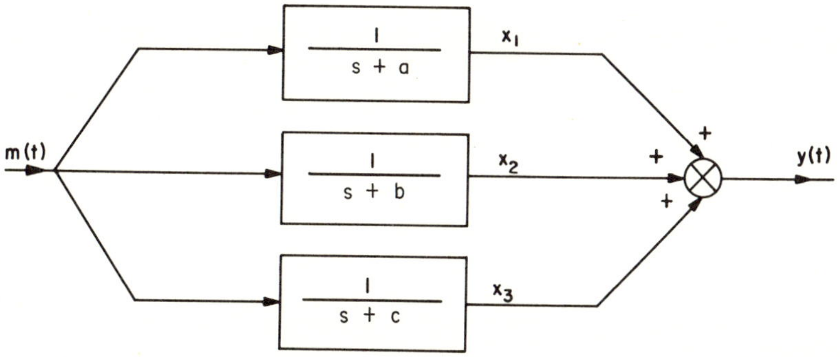

Fig. P1-5.

(a) Show that one state space representation for this system is:

$$\begin{pmatrix} \dot{x}_1 \\ \dot{x}_2 \\ \dot{x}_3 \end{pmatrix} = \begin{pmatrix} -a & 0 & 0 \\ 0 & -b & 0 \\ 0 & 0 & -c \end{pmatrix} \begin{pmatrix} x_1 \\ x_2 \\ x_3 \end{pmatrix} + \begin{pmatrix} 1 \\ 1 \\ 1 \end{pmatrix} m$$

where

$$y(t) = x_1(t) + x_2(t) + x_3(t) \quad .$$

(b) Obtain an approximate discrete-time state space representation for this system (see Appendix B). In what ways does this representation differ from the representation described in Appendix B, Section B.2?

(c) Obtain the finite-difference equation associated with the discrete-time state space representation obtained in (b).

(d) Explain why it is not possible to extract the system's basic parameters from an identification of the difference equation's coefficients.

1-6. A different approach to identifying parameters for the Fig. P1-5 system is to view that system from an input/output vantage.

(a) Show that the differential equation relating y(t) to m(t) has the following structure:

$$y^{(3)}(t) + a_1 y^{(2)}(t) + a_2 y^{(1)}(t) + a_3 y(t)$$
$$= b_1 m^{(2)}(t) + b_2 m^{(1)}(t) + b_3 m(t) \quad .$$

(b) Obtain expressions relating a_1, a_2, a_3, b_1, b_2, and b_3 to a, b, and c. Observe that, given values of a, b, and c, it is always possible to compute a_i and b_i (i = 1, 2, and 3); however, the converse is not true.

(c) Obtain an equivalent finite-difference equation for the differential equation in (a), using results in Appendix B.

(d) Compare the finite-difference equations obtained in this problem and in Problem 1-5. How are they alike, and how do they differ?

1-7. In this problem we shall investigate a technique for identifying the *matrices* of a vector differential equation using a *reference model* (A. B. Gates, *Experimental Modelling of Network-Like Systems*, Ph.D. Thesis, Case Western Reserve Univ., June 1971). The system to be identified is

$$\dot{\underline{x}}(t) = F\underline{x}(t) + C\underline{m}(t) \quad ,$$

where $\underline{x}$ is n × 1, $\underline{m}$ is p × 1, F is n × n, C is n × p, and F and C are constant matrices. The reference model, also of order n, is described by the expression

$$\dot{\underline{x}}_m(t) = F_m \underline{x}_m(t) + C_m \underline{m}(t) \quad .$$

F_m is prespecified by the designer so that the reference model is stable; $\underline{m}(t)$ is a *test signal* that is applied to both the system and reference model.

The error $\underline{e}(t)$ between $\underline{x}(t)$ and $\underline{x}_m(t)$ is

$$\underline{e}(t) = \underline{x}(t) - \underline{x}_m(t) \quad .$$

(a) Show that the differential equation for $\underline{e}(t)$ is

$$\dot{\underline{e}}(t) = F_m \underline{e}(t) + \Delta F \underline{x}(t) + \Delta C \underline{m}(t)$$

where $\Delta F = F - F_m$ and $\Delta C = C - C_m$.

(b) Show that, if $\underline{x}(t) \cong \underline{x}(t_k)$ and $\underline{m}(t) \cong \underline{m}(t_k)$ for $t_k \leq t \leq t_{k+1}$, then

$$\underline{e}(k+1) = M_1 \underline{e}(k) + M_2[\Delta F \underline{x}(k) + \Delta C \underline{m}(k)]$$

where

$$M_1 = \exp(F_m T) \quad ,$$

$$M_2 = \int_{t_k}^{t_{k+1}} \exp[F_m(t_{k+1} - \tau)]d\tau = -\int_T^0 \exp(F_m \xi)d\xi \quad ,$$

and

$$T = t_{k+1} - t_k \quad .$$

(c) Letting

$$\underline{y}(k) = M_2^{-1}[\underline{e}(k+1) - M_1 \underline{e}(k)]$$

we see, that

$$\underline{y}(k) = (\Delta F \,\vdots\, \Delta C)\begin{pmatrix} \underline{x}(k) \\ \hline \underline{m}(k) \end{pmatrix} \quad .$$

Show how the elements of ΔF and ΔC can be identified from n independent scalar equation-error identification systems.

(d) In this formulation, $\underline{y}(k)$, $\underline{x}(k)$, and $\underline{m}(k)$ must all be available for the identification of ΔF and ΔC. Show that the model reference formulation can be represented schematically as in Fig. P1-7.

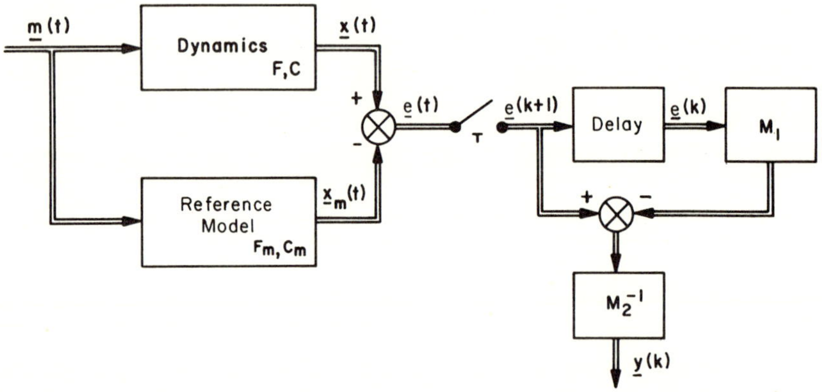

Fig. P1-7.

(e) Why do we need the reference model at all? Why not just integrate the system's differential equation?

1-8. Using the model reference technique described in Problem 1-7, show how to identify the basic parameters a, b, and c in the Fig. P1-5 system.

1-9. Starting with Eqs. (1.3-94) and (1.3-95), obtain general expressions for a_iT^i and b_iT^i, $i = 1, 2, \ldots, n$.

1-10. A system that can store energy in only one form and location is called a first-order dynamic system because the mathematical equation describing its motion can be written in terms of a single variable and its first derivative only. Some examples of first-order dynamic systems are: a single mass moving against friction, a single electrical capacitance with resistors, a single mechanical spring with friction, and a single thermal capacitance with thermal resistance.

The equation of motion for a first-order system is

$$\dot{y}(t) + ay(t) = bm(t) \quad .$$

(a) Obtain the equivalent finite-difference equation for y(k).

(b) Explain how a and b can be identified from the finite-difference equation, if: (i) y(t) and m(t) can be measured perfectly, and (ii) y(t) and m(t) cannot be measured perfectly.

1-11. A system having two separate energy-storage elements is called a second-order system. The motion of such a system can be described by a second-order differential equation. The equation of motion for an undamped, second-order system has the form

$$\ddot{y}(t) + \omega^2 y(t) = bm(t)$$

where ω is known as the *natural frequency* of the system. The solution of the homogeneous second-order system [m(t) = 0] is sinusoidal and is called simple harmonic motion. Undamped systems having such natural motion are called *harmonic oscillators*. Some examples of harmonic oscillators are: mass and spring, free piston on air column, floating buoy, mercury in a U tube, electrical inductor/capacitance circuit, simple pendulum, two-stage hydraulic servo, cantilever, wheel on hilly rail, and unbalanced cone.

(a) Obtain the equivalent finite-difference equation for y(k).

(b) Explain how ω^2 and b can be identified from the finite-difference equation, if: (i) y(t) and m(t) can be measured perfectly, and (ii) y(t) and m(t) cannot be measured perfectly.

1-12. A system having two separate energy-storage elements as well as a dissipative element is called a damped second-order system. The motion of such a system can be described by any one of the following second-order differential equations:

(i) $\ddot{y}(t) + a_1\dot{y}(t) + a_2 y(t) = b_2 m(t)$

(ii) $\ddot{y}(t) + a_1\dot{y}(t) + a_2 y(t) = b_1 \dot{m}(t)$

(iii) $\ddot{y}(t) + a_1\dot{y}(t) + a_2 y(t) = b_1 \dot{m}(t) + b_2 m(t)$.

Clearly, the equations in (i) and (ii) are special cases of the equation in (iii); hence, we shall direct our attention to (iii).

(a) Obtain the equivalent finite-difference equation for y(k).

(b) Explain how a_1, a_2, b_1, and b_2 can be identified from the finite-difference equation, if: (i) y(t) and m(t) can be measured perfectly, and (ii) y(t) and m(t) cannot be measured perfectly.

(c) What simplifications occur for the equations in (i) and (ii)?

1-13. For the first-order system in Problem 1-10, show how a and b can be identified from the *solution* to the equivalent first-order finite-difference equation. What is $\Lambda(\hat{\underline{\theta}})$?

1-14. For the harmonic oscillator in Problem 1-11, show how ω^2 and b can be identified from the *solution* to the equivalent second-order finite-difference equation. What is $\Lambda(\hat{\underline{\theta}})$?

1-15. For the damped second-order system (iii) in Problem 1-12, show how a_1, a_2, b_1, and b_2 can be identified from the *solution* to the equivalent second-order finite-difference equation. What is $\Lambda(\hat{\underline{\theta}})$?

1-16. Suppose we know that a relationship exists between y and x_1, x_2, ..., and x_n of the form

$$y = \exp(a_1x_1 + a_2x_2 + \cdots + a_nx_n) \quad .$$

We desire to estimate a_1, a_2, ..., and a_n from measurements of y and $\underline{x} = (x_1, x_2, \ldots, x_n)'$.

(a) Explain how $\underline{a}$ can be estimated from perfect measurements of y and $\underline{x}$.

(b) Explain how $\underline{a}$ can be estimated from perfect measurements of y and noisy measurements of $\underline{x}$.

(c) Can $\underline{a}$ be estimated, using the techniques of this chapter, from noisy measurements of y and $\underline{x}$? Explain.

1-17. The efficiency of a jet engine may be viewed as a linear combination of functions of inlet pressure p(t) and operating temperature T(t); that is to say,

$$E(t) = C_1 + C_2f_1[p(t)] + C_3f_2[T(t)]$$

$$+ C_4f_3[p(t), T(t)] + \nu(t)$$

where the structures of f_1, f_2, and f_3 are known a priori and $\nu(t)$ represents modeling error of known mean and variance. From tests on the engine a table of values of E(t), p(t), and T(t) are given at discrete values of t. Explain how C_1, C_2, C_3, and C_4 are estimated from these data.

1-18. Is it possible to estimate ω from noisy measurements of x(t), when

$$x(t) = \cos \omega t \quad ?$$

Construct a differential equation whose motion is $\cos \omega t$. Explain how ω can be estimated from the differential equation and the noisy measurements.

1-19. A model of a missile that is slightly more complicated than the model in Fig. 1.3-2 is depicted in Fig. P1-19. Approximate discrete-time equations of motion for the missile are:

$$\alpha(k+1) = \left[1 - \left(\frac{1845}{\mu} N_\alpha\right) T\right] \alpha(k) + T\dot{\theta}(k) - \frac{1845}{\mu} N_\delta T \delta(k) \quad ,$$

$$\dot{\theta}(k+1) = M_\alpha T \alpha(k) + \dot{\theta}(k) + M_\delta T \delta(k) + \omega(k) \quad ,$$

$$\delta(k+1) = -aT\delta(k) + aTu(k)$$

where $\omega(k)$ is a zero mean random disturbance and u(k) is the command signal that is applied to the fin (the missile is aerodynamically controlled).

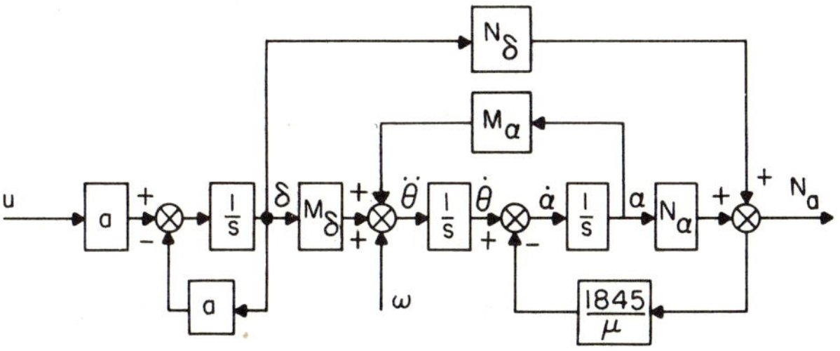

Fig. P1-19.

(a) Show that $\dot{\theta}$ satisfies the following finite-difference equation:

$$\dot{\theta}(k+2) = \phi_1\dot{\theta}(k+1) + \phi_2\dot{\theta}(k) + \phi_3\delta(k) + \phi_4\delta(k+1) + \omega(k+1) - p_3\omega(k)$$

where

$$\phi_1 = 1 + p_3 \quad , \qquad \phi_2 = p_1T - p_3 \quad ,$$

$$\phi_3 = p_1p_4 - p_2p_3 \quad , \qquad \phi_4 = p_2 \quad ,$$

$$p_1 = M_\alpha T \quad , \qquad p_2 = M_\delta T \quad ,$$

$$p_3 = 1 - \frac{1845}{\mu} TN_\alpha \quad , \qquad p_4 = -\frac{1845}{\mu} TN_\delta \quad .$$

(b) Show that the four basic parameters, M_α, M_δ, N_α, and N_δ, can be extracted from ϕ_1, ϕ_2, ϕ_3, and ϕ_4.

(c) Explain how ϕ_1, ϕ_2, ϕ_3, and ϕ_4 can be identified from the structure of the finite-difference equation, if: (i) $\dot{\theta}$ and δ can be measured perfectly, and (ii) $\dot{\theta}$ and δ cannot be measured perfectly.

(d) Explain how ϕ_1, ϕ_2, ϕ_3, and ϕ_4 can be identified from the solution to the finite-difference equation, if: (i) $\dot{\theta}$ and δ can be measured perfectly, and (ii) $\dot{\theta}$ and δ cannot be measured perfectly.

1-20. For the system in Fig. 1.3-2, explain how Z_α can be identified, if: (a) both $N_a(t)$ and $\alpha(t)$ can be measured, but with measurement errors; and (b) only $N_a(t)$ can be measured. (*Hint*: Use the structure of the solution to a certain finite-difference equation.)

1-21. Suppose we have an abstract digital system that is characterized by the finite-difference equation in Eq. (1.3-16); but, *in the present situation,* b_1^o, ..., *and* b_n^o *are known*. Assume perfect measurements of y(t) and m(t) are available.

(a) Explain how the unknown a parameters are identified from the formulation that is based on the structure of the finite-difference equation.

(b) Explain how the a parameters are identified from the formulation that is based on the solution to the finite-difference equation.

1-22. Many of the estimation techniques that are described in this book in the context of a linear representation can also be applied to the estimation of parameters in nonlinear representations, when such representations are suitably linearized.

(a) Suppose $z(k) = f(\theta) + v(k)$, where $v(k)$ is random error associated with the measurement of $f(\theta)$, and the structure of the nonlinear function of θ, $f(\theta)$, is known explicitly. Let θ^* denote a nominal value of θ, $\delta\theta = \theta - \theta^*$, and $\delta z = z - z^*$.

(i) Show, by means of a first-order Taylor series expansion of $f(\theta)$, that

$$\delta z(k) = \frac{\partial f}{\partial \theta^*}\,\delta\theta + v(k)$$

where $\partial f/\partial\theta^*$ is shorthand for "$\partial f/\partial\theta$ evaluated at $\theta = \theta^*$."

(ii) Explain how θ can be identified from the linear perturbation equation in (i).

(b) Generalize the results in (a) to the vector situation, when $z(k) = f(\underline{\theta}) + v(k)$.

(c) In both (a) and (b) explain two ways in which θ^* ($\underline{\theta}^*$) can be obtained. Observe that the identification of θ (or $\underline{\theta}$) requires θ^* (or $\underline{\theta}^*$).

1-23. *Quasilinearization* is an approximation technique for obtaining solutions to nonlinear differential equations (or difference equations) that are subject to multipoint boundary conditions [E. S. Lee, *Quasilinearization and Invariant Imbedding*, Academic Press, New York, 1968; R. Bellman and R. Kalaba, *Quasilinearization and Nonlinear Boundary-Value Problems*, American Elsevier, New York, 1965]. By quasilinearization, a nonlinear differential equation is linearized about a function (or functions) that is defined over the entire time interval of interest. A sequence of functions which converges quadratically to the solution of the original nonlinear equation is obtained. A rough initial approximation for the unknown solution can lead to the actual solution of the nonlinear equation through a sequence of solutions.

The linear equation is obtained by using the first and second terms in the Taylor's series expansion of the original nonlinear equation.

Consider the $n \times 1$ nonlinear and time-varying differential equation

$$\dot{\underline{x}}(t) = \underline{f}[\underline{x}(t), t]$$

which is assumed to have a unique solution over the interval $t \in [t_0, t_f]$, and is subject to the multipoint boundary conditions

$$\sum_{j=1}^{n} C_j(t_i)x_j(t_i) = b_i \quad , \qquad i = 1, 2, \ldots, n$$

where $t_i \in [t_0, t_f]$, and b_i, C_1, C_2, ..., and C_n are specified ahead of time. Let $\underline{x}^*(t)$ denote a nominal value of $\underline{x}(t)$; then

$$\underline{f}(\underline{x}, t) \simeq \underline{f}(\underline{x}^*, t) + F_x(\underline{x}^*, t)(\underline{x} - \underline{x}^*)$$

where the Jacobian matrix $F_x(\underline{x}, t)$ is given by the expression

$$F_x(\underline{x}, t) = \begin{pmatrix} \partial f_1(\underline{x}, t)/\partial x_1 & \cdots & \partial f_1(\underline{x}, t)/\partial x_n \\ \vdots & \ddots & \vdots \\ \partial f_n(\underline{x}, t)/\partial x_1 & \cdots & \partial f_n(\underline{x}, t)/\partial x_n \end{pmatrix}$$

and $F_x(\underline{x}^*, t)$ is shorthand for "$F_x(\underline{x}, t)$ evaluated at $\underline{x} = \underline{x}^*$."

By quasilinearization, $\underline{f}(\underline{x}, t)$ is linearized about $\underline{x}^*(t) = \underline{x}_\ell(t)$, where $\underline{x}_\ell(t)$ is the ℓth approximate solution to the nonlinear differential equation, and $\underline{x}_{\ell+1}(t)$ is obtained as the solution to the following system:

$$\dot{\underline{x}}_{\ell+1}(t) = \underline{f}[\underline{x}_\ell(t), t] + F_x[\underline{x}_\ell(t), t][\underline{x}_{\ell+1}(t) - \underline{x}_\ell(t)]$$

and

$$\sum_{j=1}^{n} C_j(t_i)x_{\ell+1,j}(t_i) = b_i \quad , \qquad i = 1, 2, \ldots, n \quad .$$

The general solution of this linear, time-varying system is

$$\underline{x}_{\ell+1}(t) = \Phi_{\ell+1}(t, t_0)\underline{x}_{\ell+1}(t_0) + \underline{p}_{\ell+1}(t)$$

where $\Phi(t, t_0)$ is the fundamental solution matrix of

$$\dot{\Phi}_{\ell+1}(t, t_0) = F_x[\underline{x}_\ell(t), t]\Phi_{\ell+1}(t, t_0)$$

$$\Phi_{\ell+1}(t_0, t_0) = I \quad .$$

$\underline{p}(t)$, which is the particular solution for the linearized differential equation, satisfies the following differential equation:

$$\dot{\underline{p}}_{\ell+1}(t) = \underline{f}[\underline{x}_\ell(t), t] + F_x[\underline{x}_\ell(t), t][\underline{p}_{\ell+1}(t) - \underline{x}_\ell(t)]$$

$$\underline{p}_{\ell+1}(t_0) \triangleq \underline{0} \quad .$$

The initial condition vector $\underline{x}_{\ell+1}(t_0)$ is determined from the n boundary conditions

$$b_i = \sum_{j=1}^{n} C_j(t_i)x_{\ell+1,j}(t_i) = \underline{C}'(t_i)\underline{x}_{\ell+1}(t_i)$$

$$= \underline{C}'(t_i)[\Phi_{\ell+1}(t_i, t_0)\underline{x}_{\ell+1}(t_0) + \underline{p}_{\ell+1}(t_i)]$$

where $i = 1, 2, \ldots, n$. This is a collection of n algebraic equations, which can be solved for the n unknowns, $x_{\ell+1,1}(t_0)$, $x_{\ell+1,2}(t_0)$, and $x_{\ell+1,n}(t_0)$.

(a) we are given the following first-order system, including measurements of $x(t)$ at $t = t_i$, $i = 0, 1, \ldots, M$:

$$\dot{x}(t) = f[x(t), t] \quad , \quad x(0) \text{ unknown,} \quad \text{and} \quad 0 \leq t \leq t_f \quad .$$

Show that, by quasilinearization and superposition, $x(t)$ can be obtained sequentially as

$$x_{\ell+1}(t) = x_{\ell+1}(0)h_{\ell+1}(t) + p_{\ell+1}(t) \qquad (!)$$

where

$$\dot{h}_{\ell+1}(t) = f_x[x_\ell(t), t]h_{\ell+1}(t)$$

$$h_{\ell+1}(0) = 1$$

and

$$\dot{p}_{\ell+1}(t) = f[x_\ell(t), t] + f_x[x_\ell(t), t][p_{\ell+1}(t) - x_\ell(t)]$$

$$p_{\ell+1}(0) = 0 \quad .$$

(b) Explain how x(0) can be estimated from (!), if: (i) $x(t_i)$ is measured perfectly and h(t) and p(t) can be obtained by numerical integrations without errors; (ii) $x(t_i)$ is measured with error and h(t) and p(t) as in (i); and, (iii) $x(t_i)$ as in (ii) but now there are random errors associated with the intergrations required to obtain h(t) and p(t).

1-24. We are given the following first-order system, including measurements of x(t) at $t = t_i$, i = 0, 1, ..., M:

$$\dot{x}(t) = f[x(t), a] \quad ,$$

$$x(0) = x_0 \quad \text{where } x_0 \text{ is known,}$$

$0 \le t \le t_f$, and a is unknown.

(a) Show, by quasilinearization [linearize about $x_\ell(t)$ and a_ℓ] and superposition, that x(t) can be obtained sequentially, as

$$x_{\ell+1}(t) = v_{\ell+1}(t) + a_{\ell+1}q_{\ell+1}(t) \qquad (!)$$

where

$$\dot{v}_{\ell+1}(t) = f(x_\ell, a_\ell) + f_x(x_\ell, a_\ell)[v_{\ell+1}(t) - x_\ell]$$

$$- a_\ell f_a(x_\ell, a_\ell)$$

$$v_{\ell+1}(0) \triangleq x_0$$

and

$$\dot{q}_{\ell+1}(t) = f_x(x_\ell, a_\ell)q_{\ell+1}(t) + f_a(x_\ell, a_\ell)$$

$$q_{\ell+1}(0) = 0 \quad .$$

(b) Explain how a can be estimated from (!), if: (i) $x(t_i)$ is measured perfectly and v(t) and q(t) can be obtained by numerical integrations without errors; (ii) $x(t_i)$ is measured with error and v(t) and q(t) as in (i); and, (iii) $x(t_i)$ as in (ii) but now there are random errors associated with the integrations required to obtain v(t) and q(t).

1-25. Now let us combine the estimation problems studied in Problems 1-23 and 1-24. Given

$$\dot{x}(t) = f[x(t), a] \quad , \qquad (!)$$

$$x(0) = x_0 \quad \text{where } x_0 \text{ is unknown,}$$

$0 \leq t \leq t_f$, and a is unknown. In addition, measurements of x(t) at $t = t_i$, i = 0, 1, ..., M, are available. Explain how a and x_0 can be identified when (!) is solved by means of quasilinearization.

1-26. Using quasilinearization, explain how the parameters which appear in each of the following nonlinear differential equations can be identified:

(a) Equation for the unsteady operation of a synchronous motor -

$$\ddot{x}(t) + C\dot{x}(t) + p \sin x(t) = L(t)$$

(b) Duffing's equation (F and ω are known) -

$$\ddot{x}(t) + C\dot{x}(t) + \alpha x(t) + \beta x^3(t) = F \cos \omega t$$

(c) Van der Pol's equation -

$$\ddot{x}(t) - \varepsilon\dot{x}(t)\left[1 - \frac{1}{3}\dot{x}^2(t)\right] + x(t) = m(t)$$

(d) Hill's equation -

$$\ddot{x}(t) - ax(t) + bp(t)x(t) = m(t)$$

where p(t) is a known periodic function.

In each case, show the equation from which the parameters will be identified and indicate which signals must be measured and/or computed.

REFERENCES

1-1. J. S. Meditch, *Stochastic Optimal Linear Estimation and Control*, McGraw-Hill, New York, 1969.

1-2. M. R. Siegel, *Schaum's Outline of Theory and Problems of Statistics*, Schaum, New York, 1961.

1-3. R. J. Schwarz and B. Friedland, *Linear Systems,* McGraw-Hill, New York, 1965.

1-4. G. N. Saridis and G. Stein, A new algorithm for linear system identification, *IEEE Trans. Auto. Control,* AC-13, 592-594 (1968).

1-5. P. C. Young, Process parameter estimation and self adaptive control, *Proc. 1965 IFAC Conf.,* Teddington, England. Also in *Theory of Self Adaptive Control System* (P. H. Hammond, ed.), pp. 118-139, Plenum Press, New York, 1966.

1-6. P. M. Lion, Rapid identification of linear and nonlinear systems, *AIAA Journal,* 5, 1835-1842 (1967).

1-7. R. A. Rucker, Real time system identification in the presence of noise, *1963 WESCON,* Paper No. 2.3, San Francisco, August, 1963.

1-8. J. J. D'Azzo and C. H. Houpis, *Feedback Control System Analysis and Synthesis* (2nd Ed.), McGraw-Hill, New York, 1966.

1-9. V. W. Eveleigh, *Adaptive Control and Optimization Techniques,* McGraw-Hill, New York, 1967.

1-10. M. Athans and P. L. Falb, *Optimal Control: An Introduction to the Theory and Its Applications,* McGraw-Hill, New York, 1966.

1-11. G. R. Cooper, J. E. Gibson, V. W. Eveleigh, J. C. Lindenlaub, J. S. Meditch, and R. H. Raible, *A Survey of the Philosophy and State of the Art of Adaptive Systems,* Rept. No. PRF2358, Purdue Univ., Lafayette, Indiana, July 1960.

1-12. R. T. Stefani, *Design and Simulation of a High Performance, Digital, Adaptive, Normal Acceleration Control System Using Modern Parameter Estimation Techniques,* Rept. No. DAC-60637, Douglas Aircraft Co., Santa Monica, Cal., May 1967.

1-13. B. A. Trakhtenbrot, *Algorithms and Automatic Computing Machines,* D. C. Heath, Boston, 1963.

2

Least–Squares Parameter Estimation

2.1 INTRODUCTION

The technique we are about to study dates back to Karl Gauss, circa 1795, and is the cornerstone for most estimation theory, both classical and modern. Least-squares estimation was invented by Gauss at a time when he was interested in predicting the motions of planets and comets using telescopic measurements. The motions of these bodies can be completely characterized by six parameters. The estimation problem that Gauss considered was one of inferring the values of these parameters from the measurement data.

We shall study least-squares estimation from two points of view: the classical batch-processing approach, and the modern sequential-processing approach. The sequential approach has been motivated by today's high-speed digital computers. Sequential least-squares estimation algorithms are computationally simpler to implement using digital computers than the classical batch algorithm; however, as we shall see, the sequential algorithms are outgrowths of the classical batch algorithm.

2.2 FORMULATION AND STATEMENT OF THE ESTIMATION PROBLEM

Suppose that L + 1 scalar measurements† of a signal $y(t)$ (e. g., attitude of a spacecraft, voltage in an electrical power system, temperature at the tip of a machine tool bit) are made at times

$$t_{k-L}, t_{k-L+1}, \ldots, t_{k-1}, t_k \quad ,$$

where

$$t_{k-L} < t_{k-L+1} < \cdots < t_{k-1} < t_k \quad , \tag{2.2-1}$$

$t_{k-L}, \ldots, t_{k-1}$, and t_k do not have to be uniformly spaced, and

$$k \geq L \quad , \tag{2.2-2}$$

which means, of course, that the first possible measurement of $y(t)$ is made at t_0.

In this notation, L controls the total number of measurements we have available (or care to use), and k controls the times at which these measurements can occur. For example, if $L = 2$ and $k = 2$, then we have three measurements made at t_0, t_1, and t_2. On the other hand, if $L = 2$ but $k = 5$, we have three measurements made at t_3, t_4, and t_5.

It is often convenient to view t_k as present time, values of $t < t_k$ as past time, and values of $t > t_k$ as future time. In this case, the measured value of y at t_k is viewed as the present measurement of y, whereas those measured values of y at $t_{k-1}, \ldots, t_{k-L+1}$, and t_{k-L} are viewed as past measurements of y. We shall adopt this viewpoint throughout the rest of the book.

Measurements of $y(t)$ are denoted $z(t)$, and

$$z(t) = y(t) + v(t) \tag{2.2-3}$$

where $v(t)$ represents a random measurement-error process. We may or may not know much about $v(t)$. For the time being, let us assume only that

$$E\{v(t)\} = 0 \quad \text{for all } t \quad . \tag{2.2-4}$$

† Vector measurements are discussed in Section 2.9.

The signal y(t) is assumed to be a linear combination of n parameters θ_1, θ_2, ..., and θ_n; that is to say,

$$y(t) = h_1(t)\theta_1 + h_2(t)\theta_2 + \cdots + h_n(t)\theta_n \quad . \tag{2.2-5}$$

Letting

$$\underline{h}(t) = [h_1(t), h_2(t), \ldots, h_n(t)]' \tag{2.2-6}$$

and

$$\underline{\theta} = (\theta_1, \theta_2, \ldots, \theta_n)' \quad , \tag{2.2-7}$$

y(t) can be written more compactly as

$$y(t) = \underline{h}'(t)\underline{\theta} \tag{2.2-8}$$

and, therefore,

$$z(t) = \underline{h}'(t)\underline{\theta} + v(t) \quad . \tag{2.2-9}$$

Let us now collect our L + 1 measurements $z(t_{k-L})$, $z(t_{k-L+1})$, ..., $z(t_{k-1})$, and $z(t_k)$:

$$\begin{aligned} z(t_k) &= \underline{h}'(t_k)\underline{\theta} + v(t_k) \\ z(t_{k-1}) &= \underline{h}'(t_{k-1})\underline{\theta} + v(t_{k-1}) \\ &\vdots \\ z(t_{k-L}) &= \underline{h}'(t_{k-L})\underline{\theta} + v(t_{k-L}) \end{aligned} \quad . \tag{2.2-10}$$

For notational convenience, we shall replace time t_k by k:

$$\begin{aligned} z(k) &= \underline{h}'(k)\underline{\theta} + v(k) \\ z(k-1) &= \underline{h}'(k-1)\underline{\theta} + v(k-1) \\ &\vdots \\ z(k-L) &= \underline{h}'(k-L)\underline{\theta} + v(k-L) \end{aligned} \quad . \tag{2.2-11}$$

Recognizing that $\underline{\theta}$ is common to each measurement we form the following *concatenation* of the L + 1 measurement equations:

$$\begin{pmatrix} z(k) \\ z(k-1) \\ \vdots \\ z(k-L) \end{pmatrix} = \begin{pmatrix} \underline{h}'(k) \\ \underline{h}'(k-1) \\ \vdots \\ \underline{h}'(k-L) \end{pmatrix} \underline{\theta} + \begin{pmatrix} v(k) \\ v(k-1) \\ \vdots \\ v(k-L) \end{pmatrix} . \quad (2.2\text{-}12)$$

Letting

$$\underline{Z}(k) = [z(k),\ z(k-1),\ \ldots,\ z(k-L)]' \quad , \quad (2.2\text{-}13)$$

$$\underline{V}(k) = [v(k),\ v(k-1),\ \ldots,\ v(k-L)]' \quad , \quad (2.2\text{-}14)$$

and

$$H(k) = \begin{pmatrix} \underline{h}'(k) \\ \underline{h}'(k-1) \\ \vdots \\ \underline{h}'(k-L) \end{pmatrix} \quad , \quad (2.2\text{-}15)$$

the *concatenated measurement equation*, Eq. (2.2-12), becomes

$$\underline{Z}(k) = H(k)\underline{\theta} + \underline{V}(k) \quad (2.2\text{-}16)$$

for $k \geq L$. We will refer to $\underline{Z}(k)$, $\underline{V}(k)$, and $H(k)$ as the *concatenated measurement vector*, *concatenated measurement-error vector*, and *concatenated observation matrix*, respectively. $\underline{Z}(k)$ is an $(L + 1) \times 1$ vector, $\underline{V}(k)$ is an $(L + 1) \times 1$ vector, and $H(k)$ is an $(L + 1) \times n$ matrix. Equation (2.2-16) is the starting point for all the analyses both in this chapter and the next.

In order to illustrate the concatenation process in more detail, let us consider a number of examples.

Example 1. Consider a signal y which is known to be related to another parameter ζ by an $(n - 1)$-th degree polynomial in ζ; that is to say,

$$y(\zeta) = \theta_1 + \theta_2\zeta + \theta_3\zeta^2 + \cdots + \theta_n\zeta^{n-1} \quad (2.2\text{-}17)$$

where $\theta_1, \theta_2, \ldots, \theta_n$ are unknown coefficients. $y(\zeta)$ is measured for $L + 1$ different values of ζ, say $\zeta_0, \zeta_1, \ldots, \zeta_L$.

For $\zeta = \zeta_i$, the measurement $z(\zeta_i)$ is given by

$$z(\zeta_i) = \theta_1 + \theta_2\zeta_i + \theta_3\zeta_i^2 + \cdots + \theta_n\zeta_i^{n-1} + v(\zeta_i) \quad . \tag{2.2-18}$$

It is easy to see that in this case, for k = L,

$$H(L) = \begin{pmatrix} 1 & \zeta_L & \zeta_L^2 & \cdots & \zeta_L^{n-1} \\ \vdots & \vdots & \vdots & \ddots & \vdots \\ 1 & \zeta_1 & \zeta_1^2 & \cdots & \zeta_1^{n-1} \\ 1 & \zeta_0 & \zeta_0^2 & \cdots & \zeta_0^{n-1} \end{pmatrix} . \tag{2.2-19}$$

The ideas of this example extend to much more general functions of ζ_i; thus, if

$$y(\zeta) = \theta_1 + \theta_2 \sin \zeta + \theta_3 e^{-\zeta} \tag{2.2-20}$$

then

$$H(L) = \begin{pmatrix} 1 & \sin \zeta_L & e^{-\zeta_L} \\ \vdots & \vdots & \vdots \\ 1 & \sin \zeta_1 & e^{-\zeta_1} \\ 1 & \sin \zeta_0 & e^{-\zeta_0} \end{pmatrix} . \tag{2.2-21}$$

▲

Example 2. Consider identifying the n × 1 initial condition vector $\underline{x}(0)$ of the linear time-varying discrete system

$$\underline{x}(k+1) = \Phi(k+1, k)\underline{x}(k) \tag{2.2-22}$$

from L + 1 measurements $z(k - L)$, $z(k - L + 1)$, ..., $z(k)$, where

$$z(k) = \underline{h}'(k)\underline{x}(k) + v(k) \quad . \tag{2.2-23}$$

From Eq. (1.3-43), we know that

$$\underline{x}(k) = \Phi(k, 0)\underline{x}(0) \quad , \tag{2.2-24}$$

which means that the concatenated measurement equation is

$$\begin{pmatrix} z(k) \\ z(k-1) \\ \vdots \\ z(k-L) \end{pmatrix} = \begin{pmatrix} \underline{h}'(k)\Phi(k, 0) \\ \underline{h}'(k-1)\Phi(k-1, 0) \\ \vdots \\ \underline{h}'(k-L)\Phi(k-L, 0) \end{pmatrix} \underline{x}(0) + \begin{pmatrix} v(k) \\ v(k-1) \\ \vdots \\ v(k-L) \end{pmatrix} . \tag{2.2-25}$$

In this example, $\underline{x}(0)$ plays the role of the unknown parameter, vector $\underline{\theta}$. ▲

Example 3. We wish to estimate the weights $f_1, f_2, \ldots, f_N$ in the superposition summation

$$g(k) = \sum_{i=1}^{N} f_i w(k - i) \tag{2.2-26}$$

that relates the input sequence $w(k - 1), w(k - 2), \ldots, w(k - N)$ to the output $g(k)$ of a linear system (see Section 1.3.2). Measurements of $g(k)$ are made at $t_{k-L}, t_{k-L+1}, \ldots, t_k$; that is to say,

$$z(t_j) = g(t_j) + v(t_j) \tag{2.2-27}$$

for $j = k - L, k - L + 1, \ldots, k$. Concatenating these $L + 1$ measurements, we find that

$$\begin{pmatrix} z(k) \\ z(k-1) \\ \vdots \\ z(k-L) \end{pmatrix} = \begin{pmatrix} w(k-1) & w(k-2) & \cdots & w(k-N) \\ w(k-2) & w(k-3) & \cdots & w(k-N-1) \\ \vdots & \vdots & \ddots & \vdots \\ w(k-L-1) & w(k-L-2) & \cdots & w(k-N-L) \end{pmatrix}$$

$$\times \begin{pmatrix} f_1 \\ f_2 \\ \vdots \\ f_N \end{pmatrix} + \begin{pmatrix} v(k) \\ v(k-1) \\ \vdots \\ v(k-L) \end{pmatrix} . \tag{2.2-28}$$

▲

At this point, some discussion on the parameter L is in order. Equation (2.2-16) is a system of $L + 1$ equations in n unknowns. *If*

$(L + 1) < n$, we have fewer equations than unknowns. Such an underdetermined system of equations does not lead to unique or very meaningful values for $\theta_1, \theta_2, \ldots, \theta_n$. *If* $(L + 1) = n$, we have exactly as many equations as unknowns and, *if measurement errors are not present*, we could solve for $\underline{\theta}$ exactly, once and for all, as

$$\underline{\theta} = \mathcal{H}^{-1}(k)\underline{Z}(k) \quad . \tag{2.2-29}$$

$\mathcal{H}^{-1}(k)$ will exist as long as the n measurements are linearly independent. This case is of marginal interest to us, since it leaves us no freedom to attempt to do something about measurement errors. For, if measurement errors are not exactly zero, then

$$\underline{\theta} = \mathcal{H}^{-1}(k)\underline{Z}(k) - \mathcal{H}^{-1}(k)\underline{V}(k) \tag{2.2-30}$$

and it is clear that the characteristics of the random measurement-error process $v(t)$ will greatly affect $\underline{\theta}$, which is not desirable.

If $(L + 1) > n$, we have more equations than unknowns, and within this overdetermined structure, we will, as Morrison states (2-2), "be free to maneuver," in an attempt to offset the measurement errors. This is the case of real interest to us.

We denote the estimate of $\underline{\theta}$ based on the $L + 1$ data samples, $z(k), z(k - 1), \ldots, z(k - L)$ - the components of $\underline{Z}(k)$ - as $\hat{\underline{\theta}}(k)$. If we knew $\hat{\underline{\theta}}(k)$ was best (optimum) in some sense, then the best estimate of $\underline{Z}(k)$, in Eq. (2.2-16), denoted $\hat{\underline{Z}}(k)$, would be

$$\hat{\underline{Z}}(k) = \mathcal{H}(k)\hat{\underline{\theta}}(k) \tag{2.2-31}$$

for $k \geq L$. This estimate discounts the measurement-error process, since it has been assumed to be of zero mean. The symbol $\tilde{\underline{\theta}}(k)$ denotes the error between $\hat{\underline{\theta}}(k)$ and $\underline{\theta}$, and is given by

$$\tilde{\underline{\theta}}(k) = \underline{\theta} - \hat{\underline{\theta}}(k) \quad . \tag{2.2-32}$$

Letting $\tilde{\underline{Z}}(k)$ denote the error between $\hat{\underline{Z}}(k)$ and $\underline{Z}(k)$, that is to say $\tilde{\underline{Z}}(k) = \underline{Z}(k) - \hat{\underline{Z}}(k)$, it follows from Eq. (2.2-16), (2.2-31), and (2.2-32) that

$$\tilde{\underline{Z}}(k) = \mathcal{H}(k)\tilde{\underline{\theta}}(k) + \underline{V}(k) \quad . \tag{2.2-33}$$

Important observation: $\tilde{\underline{Z}}(k)$ is a vector *equation error*, of exactly the type that we discussed in Chapter 1, Section 1.2.4. $\underline{Z}(k)$ in Eq. (2.2-16) represents the measured output of the identification system, and $\hat{\underline{Z}}(k)$ represents the output of the model of the identification system. We depict these relationships in Fig. 2.2-1. ▲

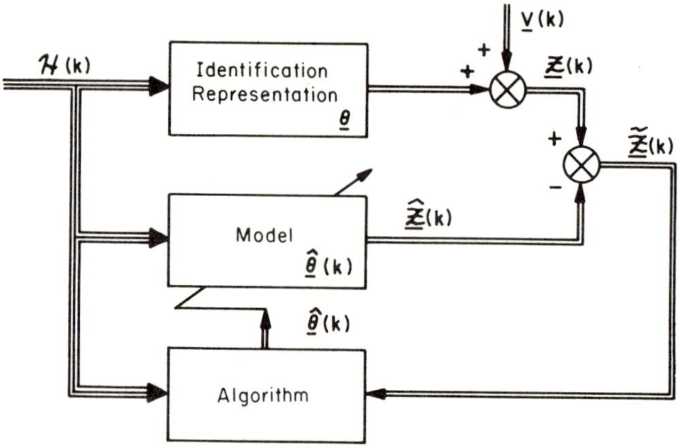

Fig. 2.2-1. Interpretation of concatenated measurement equation and estimation equation as a vector equation error identification system.

We now direct our attention to a method for obtaining $\hat{\underline{\theta}}(k)$ that is based on minimizing a specific measure of the vector equation error $\tilde{\underline{Z}}(k)$. $\hat{\underline{\theta}}(k)$ will be chosen so as to minimize a weighted sum of the squares of the components of $\tilde{\underline{Z}}(k)$.

Let

$$J[\hat{\underline{\theta}}(k)] = w(k)\tilde{z}^2(k) + w(k-1)\tilde{z}^2(k-1) + \cdots + w(k-L)\tilde{z}^2(k-L) \quad , \qquad (2.2\text{-}34)$$

which can also be written as

$$J[\hat{\underline{\theta}}(k)] = \tilde{\underline{Z}}'(k)W(k)\tilde{\underline{Z}}(k) \quad , \qquad (2.2\text{-}35)$$

where

$$W(k) = \begin{pmatrix} w(k) & 0 & \cdots & 0 \\ 0 & w(k-1) & \cdots & 0 \\ \vdots & \vdots & \ddots & \vdots \\ 0 & 0 & \cdots & w(k-L) \end{pmatrix} . \quad (2.2\text{-}36)$$

We shall choose $\hat{\underline{\theta}}(k)$ so that $J[\hat{\underline{\theta}}(k)]$ is minimized. This technique for choosing $\hat{\underline{\theta}}(k)$ is known as *generalized least squares*, or *weighted least squares*. When each component of $\tilde{\underline{Z}}(k)$ is weighted by the same amount, in which case

$$W(k) = wI \quad (2.2\text{-}37)$$

where I is the $(L + 1) \times (L + 1)$ identity matrix, we say that we are choosing $\hat{\underline{\theta}}(k)$ by the method of *least squares*. Obviously, weighted least squares is more general than least squares, as the former includes the latter as a special case; hence, our derivations that follow will, for the most part, be for the weighted least-squares technique.

We shall assume that W(k) is symmetric and positive definite. Recall from matrix theory that a *matrix* W *is said to be positive definite* if

$$\underline{a}'W\underline{a} > 0 \quad (2.2\text{-}38)$$

for $\underline{a} \neq \underline{0}$. We shall see, in Section 2.3, that in order to compute the weighted least-squares estimate of $\underline{\theta}$, we shall have to compute $[\mathcal{H}'(k)W(k)\mathcal{H}(k)]^{-1}$. This inverse will always exist when W(k) is positive definite and $\mathcal{H}(k)$ is of maximum rank (*1-2*).

Besides these two restrictions on W(k), how should one choose its elements? There is no general rule to cover all situations. By suitably choosing the elements of W(k) we can weight earlier measurements of y(t) more heavily than later measurements, or vice versa. A weighting matrix that can be used in both of these situations is

$$W(k) = \begin{pmatrix} \mu^{k-L} & 0 & \cdots & 0 \\ 0 & \mu^{k-L+1} & \cdots & 0 \\ \vdots & \vdots & \ddots & \vdots \\ 0 & 0 & \cdots & \mu^{k} \end{pmatrix} \qquad (2.2\text{-}39)$$

for $k \geq L$. When $0 < \mu < 1$, we shall be weighting recent measurements more heavily than past ones. For example, if $k = L$, then

$$J[\hat{\underline{\theta}}(L)] = \tilde{z}^2(L) + \mu\tilde{z}^2(L - 1) + \cdots + \mu^L\tilde{z}^2(0) \qquad (2.2\text{-}40)$$

and it is clear that we are discounting earlier measurements in favor of later ones. It is also clear that when $\mu > 1$ we shall be weighting earlier measurements more heavily than later ones.

2.3 SOLUTION FOR THE GENERALIZED LEAST-SQUARES ESTIMATE

We shall obtain the estimate $\hat{\underline{\theta}}(k)$ that minimizes $J[\hat{\underline{\theta}}(k)]$ in Eq. (2.2-35) by using vector calculus. Before proceeding to the formal minimization of $J[\hat{\underline{\theta}}(k)]$ with respect to $\hat{\underline{\theta}}(k)$, we pause to derive two very useful vector derivatives.

2.3.1 Two Useful Vector Derivatives

Here we shall derive the vector derivatives of two scalars. *By definition, a vector derivative of a scalar will be a vector* (a column matrix).

Let $\underline{m}$ and $\underline{b}$ be two $n \times 1$ nonzero vectors, and A be an $n \times n$ symmetric matrix.

First, let us show that

$$\frac{d}{d\underline{m}} (\underline{b}'\underline{m}) = \underline{b} \quad . \qquad (2.3\text{-}1)$$

Proof. Letting $\underline{b} = (b_1, b_2, \ldots, b_n)'$ and $\underline{m} = (m_1, m_2, \ldots, m_n)'$, it is clear that

$$\underline{b}'\underline{m} = \sum_{i=1}^{n} b_i m_i \quad ; \qquad (2.3\text{-}2)$$

hence,

$$\frac{d}{d\underline{m}}(\underline{b}'\underline{m}) = \begin{pmatrix} \frac{\partial}{\partial m_1}\sum b_i m_i \\ \frac{\partial}{\partial m_2}\sum b_i m_i \\ \vdots \\ \frac{\partial}{\partial m_n}\sum b_i m_i \end{pmatrix} = \begin{pmatrix} b_1 \\ b_2 \\ \vdots \\ b_n \end{pmatrix} = \underline{b} \quad . \tag{2.3-3}$$

▲

Next, let us show that

$$\frac{d}{d\underline{m}}(\underline{m}'A\underline{m}) = 2A\underline{m} \quad . \tag{2.3-4}$$

Proof. Letting the ijth component of A be denoted as a_{ij}, it is clear that the *quadratic form* $\underline{m}'A\underline{m}$ can also be written as

$$\underline{m}'A\underline{m} = \sum_{i=1}^{n}\sum_{j=1}^{n} m_i a_{ij} m_j \quad ; \tag{2.3-5}$$

hence,

$$\frac{d}{d\underline{m}}(\underline{m}'A\underline{m}) = \begin{pmatrix} \frac{\partial}{\partial m_1}\sum\sum m_i a_{ij} m_j \\ \frac{\partial}{\partial m_2}\sum\sum m_i a_{ij} m_j \\ \vdots \\ \frac{\partial}{\partial m_n}\sum\sum m_i a_{ij} m_j \end{pmatrix} = \begin{pmatrix} \sum_{j=1}^{n} a_{1j}m_j + \sum_{i=1}^{n} a_{i1}m_i \\ \sum_{j=1}^{n} a_{2j}m_j + \sum_{i=1}^{n} a_{i2}m_i \\ \vdots \\ \sum_{j=1}^{n} a_{nj}m_j + \sum_{i=1}^{n} a_{in}m_i \end{pmatrix} . \tag{2.3-6}$$

Using the fact that A is symmetric, which means that

$$a_{ij} = a_{ji} \qquad \text{for all} \quad i \neq j \quad , \tag{2.3-7}$$

it is easy to see that

$$\frac{d}{d\underline{m}}(\underline{m}'A\underline{m}) = 2\begin{pmatrix} \sum a_{1j}m_j \\ \sum a_{2j}m_j \\ \vdots \\ \sum a_{nj}m_j \end{pmatrix} = 2A\underline{m} \quad . \tag{2.3-8}$$

▲

2.3.2 Obtaining $\hat{\underline{\theta}}(k)$

To begin, we shall express the right-hand side of Eq. (2.2-35) as a function of $\hat{\underline{\theta}}(k)$. This is easily accomplished as follows:

$$\begin{aligned} J[\hat{\underline{\theta}}(k)] &= \tilde{\underline{Z}}'(k)W(k)\tilde{\underline{Z}}(k) \\ &= [\underline{Z}(k) - \hat{\underline{Z}}(k)]'W(k)[\underline{Z}(k) - \hat{\underline{Z}}(k)] \\ &= [\underline{Z}(k) - H(k)\hat{\underline{\theta}}(k)]'W(k)[\underline{Z}(k) - H(k)\hat{\underline{\theta}}(k)] \quad . \end{aligned} \tag{2.3-9}$$

The last relationship was obtained with the help of Eq. (2.2-31). Next we expand this quadratic form to give

$$\begin{aligned} J[\hat{\underline{\theta}}(k)] = {} & \underline{Z}'(k)W(k)\underline{Z}(k) - 2\underline{Z}'(k)W(k)H(k)\hat{\underline{\theta}}(k) \\ & + \hat{\underline{\theta}}'(k)H'(k)W(k)H(k)\hat{\underline{\theta}}(k) \quad . \end{aligned} \tag{2.3-10}$$

Finally, we take the vector derivative of $J[\hat{\underline{\theta}}(k)]$ with respect to $\hat{\underline{\theta}}(k)$, and set the result equal to zero. Using the vector derivative formulas in Eqs. (2.3-1) and (2.3-4), it is clear that

$$\begin{aligned} \frac{d}{d\hat{\underline{\theta}}(k)} J[\hat{\underline{\theta}}(k)] = {} & -2[\underline{Z}'(k)W(k)H(k)]' \\ & + 2H'(k)W(k)H(k)\hat{\underline{\theta}}(k) = \underline{0} \ . \end{aligned} \tag{2.3-11}$$

Hence,

$$\hat{\underline{\theta}}(k) = [H'(k)W(k)H(k)]^{-1}H'(k)W(k)\underline{Z}(k) \tag{2.3-12}$$

for $k \geq L$. That $\hat{\underline{\theta}}(k)$ in Eq. (2.3-12) does indeed minimize (and not maximize) $J[\hat{\underline{\theta}}(k)]$ is clear from the fact that

$$\frac{d^2}{d\hat{\underline{\theta}}^2(k)} J[\hat{\underline{\theta}}(k)] = 2H'(k)W(k)H(k) > 0 \quad , \tag{2.3-13}$$

which is the vector calculus analog to the scalar calculus requirement that $\hat{\theta}$ minimizes $J(\hat{\theta})$ if $dJ(\hat{\theta})/d\hat{\theta} = 0$ and $d^2J(\hat{\theta})/d\hat{\theta}^2$ is positive. Equation (2.3-13) means that $\mathcal{H}'(k)W(k)\mathcal{H}(k)$ must be positive definite. This is guaranteed by our earlier requirement that W(k) be positive definite, assuming that $\mathcal{H}(k)$ is of maximum rank *(2-1)*.

$\hat{\underline{\theta}}(k)$ in Eq. (2.3-12) is the *generalized least-squares* estimate of $\underline{\theta}$. When W(k) = wI, we have the *least squares* estimate of $\underline{\theta}$, given by the expression

$$\hat{\underline{\theta}}(k) = [\mathcal{H}'(k)\mathcal{H}(k)]^{-1}\mathcal{H}'(k)\underline{Z}(k) \tag{2.3-14}$$

for $k \geq L$. We summarize these results in the following:

Theorem 2-0. *The* generalized least-squares *estimate of* $\underline{\theta}$ *is given in Eq. (2.3-12), and, the* least-squares *estimate of* $\underline{\theta}$ *is given in Eq. (2.3-14).*

Equation (2.3-11), which we rewrite here for convenience as

$$[\mathcal{H}'(k)W(k)\mathcal{H}(k)]\hat{\underline{\theta}}(k) = \mathcal{H}'(k)W(k)\underline{Z}(k) \quad , \tag{2.3-15}$$

is referred to by statisticians as the *normal equation*. It is a system of n linear equations in the n unknowns $\hat{\theta}_1(k)$, $\hat{\theta}_2(k)$, ..., $\hat{\theta}_n(k)$. The difficult part about computing $\hat{\underline{\theta}}(k)$ by means of either Eq. (2.3-12) or (2.3-14) is computing the inverse of $\mathcal{H}'(k)W(k)\mathcal{H}(k)$ or $\mathcal{H}'(k)\mathcal{H}(k)$. These matrices are both of dimension $n \times n$. The more unknown parameters there are, the larger n becomes, and the more computing time is required to evaluate these inverses.

One way to circumvent this difficulty is to solve for $\hat{\underline{\theta}}(k)$ directly from the normal equation, using a technique such as Gaussian elimination. Graybill *(2-1)* has an entire chapter devoted to various ways of solving the normal equation. We shall not dwell on solutions of the normal equation in this book because, as we shall see in Section 2.5, a very useful measure of parameter identification error, which is usually needed to provide us with some measure of confidence in our estimate, depends explicitly on $[\mathcal{H}'(k)W(k)\mathcal{H}(k)]^{-1}$; hence, this inverse must often be computed, regardless of how $\hat{\underline{\theta}}(k)$ is computed.

2.4 APPLICATIONS

2.4.1 Curve Fitting by Arbitrary Functions

For an explanation of the curve fitting problem, the reader is referred to Section 1.3.1. Briefly, we wish to fit a given set of data $z(\zeta_1)$, $z(\zeta_2)$, ..., $z(\zeta_q)$ by the approximating function $\hat{z}(\zeta)$, where

$$\hat{z}(\zeta) = \sum_{j=1}^{N} \hat{\theta}_j \phi_j(\zeta) \quad . \tag{2.4-1}$$

Following the concatenation procedure displayed in Example 1, which was for the special case when $\phi_1 = 1$, $\phi_2 = \zeta$, $\phi_3 = \zeta^2$, ..., $\phi_N = \zeta^{N-1}$, it is clear that

$$\underline{Z}(q) = [z(\zeta_q), z(\zeta_{q-1}), \ldots, z(\zeta_1)]' \tag{2.4-2}$$

and

$$\mathcal{H}(q) = \begin{pmatrix} \phi_1(\zeta_q) & \phi_2(\zeta_q) & \cdots & \phi_N(\zeta_q) \\ \phi_1(\zeta_{q-1}) & \phi_2(\zeta_{q-1}) & \cdots & \phi_N(\zeta_{q-1}) \\ \vdots & \vdots & \ddots & \vdots \\ \phi_1(\zeta_1) & \phi_2(\zeta_1) & \cdots & \phi_N(\zeta_1) \end{pmatrix} \quad . \tag{2.4-3}$$

Hence, the *least-squares estimate* of θ_j ($j = 1, 2, \ldots, N$) based on q data $\hat{\theta}_j(q)$ is given by Eq. (2.4-4). This estimate is complicated by the inversion of a matrix whose elements require a significant number of computations. There are several ways to simplify these results. Let us explore some of these next.

2.4.2 Curve Fitting by Discrete Orthonormal Polynomials

The first way in which we can simplify the estimate in Eq. (2.4-4) is to choose a clever set of basis functions $\phi_1(\zeta)$, $\phi_2(\zeta)$, ..., $\phi_N(\zeta)$. The discrete orthonormal polynomials constitute such a choice. *These polynomials are orthonormal over a discrete set of equidistant integer values for* ζ; that is to say,

$$\hat{\underline{\theta}}(q) = \begin{pmatrix} \sum_{j=0}^{q-1} \phi_1^2(q - j) & \sum_{j=0}^{q-1} \phi_1(q - j)\phi_2(q - j) & \cdots & \sum_{j=0}^{q-1} \phi_1(q - j)\phi_N(q - j) \\ \sum_{j=0}^{q-1} \phi_1(q - j)\phi_2(q - j) & \sum_{j=0}^{q-1} \phi_2^2(q - j) & \cdots & \sum_{j=0}^{q-1} \phi_2(q - j)\phi_N(q - j) \\ \vdots & \vdots & \ddots & \vdots \\ \sum_{j=0}^{q-1} \phi_1(q - j)\phi_N(q - j) & \sum_{j=0}^{q-1} \phi_2(q - j)\phi_N(q - j) & \cdots & \sum_{j=0}^{q-1} \phi_N^2(q - j) \end{pmatrix}^{-1}$$

$$\times \begin{pmatrix} \sum_{j=0}^{q-1} \phi_1(q - j)z(q - j) \\ \sum_{j=0}^{q-1} \phi_2(q - j)z(q - j) \\ \vdots \\ \sum_{j=0}^{q-1} \phi_N(q - j)z(q - j) \end{pmatrix} \qquad (2.4\text{-}4)^{\dagger}$$

† For notational convenience, and in keeping with earlier examples, we write $\phi_i(q - j)$ in place of $\phi_i(\zeta_{q-j})$.

$$\sum_{j=0}^{N} \phi_i(\zeta = j)\phi_\ell(\zeta = j)\mu(j) = 0 \tag{2.4-5}$$

for all $i \neq \ell$, and

$$\sum_{j=0}^{N} \phi_i^2(\zeta = j)\mu(j) = 1 \tag{2.4-6}$$

where N and the weighting function $\mu(j)$ must be chosen ahead of time. It is important to observe that, while Eqs. (2.4-5) and (2.4-6) involve the argument ζ only at the integral values 0, 1, 2, ..., N, the polynomials $\phi(\zeta)$ are nevertheless continuous functions of ζ (*2-2*). Discrete Chebyshev, Krawtchouk, Charlier, Meixner, and Hahn polynomials are given in Abromowitz and Stegun (*2-3*). Discrete Legendre and Laguerre polynomials are derived in Morrison (*2-2*). These polynomials are summarized in Tables 2-1 and 2-2, respectively. For the *discrete Legendre polynomials*, $\mu(j) = 1$ for all admissible values of j; whereas, for the *discrete Laguerre polynomials*, $N = \infty$ and $\mu(j) = \mu^j$, where $|\mu| < 1$, for all admissible values of j (j = 0, 1, ...).

Now, let us investigate the effects of choosing discrete Legendre polynomials as the basis functions in Eq. (2.4-4). It is clear from Eqs. (2.4-5) and (2.4-6) that the first matrix on the right-hand side of Eq. (2.4-4) reduces to the identity matrix; hence

$$\hat{\theta}_i(q) = \sum_{j=0}^{q-1} \phi_i(q - j)z(q - j) \tag{2.4-7}$$

for i = 1, 2, ..., N. The computational advantages for using discrete orthonormal basis functions is clear. In this example, the tedious matrix inversion in Eq. (2.4-4) completely disappears when such functions are used.

2.4.3 Curve Fitting by Simple Functions

The simplest approximating function to a set of data is the *straight line*. In this case, great simplifications of Eq. (2.4-4) occur in two ways: (1) N, the number of unknown parameters in

TABLE 2-1

Discrete Legendre Polynomials[a]

$$\phi_i(\zeta) = \frac{p_i(\zeta)}{C_i(N)} \quad \text{where} \quad p_i(\zeta) = \sum_{\nu=0}^{i} (-1)^\nu \binom{i}{\nu}\binom{i+\nu}{\nu}\frac{\zeta^{(\nu)}}{N^{(\nu)}} \quad \text{and} \quad C_i^2(N) = \frac{(N+i+1)^{(i+1)}}{(2i+1)N^{(i)}}$$

i	$p_i(\zeta)$	$C_i^2(N)$
0	1	$N + 1$
1	$1 - 2\frac{\zeta}{N}$	$\frac{(N+2)(N+1)}{3N}$
2	$1 - 6\frac{\zeta}{N} + 6\frac{\zeta(\zeta-1)}{N(N-1)}$	$\frac{(N+3)(N+2)(N+1)}{5N(N-1)}$
3	$1 - 12\frac{\zeta}{N} + 30\frac{\zeta(\zeta-1)}{N(N-1)} - 20\frac{\zeta(\zeta-1)(\zeta-2)}{N(N-1)(N-2)}$	$\frac{(N+4)(N+3)(N+2)(N+1)}{7N(N-1)(N-2)}$
4	$1 - 20\frac{\zeta}{N} + 90\frac{\zeta(\zeta-1)}{N(N-1)} - 140\frac{\zeta(\zeta-1)(\zeta-2)}{N(N-1)(N-2)}$ $+ 70\frac{\zeta(\zeta-1)(\zeta-2)(\zeta-3)}{N(N-1)(N-2)(N-3)}$	$\frac{(N+5)(N+4)(N+3)(N+2)(N+1)}{9N(N-1)(N-2)(N-3)}$

[a] $\binom{i}{\nu} = \frac{i!}{\nu!(i-\nu)!}$ and $\zeta^{(\nu)} = \zeta(\zeta-1)\ldots(\zeta-\nu+1)$.

TABLE 2-2

Discrete Laguerre Polynomials

$$\phi_i(\zeta) = \frac{p_i(\zeta)}{C_i(\mu)} \quad \text{where} \quad p_i(\zeta) = \mu^i \sum_{\nu=0}^{i} (-1)^\nu \binom{i}{\nu}\left(\frac{1-\mu}{\mu}\right)^\nu \binom{\zeta}{\nu}, \quad C_i(\mu) = \frac{\mu^i}{1-\mu}, \quad \text{and} \quad |\mu| < 1$$

i	$p_i(\zeta)$	$C_i^2(\mu)$
0	1	$\frac{1}{1-\mu}$
1	$\mu\left[1 - \left(\frac{1-\mu}{\mu}\right)\zeta\right]$	$\frac{\mu}{1-\mu}$
2	$\mu^2\left[1 - 2\left(\frac{1-\mu}{\mu}\right)\zeta + \left(\frac{1-\mu}{\mu}\right)^2 \frac{\zeta(\zeta-1)}{2!}\right]$	$\frac{\mu^2}{1-\mu}$
3	$\mu^3\left[1 - 3\left(\frac{1-\mu}{\mu}\right)\zeta + 3\left(\frac{1-\mu}{\mu}\right)^2 \frac{\zeta(\zeta-1)}{2!} - \left(\frac{1-\mu}{\mu}\right)^3 \frac{\zeta(\zeta-1)(\zeta-2)}{3!}\right]$	$\frac{\mu^3}{1-\mu}$
4	$\mu^4\left[1 - 4\left(\frac{1-\mu}{\mu}\right)\zeta + 6\left(\frac{1-\mu}{\mu}\right)^2 \frac{\zeta(\zeta-1)}{2!} - 4\left(\frac{1-\mu}{\mu}\right)^3 \frac{\zeta(\zeta-1)(\zeta-2)}{3!} + \left(\frac{1-\mu}{\mu}\right)^4 \frac{\zeta(\zeta-1)(\zeta-2)(\zeta-3)}{4!}\right]$	$\frac{\mu^4}{1-\mu}$

Eq. (2.4-1), is set equal to 2; and (2) $\phi_1(\zeta) = 1$, whereas $\phi_2(\zeta) = \zeta$. Statisticians refer to the equation

$$\hat{z}(\zeta) = \hat{\theta}_1 + \hat{\theta}_2\zeta \tag{2.4-8}$$

as the *least-squares line* or *regression line* (2-4). Its properties have been studied extensively by them.

Let us proceed to evaluate $\hat{\underline{\theta}}(q)$ for the regression line. Substituting for $\phi_1(\zeta)$ and $\phi_2(\zeta)$ in Eq. (2.4-4), we find

$$\hat{\underline{\theta}}(q) = \begin{pmatrix} q & \sum_{j=0}^{q-1} \zeta_{q-j} \\ \sum_{j=0}^{q-1} \zeta_{q-j} & \sum_{j=0}^{q-1} \zeta_{q-j}^2 \end{pmatrix}^{-1} \begin{pmatrix} \sum_{j=0}^{q-1} z(q-j) \\ \sum_{j=0}^{q-1} \zeta_{q-j} z(q-j) \end{pmatrix} \tag{2.4-9}$$

from which it is easy to show that

$$\hat{\theta}_1(q) = \frac{\left(\sum z\right)\left(\sum \zeta^2\right) - \left(\sum \zeta\right)\left(\sum \zeta z\right)}{q\sum \zeta^2 - \left(\sum \zeta\right)^2} \tag{2.4-10}$$

and

$$\hat{\theta}_2(q) = \frac{q\sum \zeta z - \left(\sum \zeta\right)\left(\sum z\right)}{q\sum \zeta^2 - \left(\sum \zeta\right)^2} \tag{2.4-11}$$

where we are using the following shorthand notation:

$$\sum = \sum_{j=0}^{q-1} \quad , \quad z = z(q-j) \quad , \quad \text{and} \quad \zeta = \zeta_{q-j} \quad .$$

2.4.4 Identification of Aerodynamic Parameters

Let us consider the identification of the two aerodynamic parameters M_α and M_δ that are associated with the high performance aerospace control system that was described in Section 1.3.5. Our approach will be to estimate M_α and M_δ from the equation

$$\ddot{\theta}_m(k) = M_\alpha \alpha(k) + M_\delta \delta(k) + v_{\ddot{\theta}}(k) \tag{2.4-12}$$

where $\ddot{\theta}_m(k)$ denotes the *measured value* of $\ddot{\theta}(k)$, that is corrupted by measurement noise $v_{\ddot{\theta}}(k)$. For the present, we shall assume that $\alpha(k)$

and $\delta(k)$ can both be measured perfectly; however, we shall reexamine this application in Section 2.5 for the situation when this assumption cannot be made. The concatenated measurement equation for L + 1 measurements is

$$\begin{pmatrix} \ddot{\theta}_m(k) \\ \ddot{\theta}_m(k-1) \\ \vdots \\ \ddot{\theta}_m(k-L) \end{pmatrix} = \begin{pmatrix} \alpha(k) & \delta(k) \\ \alpha(k-1) & \delta(k-1) \\ \vdots & \vdots \\ \alpha(k-L) & \delta(k-L) \end{pmatrix} \begin{pmatrix} M_\alpha \\ M_\delta \end{pmatrix} + \begin{pmatrix} v_{\ddot{\theta}}(k) \\ v_{\ddot{\theta}}(k-1) \\ \vdots \\ v_{\ddot{\theta}}(k-L) \end{pmatrix} . \tag{2.4-13}$$

Hence, the least-squares estimates of M_α and M_δ are

$$\begin{pmatrix} \hat{M}_\alpha(k) \\ \hat{M}_\delta(k) \end{pmatrix} = \begin{pmatrix} \sum_{j=0}^{L} \alpha^2(k-j) & \sum_{j=0}^{L} \alpha(k-j)\delta(k-j) \\ \sum_{j=0}^{L} \alpha(k-j)\delta(k-j) & \sum_{j=0}^{L} \delta^2(k-j) \end{pmatrix}^{-1} \times \begin{pmatrix} \sum_{j=0}^{L} \alpha(k-j)\ddot{\theta}_m(k-j) \\ \sum_{j=0}^{L} \delta(k-j)\ddot{\theta}_m(k-j) \end{pmatrix} . \tag{2.4-14}$$

The expressions for $\hat{M}_\alpha(k)$ and $\hat{M}_\delta(k)$ are only slightly more complicated than those obtained in the preceding subsection for $\hat{\theta}_1(q)$ and $\hat{\theta}_2(q)$. This is clear when Eqs. (2.4-14) and (2.4-9) are compared.

2.5 SOME PROPERTIES OF GENERALIZED LEAST SQUARES AND LEAST-SQUARES PARAMETER ESTIMATES

In this section we shall examine the bias and variance of generalized least squares and/or least-squares estimators.

2.5.1 Bias

In some applications, measurement errors may be so small that for all practical purposes $z(t) = y(t)$. In these cases, $\hat{\underline{\theta}}(k)$ is a deterministic quantity and it is clear, from Eqs. (2.3-12) and (2.2-16), that

$$\hat{\underline{\theta}}(k) = [H'(k)W(k)H(k)]^{-1}H'(k)W(k)H(k)\underline{\theta} = \underline{\theta} \quad . \qquad (2.5\text{-}1)$$

The estimate is exactly equal to the unknown parameter. Naturally, we could have obtained $\underline{\theta}$ much more directly by inverting $H'(k)$, as in Eq. (2.2-29), than by computing it from Eq. (2.3-12).

When measurement errors are not negligible, $\hat{\underline{\theta}}(k)$ is random, because $\underline{Z}(k)$ is random. Of course it is no longer possible to have $\hat{\underline{\theta}}(k) = \underline{\theta}$ for all $k \geq L$; but it is possible for $\hat{\underline{\theta}}(k)$ to be an *unbiased estimate* of $\underline{\theta}$. By an unbiased estimate, we mean one for which

$$E\{\hat{\underline{\theta}}(k)\} = \underline{\theta} \qquad (2.5\text{-}2)$$

for all $k \geq L$. We shall now study whether or not $\hat{\underline{\theta}}(k)$ is an unbiased estimate of $\underline{\theta}$.

Theorem 2-1. *The generalized least-squares estimate of* $\underline{\theta}$, *given in Eq. (2.3-12), is unbiased if* $\underline{V}(k)$ *has zero mean and if* $\underline{V}(k)$ *and* $H(k)$ *are statistically independent.*

Proof. From Eqs. (2.3-12) and (2.2-16), we find that

$$\hat{\underline{\theta}}(k) = [H'(k)W(k)H(k)]^{-1}H'(k)W(k)[H(k)\underline{\theta} + \underline{V}(k)]$$

$$= \underline{\theta} + [H'(k)W(k)H(k)]^{-1}H'(k)W(k)\underline{V}(k) \quad . \qquad (2.5\text{-}3)$$

Taking the expectation on both sides of this equation, it follows that

$$E\{\hat{\underline{\theta}}(k)\} = \underline{\theta} + E\{[H'(k)W(k)H(k)]^{-1}H'(k)\}W(k)E\{\underline{V}(k)\} \qquad (2.5\text{-}4)$$

where we have used the fact that $H(k)$ and $\underline{V}(k)$ are statistically independent.† The second term in Eq. (2.5-4) is zero, since $E\{\underline{V}(k)\} = \underline{0}$, and therefore

$$E\{\hat{\underline{\theta}}(k)\} = \underline{\theta} \quad . \tag{2.5-5}$$

▲

It is important to observe that the statistical independence of $H(k)$ and the zero mean $\underline{V}(k)$ is sufficient but not necessary for the unbiasedness of $\hat{\underline{\theta}}(k)$. In Eq. (2.5-4), if $E\{\hat{\underline{\theta}}(k)\} = \underline{\theta}$, then

$$E\{[H'(k)W(k)H(k)]^{-1}H'(k)W(k)\underline{V}(k)\} = \underline{0} \quad ,$$

which is a vector *orthogonality* condition.‡ Orthogonality is a weaker condition than independence, but is more difficult to verify ahead of time than independence. What this means is that, if $H(k)$ and $\underline{V}(k)$ are not statistically independent, conclusions about the bias of $\hat{\underline{\theta}}(k)$ cannot be drawn. Obviously, such uncertainty is undesirable and is often intolerable.¶

In order to illustrate some applications of this very important property, let us first take another look at Examples 1, 2, and 3 in Section 2.2.

---

† If two random variables, a and b, are statistically independent, $f(a, b) = f(a)f(b)$ (Appendix E); thus, $E\{ab\} = E\{a\}E\{b\}$ and $E\{g(a)h(b)\} = E\{g(a)\}E\{g(b)\}$.

‡ Two random variables a and b are orthogonal if $E\{ab\} = 0$.

¶ An approach to this source of difficulty, which is well known in the statistical literature (*2-21*) and has also received some attention in the engineering literature (*2-22*), is to use the *method of instrumental variables*. In this method $H(k)$ is replaced by $H^*(k)$ where $H^*(k)$ is chosen so that it is uncorrelated with $\underline{V}(k)$. There can be, and in general there will be, many choices of $H^*(k)$ that may qualify as instrumental variables. The difficulty of checking that the chosen $H^*(k)$ is indeed uncorrelated with $\underline{V}(k)$ makes the method somewhat unattractive.

Example 1 (Continued). Symbol ζ is often associated with time. ζ_i is not a measurement of ζ, but is rather a prespecified value of ζ; hence, there is no randomness associated with ζ_i, and $H(L)$ in Eq. (2.2-19) is deterministic. In this case, $E\{H(k)\underline{V}(k)\} = H(L)E\{\underline{V}(L)\}$, and it is clear that $\hat{\underline{\theta}}(k)$ is an unbiased estimate of $\underline{\theta}$. ▲

Example 2 (Continued). Here, as in Example 1, $H(k)$ is deterministic [see Eq. (2.2-25)]; thus, $\hat{\underline{x}}(k)$ computed from Eq. (2.3-12) will be an unbiased estimate of $\underline{x}(0)$. ▲

Example 3 (Continued). If the input sequence w(k - 1), w(k - 2), ..., w(k - N) is deterministic and is known ahead of time without error, then $H(k)$ in Eq. (2.2-28) is deterministic; and, therefore, $\hat{\underline{f}}(k)$ computed from Eq. (2.3-12) will be an unbiased estimate of $\underline{f}$.

Often, however, one uses a random input sequence for w(k - 1), w(k - 2), ..., w(k - N); however, this random sequence is in no way related to the measurement error process, which means, of course, that they are statistically independent, and again, $\hat{\underline{f}}(k)$ will be an unbiased estimate of $\underline{f}$. ▲

As a further illustration of an application of Theorem 2-1, let us take a look at the problem of identifying the coefficients of a finite-difference equation that is excited by a random sequence.

Example 4. Here we shall be interested in identifying the coefficients $a_1, a_2, \ldots, a_n$ in the finite-difference equation

$$y(k + n) + a_1 y(k + n - 1) + \cdots + a_n y(k) = \omega(k) \quad . \quad (2.5\text{-}6)$$

We showed in Section 1.3.3 that one equation-error approach to the identification of these coefficients is to first solve for y(k + n) and to then view y(k + n) as the output of the "Identification representation" block in Fig. 1.2-2c. We shall assume that it is possible to measure y(k), y(k + 1), ..., y(k + n) perfectly. The situation when this is not true is discussed in Chapter 5.

Equation (2.5-6) is concatenated as follows:

$$\begin{pmatrix} y(k+n) \\ y(k+n-1) \\ \vdots \\ y(k+n-L) \end{pmatrix} = \begin{pmatrix} y(k+n-1) & \cdots & y(k) \\ y(k+n-2) & \cdots & y(k-1) \\ \vdots & \ddots & \vdots \\ y(k+n-L-1) & \cdots & y(k-L) \end{pmatrix} \begin{pmatrix} -a_1 \\ -a_2 \\ \vdots \\ -a_n \end{pmatrix}$$

$$+ \begin{pmatrix} \omega(k) \\ \omega(k-1) \\ \vdots \\ \omega(k-L) \end{pmatrix} . \tag{2.5-7}$$

We shall now demonstrate that H and $\underline{V}$ in Eq. (2.5-7) are dependent. Let us represent the explicit dependence of H on its elements and $\underline{V}$ on its elements in the following manner:

$$H = H[y(k-L), \ldots, y(k-1), y(k), \ldots, y(k+n-1)] \tag{2.5-8}$$

and

$$\underline{V} = \underline{V}[\omega(k-L), \ldots, \omega(k-1), \omega(k)] . \tag{2.5-9}$$

Recall, from Section 1.3.3, that Eq. (2.5-6) is equivalent to the following first-order vector difference equation:

$$\begin{pmatrix} x_1(k+1) \\ x_2(k+1) \\ \vdots \\ x_n(k+1) \end{pmatrix} = \left(\begin{array}{c|ccc} 0 & & & \\ 0 & & I_{n-1} & \\ \vdots & & & \\ \hline -a_n & -a_{n-1} & \cdots & -a_1 \end{array}\right) \begin{pmatrix} x_1(k) \\ x_2(k) \\ \vdots \\ x_n(k) \end{pmatrix} + \begin{pmatrix} d_1 \\ d_2 \\ \vdots \\ d_n \end{pmatrix} \omega(k) \tag{2.5-10}$$

where $y(k) = x_1(k)$. This equation can also be written as

$$\underline{x}(k+1) = \Phi\underline{x}(k) + \underline{\gamma}\omega(k) , \tag{2.5-11}$$

the solution of which is [see Eq. (1.3-58)]

$$\underline{x}(k) = \Phi^k \underline{x}(0) + \sum_{i=1}^{k} \Phi^{k-i} \underline{\gamma}\omega(i - 1) \quad . \tag{2.5-12}$$

It is clear from this solution that [recall that $y(k + j) = x_1(k + j)$]

$$\begin{aligned} y(k + n - 1) &= y[\omega(0), \omega(1), \ldots, \omega(k + n - 2)] \\ y(k + n - 2) &= y[\omega(0), \omega(1), \ldots, \omega(k + n - 3)] \\ &\vdots \\ y(k) &= y[\omega(0), \omega(1), \ldots, \omega(k - 1)] \\ y(k - 1) &= y[\omega(0), \omega(1), \ldots, \omega(k - 2)] \\ &\vdots \\ y(k - L) &= y[\omega(0), \omega(1), \ldots, \omega(k - L - 1)] \end{aligned} \tag{2.5-13}$$

which means that, in Eq. (2.5-8)

$$\begin{aligned} \mathcal{H} = \mathcal{H}[&\omega(0), \omega(1), \ldots, \omega(k - L - 1), \\ &\omega(k - L), \ldots, \omega(k - 2), \omega(k - 1), \\ &\omega(k), \ldots, \omega(k + n - 3), \omega(k + n - 2)] \quad . \end{aligned} \tag{2.5-14}$$

Comparing Eqs. (2.5-14) and (2.5-9), we see that $\mathcal{H}$ and $\underline{\mathcal{V}}$ depend upon similar values of ω; hence, they are dependent, which means that estimates of $a_1, a_2, \ldots, a_n$ obtained by means of the generalized least-squares algorithm may or may not be biased. Such uncertainty is undesirable.

We shall return to this application in Chapter 5, where we will obtain an algorithm that gives unbiased estimates of $a_1, a_2, \ldots, a_n$.

▲

Example 5. As a final illustration of Theorem 2-1, let us take another look at the aerodynamic parameter identification application that we examined in Section 2.4.4.

To begin, let us examine the situation in Eq. (2.4-13). Recall that $\alpha(k)$ and $\delta(k)$ were assumed to be measured without errors; hence,

$$H(k) = \begin{pmatrix} \alpha(k) & \delta(k) \\ \alpha(k-1) & \delta(k-1) \\ \vdots & \vdots \\ \alpha(k-L) & \delta(k-L) \end{pmatrix}$$

is deterministic and is clearly independent of the concatenated measurement-error vector $\underline{V}(k) = [v_{\ddot{\theta}}(k), v_{\ddot{\theta}}(k-1), \ldots, v_{\ddot{\theta}}(k-L)]'$. In this case, $\hat{M}_\alpha(k)$ and $\hat{M}_\delta(k)$ obtained as in Eq. (2.4-14), for example, will be unbiased estimates of M_α and M_δ, respectively.

Next, let us examine the more realistic situation, when the measured values of $\alpha(k)$ and $\delta(k)$ differ from the actual quantities, because of measurement errors. Letting $\alpha_m(k)$ and $\delta_m(k)$ denote the *measured values* of $\alpha(k)$ and $\delta(k)$, we have

$$\alpha_m(k) = \alpha(k) + n_\alpha(k) \tag{2.5-15}$$

and

$$\delta_m(k) = \delta(k) + n_\delta(k) \tag{2.5-16}$$

where $n_\alpha(k)$ and $n_\delta(k)$ are zero mean measurement-error processes. In order to identify M_α and M_δ from Eq. (2.4-12), that equation must be rewritten in terms of measured signals; hence, utilizing Eqs. (2.5-15) and (2.5-16), we find that

$$\ddot{\theta}_m(k) = M_\alpha \alpha_m(k) + M_\delta \delta_m(k) + v_1(k) \tag{2.5-17}$$

where

$$v_1(k) = v_{\ddot{\theta}}(k) - M_\alpha n_\alpha(k) - M_\delta n_\delta(k) \quad . \tag{2.5-18}$$

In this case, the concatenated measurement equation is

$$\begin{pmatrix} \ddot{\theta}_m(k) \\ \ddot{\theta}_m(k-1) \\ \vdots \\ \ddot{\theta}_m(k-L) \end{pmatrix} = \begin{pmatrix} \alpha_m(k) & \delta_m(k) \\ \alpha_m(k-1) & \delta_m(k-1) \\ \vdots & \vdots \\ \alpha_m(k-L) & \delta_m(k-L) \end{pmatrix} \begin{pmatrix} M_\alpha \\ M_\delta \end{pmatrix} + \begin{pmatrix} v_1(k) \\ v_1(k-1) \\ \vdots \\ v_1(k-L) \end{pmatrix}. \tag{2.5-19}$$

Observe, first, that

$$H(k) = \begin{pmatrix} \alpha_m(k) & \delta_m(k) \\ \alpha_m(k-1) & \delta_m(k-1) \\ \vdots & \vdots \\ \alpha_m(k-L) & \delta_m(k-L) \end{pmatrix}$$

is random; and, second, that $\underline{V}(k) = [v_1(k), v_1(k-1), \ldots, v_1(k-L)]'$ and $H(k)$ are dependent. The second observation is seen to be true upon comparison of Eqs. (2.5-15), (2.5-16), *and* (2.5-18).

We must conclude, therefore, that generalized least-squares estimates of M_α and M_δ may or may not be biased when $\alpha(k)$ and $\delta(k)$ cannot be measured perfectly. We shall return to this application in Chapter 5, where we will show how to obtain unbiased estimates of these parameters. ▲

In the preceding five examples, we have seen three situations where estimates will be unbiased, and two where estimates may or may not be biased. Generalizations are possible, and we provide these in Section 2.13.

Let us also note that, in terms of the identification error vector $\tilde{\underline{\theta}}(k)$, unbiasedness means

$$E\{\tilde{\underline{\theta}}(k)\} = \underline{0} \quad . \tag{2.5-20}$$

2.5.2 Variance and Covariance

Information about the dispersion of $\tilde{\theta}_i(k)$ about its zero mean value is provided by the variance of $\tilde{\theta}_i(k)$. The variance of $\tilde{\theta}_i(k)$ is given by the ith diagonal element of the covariance matrix of $\tilde{\underline{\theta}}(k)$. We proceed to compute Cov $[\tilde{\underline{\theta}}(k)]$.

Theorem 2-2. *For the generalized least-squares estimate of* $\underline{\theta}$ *given in Eq. (2.3-12), if* $\underline{V}(k)$ *has zero mean,* $\underline{V}(k)$ *and* $H(k)$ *are statistically independent, and*

$$E\{\underline{V}(k)\underline{V}'(k)\} = R(k) \quad , \tag{2.5-21}$$

then

$$\text{Cov}\,[\underline{\tilde{\theta}}(k)] = E\{[H'(k)W(k)H(k)]^{-1}H'(k)\}W(k)R(k)W(k)$$
$$\times\, E\{H(k)[H'(k)W(k)H(k)]^{-1}\} \quad . \qquad (2.5\text{-}22)$$

Proof. From elementary probability theory (*2-5*), we know that

$$\text{Cov}\,[\underline{\tilde{\theta}}(k)] = E\{[\underline{\tilde{\theta}}(k) - E\{\underline{\tilde{\theta}}(k)\}][\underline{\tilde{\theta}}(k) - E\{\underline{\tilde{\theta}}(k)\}]'\} \quad ; \qquad (2.5\text{-}23)$$

however, since $\underline{V}(k)$ is a zero mean vector and is statistically independent of $H(k)$, we have [Eq. (2.5-20)]

$$\text{Cov}\,[\underline{\tilde{\theta}}(k)] = E\{\underline{\tilde{\theta}}(k)\underline{\tilde{\theta}}'(k)\}$$
$$= E\{[\underline{\theta} - \underline{\hat{\theta}}(k)][\underline{\theta} - \underline{\hat{\theta}}(k)]'\} \qquad (2.5\text{-}24)$$

which can be evaluated, using the expression for $\underline{\theta} - \underline{\hat{\theta}}(k)$ in Eq. (2.5-3), as follows:

$$\text{Cov}\,[\underline{\tilde{\theta}}(k)] = E\{[-(H'WH)^{-1}H'W\underline{V}][-(H'WH)^{-1}H'W\underline{V}]'\}$$
$$= E\{(H'WH)^{-1}H'W\underline{V}\,\underline{V}'WH(H'WH)^{-1}\} \qquad (2.5\text{-}25)$$

where† we have made use of the facts that W(k) is a symmetric matrix and the transpose and inverse symbols may be permuted. Because $H(k)$ and $\underline{V}(k)$ are statistically independent,

$$\text{Cov}\,[\underline{\tilde{\theta}}(k)] = E\{(H'WH)^{-1}H'\}WE\{\underline{V}\,\underline{V}'\}WE\{H(H'WH)^{-1}\}$$
$$= E\{(H'WH)^{-1}H'\}WRWE\{H(H'WH)^{-1}\} \qquad (2.5\text{-}26)$$

which is Eq. (2.5-22). ▲

The following special cases of Theorem 2-2 are important in practical applications, and treat the case when $H(k)$ is deterministic.

<u>Corollary 2-1</u>. *For the generalized least-squares estimate of* $\underline{\theta}$ *given in Eq. (2.3-12), if* $\underline{V}(k)$ *has zero mean and* $H(k)$ *is deterministic, then*

$$\text{Cov}\,[\underline{\tilde{\theta}}(k)] = [H'(k)W(k)H(k)]^{-1}H'(k)W(k)R(k)W(k)H(k)$$
$$\times\, [H'(k)W(k)H(k)]^{-1} \quad . \qquad (2.5\text{-}27)$$

---

† In proofs, we will often simplify notation by omitting functional dependences of vectors and matrices on k.

Proof. Equation (2.5-27) is a direct consequence of Eq. (2.5-26) and the fact that H is deterministic. ▲

Corollary 2-2. *If* $H(k)$ *is deterministic and if the components of* $\underline{V}(k)$ *are equally distributed with zero mean and variance* σ^2*, then for a least-squares estimator*

$$\mathrm{Cov}\,[\tilde{\underline{\theta}}(k)] = \sigma^2[H'(k)H(k)]^{-1} \quad . \tag{2.5-28}$$

Proof. For a least-squares estimator [see Eq. (2.2-37)]

$$W(k) = w(k)I \quad , \tag{2.5-29}$$

and if the components of $\underline{V}(k)$ are equally distributed with variance σ^2, then

$$R(k) = \sigma^2 I \quad . \tag{2.5-30}$$

Equation (2.5-28) follows upon substitution of Eqs. (2.5-29) and (2.5-30) into Eq. (2.5-27). ▲

When $R(k)$ is diagonal with σ^2 in place of each diagonal element, i. e., $R(k)$ as in Eq. (2.5-30), then $\underline{V}(k)$ is said to be a *homoskedastic* error process. Equation (2.5-30) is known as the condition of *homoskedasticity*. When $R(k)$ is diagonal but the diagonal elements are not necessarily the same, then $\underline{V}(k)$ is said to be a *heteroskedastic* error process. The condition of *heteroskedasticity* is $R(k) = \mathrm{diag}[\sigma^2(k), \sigma^2(k-1), \ldots, \sigma^2(k-L)]$. The variance of a homoskedastic error process is the same regardless of the amplitude of the measurement z, whereas the variance of a heteroskedastic error process can be made to be dependent upon the amplitude of the measurement. The terms homoskedasticity and heteroskedasticity are quite common in the literature on statistics and econometrics, but are not used, as yet, in the engineering literature.

Usually, when one uses a least-squares estimation algorithm, he does not know the numerical value of σ^2. If he did, then, as we show in the next chapter, he could use this information directly in the estimate for $\underline{\theta}$. Observe, in Eqs. (2.3-14) and (2.3-12), that nowhere does $R(k)$ appear. We show next that, under certain conditions, it is possible to obtain an unbiased estimate of σ^2 as a byproduct of estimating $\underline{\theta}$.

Theorem 2-3. *If* $\underline{\theta}$ *is estimated using the least-squares algorithm in Eq. (2.3-14),* $H(k)$ *is nonrandom, the components of* $\underline{V}(k)$ *are equally distributed with zero mean and variance* σ^2*, and* σ^2 *is unknown, then an unbiased estimate of* σ^2*, denoted* $\hat{\sigma}^2(k)$*, is*

$$\hat{\sigma}^2(k) = \frac{\tilde{\underline{Z}}'(k)\tilde{\underline{Z}}(k)}{\dim[\underline{V}(k)] - \dim[\underline{\theta}]} \tag{2.5-31}$$

where dim[ ] *is short for dimension of* [ ].

Proof. We shall proceed by computing $E\{\tilde{\underline{Z}}'(k)\tilde{\underline{Z}}(k)\}$ and then approximating it as $\tilde{\underline{Z}}'(k)\tilde{\underline{Z}}(k)$, since $\tilde{\underline{Z}}'(k)\tilde{\underline{Z}}(k)$ can be obtained from the measurements and estimate.

First, let us compute an expression for $\tilde{\underline{Z}}(k)$:

$$\tilde{\underline{Z}}(k) = \underline{Z}(k) - \hat{\underline{Z}}(k) = \underline{Z} - H\hat{\underline{\theta}} \quad . \tag{2.5-32}$$

Setting W(k) = w(k)I in Eq. (2.5-3), we find that

$$\hat{\underline{\theta}} = \underline{\theta} + (H'H)^{-1}H'\underline{V} \quad . \tag{2.5-33}$$

Hence

$$\begin{aligned}\tilde{\underline{Z}} &= H\underline{\theta} + \underline{V} - H[\underline{\theta} + (H'H)^{-1}H'\underline{V}] \\ &= \underline{V} - H(H'H)^{-1}H'\underline{V} \\ &= [I_{L+1} - H(H'H)^{-1}H']\underline{V}\end{aligned} \tag{2.5-34}$$

where I_{L+1} is the $(L + 1) \times (L + 1)$ identity matrix, and

$$\dim[\underline{V}] = L + 1 \quad . \tag{2.5-35}$$

Now, let us show that $M = I_{L+1} - H(H'H)^{-1}H'$ is an *idempotent matrix*; that is to say, M is a matrix for which $M' = M$ and $M^2 = M$ (*2-1*). Observe that

$$M' = [I_{L+1} - H(H'H)^{-1}H']' = I_{L+1} - H(H'H)^{-1}H' = M \tag{2.5-36}$$

and that

$$M^2 = [I_{L+1} - H(H'H)^{-1}H'][I_{L+1} - H(H'H)^{-1}H']$$

$$= I_{L+1} - 2H(H'H)^{-1}H' + H(H'H)^{-1}H'$$

$$= I_{L+1} - H(H'H)^{-1}H' = M \qquad (2.5\text{-}37)$$

and, therefore, M is indeed an idempotent matrix. It follows from Eqs. (2.5-34), (2.5-36), and (2.5-37) that

$$E\{\underline{\tilde{Z}}'\underline{\tilde{Z}}\} = E\{\underline{V}'[I_{L+1} - H(H'H)^{-1}H']'[I_{L+1} - H(H'H)^{-1}H']\underline{V}\}$$

$$= E\{\underline{V}'[I_{L+1} - H(H'H)^{-1}H']\underline{V}\} \quad . \qquad (2.5\text{-}38)$$

Next, we pause to state some well-known facts about the trace of a matrix, which we will need to evaluate $E\{\underline{\tilde{Z}}'\underline{\tilde{Z}}\}$.†

(1) $\underline{a}'B\underline{a} = \mathrm{tr}(B\underline{a}\underline{a}')$

(2) $E\{\mathrm{tr}\ A\} = \mathrm{tr}\ E\{A\}$

(3) $\mathrm{tr}\ cA = c\ \mathrm{tr}\ A$, where c is a scalar

(4) $\mathrm{tr}(A + B) = \mathrm{tr}\ A + \mathrm{tr}\ B$

(5) $\mathrm{tr}\ I_N = N$, where I_N is the $N \times N$ identity matrix

(6) $\mathrm{tr}\ AB = \mathrm{tr}\ BA$

Using these facts, we now evaluate $E\{\underline{\tilde{Z}}'\underline{\tilde{Z}}\}$, in Eq. (2.5-38), as follows:

$$E\{\underline{\tilde{Z}}'\underline{\tilde{Z}}\} = E\left\{\mathrm{tr}\left[[I_{L+1} - H(H'H)^{-1}H']\underline{V}\underline{V}'\right]\right\}$$

$$= \mathrm{tr}\ E\{[I_{L+1} - H(H'H)^{-1}H']\underline{V}\underline{V}'\}$$

$$= \mathrm{tr}\left\{[I_{L+1} - H(H'H)^{-1}H']E\{\underline{V}\underline{V}'\}\right\}$$

$$= \mathrm{tr}\{\sigma^2[I_{L+1} - H(H'H)^{-1}H']\}$$

$$= \sigma^2\ \mathrm{tr}[I_{L+1} - H(H'H)^{-1}H']$$

† The trace of a matrix is equal to the sum of the diagonal elements of the matrix.

$$= \sigma^2(L + 1) - \sigma^2 \, \mathrm{tr}[H(H'H)^{-1}H']$$

$$= \sigma^2(L + 1) - \sigma^2 \, \mathrm{tr}[H'H(H'H)^{-1}]$$

$$= \sigma^2(L + 1) - \sigma^2 \, \mathrm{tr}[I_{\dim(H'H)}]$$

$$= \sigma^2[(L + 1) - n] \quad , \tag{2.5-39}$$

where we have also made use of the fact that H is an $(L + 1) \times n$ matrix.

Finally, we see that

$$\sigma^2 = \frac{E\{\tilde{\underline{Z}}'\tilde{\underline{Z}}\}}{[(L + 1) - n]} \quad . \tag{2.5-40}$$

In practice, we do not know $E\{\tilde{\underline{Z}}'\tilde{\underline{Z}}\}$. All we have is a single sample value of the random variable $\tilde{\underline{Z}}'\tilde{\underline{Z}}$. In lieu of $E\{\tilde{\underline{Z}}'\tilde{\underline{Z}}\}$, we decide simply to use that sample value, and so we estimate σ^2 by $\hat{\sigma}^2(k)$, where

$$\hat{\sigma}^2(k) = \frac{\tilde{\underline{Z}}'(k)\tilde{\underline{Z}}(k)}{[(L + 1) - n]} \quad , \tag{2.5-41}$$

which is the result stated as Eq. (2.5-31), since $(L + 1) = \dim[\underline{V}(k)]$ and $n = \dim[\underline{\theta}]$. It is clear from Eqs. (2.5-41) and (2.5-40) that $\hat{\sigma}^2(k)$ is an unbiased estimate of σ^2. ▲

In this theorem we have shown how the vector equation error $\tilde{\underline{Z}}(k)$ can be used for calibrating an observation instrument.

Example 6.[†] Suppose we wish to calibrate an instrument by making a series of uncorrelated measurements on a constant quantity. The statistics of the instrument's errors are stationary (i. e., the variances of all observations are equal). Denoting the constant quantity as θ - a scalar - our measurement equation becomes

$$z(k) = \theta + v(k) \quad . \tag{2.5-42}$$

[†] This example is based, to a large extent, on Problem 6.2-4 in Ref. *2-2*.

The concatenated measurement equation is

$$\begin{pmatrix} z(k) \\ z(k-1) \\ \vdots \\ z(k-L) \end{pmatrix} = \begin{pmatrix} 1 \\ 1 \\ \vdots \\ 1 \end{pmatrix} \theta + \begin{pmatrix} v(k) \\ v(k-1) \\ \vdots \\ v(k-L) \end{pmatrix} \qquad (2.5\text{-}43)$$

for $k \geq L$. Clearly, $\mathcal{H} = (1, 1, \ldots, 1)'$, an $(L + 1) \times 1$ vector; hence, the least-squares estimate of θ, from Eq. (2.3-14), is

$$\hat{\theta}(k) = \frac{1}{L+1} \sum_{i=0}^{L} z(k-i) \qquad (2.5\text{-}44)$$

which is nothing more than the average (sample mean) of the $L + 1$ measurements $z(k)$, $z(k - 1)$, ..., $z(k - L)$.

From Eqs. (2.5-31) and (2.5-32), we obtain the following estimate for σ^2:

$$\hat{\sigma}^2(k) = \frac{1}{L} [\underline{Z}(k) - \mathcal{H}\hat{\theta}(k)]'[\underline{Z}(k) - \mathcal{H}\hat{\theta}(k)]$$

$$= \frac{1}{L} \sum_{j=0}^{L} [z(k-j) - \hat{\theta}(k)]^2 \qquad (2.5\text{-}45)$$

where $\hat{\theta}(k)$ is given in Eq. (2.5-44).

Let us look at some general properties of the sample mean as an estimator. Suppose $z(k)$ has any distribution with mean θ and variance σ^2 [as is implied by the nature of Eq. (2.5-42)]. By Theorem 2-1 and Corollary 2-2, it follows that the sample mean is an unbiased estimator of θ and has variance equal to $\sigma^2/(L + 1)$. Observe that the dispersion of the sample mean about θ can always be made smaller by using additional measurements. An estimator is said to be *consistent* if, as the number of measurements increases to infinity, the bias and variance both become smaller. Clearly, as $L \to \infty$ the variance of the sample mean approaches zero; hence, the sample mean is a consistent estimator of θ.

2.6 CHOICE OF L

Recall, from Section 2.2, that L controls the total number of measurements we have available, or care to use, for estimating $\underline{\theta}$. We discussed the cases when there are fewer equations than unknown parameters - $(L + 1) < n$ - exactly as many equations as unknowns - $(L + 1) = n$ - and more equations than unknowns - $(L + 1) > n$ - and decided that only the latter case is of real interest to us. In this chapter (and the next), L will be chosen so that

$$L > n - 1 \quad . \tag{2.6-1}$$

There are two basically different approaches for choosing L. In the first approach, L is fixed at some numerical value, say L*, in which case $\underline{\theta}$ is estimated from exactly L* + 1 measurements. The resulting estimate is sometimes referred to as a *fixed-memory estimate* (2-2). Regardless of how k is chosen (see Fig. 2.6-1), $\underline{\theta}$ *is estimated from exactly the same number of measurements*. In (a) of Fig. 2.6-1, $\underline{\theta}$ is estimated from $\{z(4), z(3), z(2), z(1), z(0)\}$, whereas, in (b) of Fig. 2.6-1, $\underline{\theta}$ is estimated from $\{z(6), z(5), z(4), z(3), z(2)\}$.

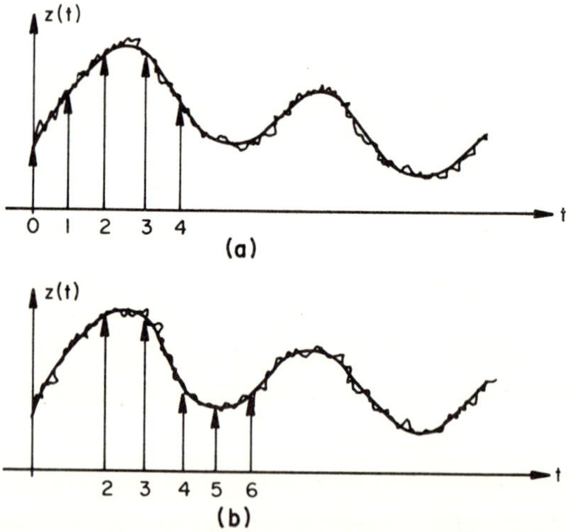

Fig. 2.6-1. Examples of data used in a fixed-memory estimate: (a) L = 4 and k = 4; (b) L = 4 and k = 6.

A second approach to choosing L is to set it equal to k, requiring, of course, that $k > n - 1$. The resulting estimate is sometimes referred to as an *expanding-memory estimate* (2-2). Increasing k from k_1^* to k_2^*, where $k_2^* > k_1^*$, increases the number of measurements used in the estimate of $\underline{\theta}$. In the first case, there will be $k_1^* + 1$ measurements, whereas in the second case there will be $k_2^* + 1$ measurements, and, $k_2^* + 1 > k_1^* + 1$. This is illustrated in Fig. 2.6-2 for $k_1^* = 4$ and $k_2^* = 6$ (compare Figs. 2.6-2 and 2.6-1).

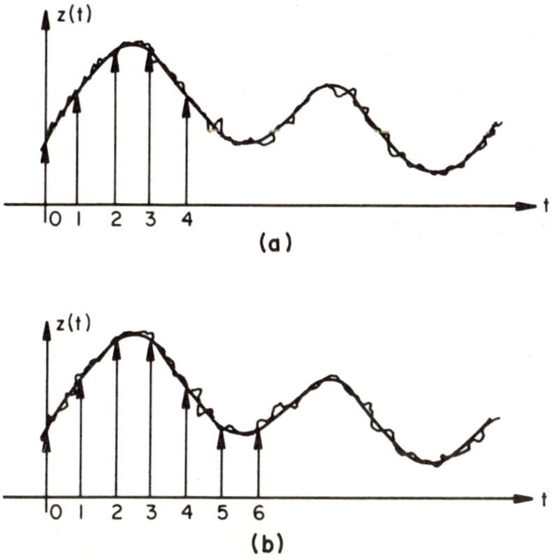

Fig. 2.6-2. Examples of data used in an expanding-memory estimate: (a) k = 4, and (b) k = 6.

The expanding-memory concept is very useful in practical applications. Suppose, for example, we are identifying 10 parameters, θ_1, θ_2, ..., θ_{10}, from 60 voltage measurements. We could, of course, estimate these 10 parameters at once, using all these measurements. Let us denote this estimate as $\hat{\underline{\theta}}_{60}$. On the other hand, many other estimates of $\underline{\theta}$ are possible. Let us begin by estimating $\underline{\theta}$ using only 11 measurements, calling this estimate $\hat{\underline{\theta}}_{11}$. Proceeding in a similar manner, we generate the following arbitrary sequence of estimates of $\underline{\theta}$:

$$\hat{\underline{\theta}}_{11}, \hat{\underline{\theta}}_{18}, \hat{\underline{\theta}}_{29}, \hat{\underline{\theta}}_{34}, \hat{\underline{\theta}}_{46}, \hat{\underline{\theta}}_{57}, \hat{\underline{\theta}}_{60} .$$

Each of these estimates (other than $\hat{\underline{\theta}}_{11}$) may be interpreted as expanding the data base associated with the one before it; thus, for example, $\hat{\underline{\theta}}_{29}$ not only uses the measurements that were used to compute $\hat{\underline{\theta}}_{18}$, but it uses the next 11 measurements as well. Intuitively, we would expect the above sequence to converge to $\hat{\underline{\theta}}_{60}$, and be closer to $\hat{\underline{\theta}}_{60}$ as more and more data are used.

The example we have just discussed is one in which the entire measurement time history is available ahead of time. This situation is often known as an *off-line* or *post-flight* situation.

Another situation in which the expanding-memory concept is important and very useful is the *on-line* situation. Now we wish to identify $\underline{\theta}$, as measurements become available. Series $\hat{\underline{\theta}}(k)$, $\hat{\underline{\theta}}(k + 1)$, $\hat{\underline{\theta}}(k + 2)$, ..., etc., and $\hat{\underline{\theta}}(k)$, $\hat{\underline{\theta}}(k + \ell)$, $\hat{\underline{\theta}}(k + 2\ell)$, ..., etc., represent two (out of many) possible sequences of on-line expanding-memory estimates of $\underline{\theta}$.

In this book, our attention will be directed almost exclusively at expanding-memory parameter identification. It is extremely useful in both off-line and on-line situations.

2.7 SEQUENTIAL GENERALIZED LEAST-SQUARES ESTIMATION

Regardless of how L is chosen (its choice leads to either a fixed-memory or expanding-memory identifier), $\underline{\theta}$ must be estimated from either Eq. (2.3-12), in the case of generalized least-squares estimation, or Eq. (2.3-14), in the case of least-squares estimation. Suppose we decide to add more measurements, increasing the total number of them from L + 1 to L' + 1. The algorithms in Eqs. (2.3-14) and (2.3-12) do not take the earlier estimate into account; that is to say, $\hat{\underline{\theta}}(k)$ based on L' + 1 measurements must be computed from the entire collection of L' + 1 measurements. No use of $\hat{\underline{\theta}}(k)$ based on L + 1 measurements is made during the calculation of $\hat{\underline{\theta}}(k)$ based on L' + 1 measurements. This seems quite wasteful. We intuitively expect that it should be possible to compute the estimate based on L' + 1 measurements from the estimate based on L + 1 measurements, and a modification of this earlier estimate to account for the L' - L new measurements. Let us proceed to justify our intuition.

2.7.1 Derivation of a Sequential Algorithm: One Additional Measurement

To begin, we consider the case when one additional measurement z(k + 1), made at t_{k+1}, becomes available:

$$z(k+1) = \underline{h}'(k+1)\underline{\theta} + v(k+1) \quad . \tag{2.7-1}$$

Let us now concatenate our L + 2 measurements z(k - L), z(k - L + 1), ..., z(k), z(k + 1), as we did in Eq. (2.2-12):

$$\begin{pmatrix} z(k+1) \\ z(k) \\ z(k-1) \\ \vdots \\ z(k-L) \end{pmatrix} = \begin{pmatrix} \underline{h}'(k+1) \\ \underline{h}'(k) \\ \underline{h}'(k-1) \\ \vdots \\ \underline{h}'(k-L) \end{pmatrix} \underline{\theta} + \begin{pmatrix} v(k+1) \\ v(k) \\ v(k-1) \\ \vdots \\ v(k-L) \end{pmatrix} \quad . \tag{2.7-2}$$

Letting

$$\underline{Z}(k+1) = [z(k+1), z(k), z(k-1), \ldots, z(k-L)]' \ , \tag{2.7-3}$$

$$\underline{V}(k+1) = [v(k+1), v(k), v(k-1), \ldots, v(k-L)]' \ , \tag{2.7-4}$$

and

$$H(k+1) = \begin{pmatrix} \underline{h}'(k+1) \\ \underline{h}'(k) \\ \underline{h}'(k-1) \\ \vdots \\ \underline{h}'(k-L) \end{pmatrix} \ , \tag{2.7-5}$$

the concatenated measurement equation, (2.7-2), becomes

$$\underline{Z}(k+1) = H(k+1)\underline{\theta} + \underline{V}(k+1) \quad . \tag{2.7-6}$$

Letting $\hat{\underline{\theta}}(k + 1)$ denote the estimate of $\underline{\theta}$ based on the L + 2 data samples z(k + 1), z(k), ..., z(k - L), it is clear that the generalized least-squares estimate, analogous to Eq. (2.3-12), is

$$\hat{\underline{\theta}}(k + 1) = [H'(k + 1)W(k + 1)H(k + 1)]^{-1}H'(k + 1)W(k + 1)\underline{Z}(k + 1) \ . \tag{2.7-7}$$

We shall now show that it is possible to determine $\hat{\underline{\theta}}(k + 1)$ from $\hat{\underline{\theta}}(k)$ and z(k + 1).

Observe, from Eqs. (2.7-3) and (2.2-13), (2.7-4) and (2.2-14), and (2.7-5) and (2.2-15) that

$$\underline{Z}(k + 1) = \begin{pmatrix} z(k + 1) \\ \hline \underline{Z}(k) \end{pmatrix} , \tag{2.7-8}$$

$$\underline{V}(k + 1) = \begin{pmatrix} v(k + 1) \\ \hline \underline{V}(k) \end{pmatrix} , \tag{2.7-9}$$

and

$$H(k + 1) = \begin{pmatrix} \underline{h}'(k + 1) \\ \hline H(k) \end{pmatrix} \tag{2.7-10}$$

In addition, W(k + 1), which is defined in a manner that is analogous to W(k) in Eq. (2.2-36), can be written as

$$W(k + 1) = \begin{pmatrix} w(k + 1) & 0 & \cdots & 0 \\ 0 & w(k) & \cdots & 0 \\ \vdots & \vdots & \ddots & \vdots \\ 0 & 0 & \cdots & w(k - L) \end{pmatrix} = \left(\begin{array}{c|c} w(k + 1) & \underline{0}' \\ \hline \underline{0} & W(k) \end{array}\right) , \tag{2.7-11}$$

where $\underline{0}$ is the (L + 1) × 1 zero vector.

We proceed by substituting Eqs. (2.7-8) through (2.7-11) into Eq. (2.7-7) (dropping the dependence upon k and k + 1, for notational simplicity):

$$\hat{\underline{\theta}}(k+1) = \left[(\underline{h} \,\vdots\, \mathcal{H}') \begin{pmatrix} w & \underline{0}' \\ \underline{0} & W \end{pmatrix} \begin{pmatrix} \underline{h}' \\ \mathcal{H} \end{pmatrix} \right]^{-1} (\underline{h}' \,\vdots\, \mathcal{H}') \begin{pmatrix} w & \underline{0}' \\ \underline{0} & W \end{pmatrix} \begin{pmatrix} z \\ \underline{Z} \end{pmatrix}$$

$$= [\underline{h}w\underline{h}' + \mathcal{H}'W\mathcal{H}]^{-1}[\underline{h}wz + \mathcal{H}'W\underline{Z}] \quad . \tag{2.7-12}$$

In order to simplify this expression, we let

$$P(k) = [\mathcal{H}'(k)W(k)\mathcal{H}(k)]^{-1} \tag{2.7-13}$$

which means, of course, that [Eq. (2.3-12)]

$$\hat{\underline{\theta}}(k) = P(k)\mathcal{H}'(k)W(k)\underline{Z}(k) \quad . \tag{2.7-14}$$

It is also clear, from our expansion of $\mathcal{H}'(k+1)W(k+1)\mathcal{H}(k+1)$ in Eq. (2.7-12), that

$$P(k+1) = [\underline{h}w\underline{h}' + \mathcal{H}'W\mathcal{H}]^{-1}$$

$$= [\underline{h}w\underline{h}' + P^{-1}(k)]^{-1} \quad ; \tag{2.7-15}$$

hence,

$$P^{-1}(k+1) = P^{-1}(k) + \underline{h}(k+1)w(k+1)\underline{h}'(k+1) \tag{2.7-16}$$

which is a sequential equation for computing $P^{-1}(k+1)$ from $P^{-1}(k)$. From Eqs. (2.7-12), (2.7-14), and (2.7-15), we find that

$$\hat{\underline{\theta}}(k+1) = P(k+1)[\underline{h}wz + P^{-1}(k)\hat{\underline{\theta}}(k)] \quad . \tag{2.7-17}$$

Solving Eq. (2.7-16) for $P^{-1}(k)$ and putting the resulting expression into Eq. (2.7-17), we then find that

$$\hat{\underline{\theta}}(k+1) = P(k+1)\{\underline{h}wz + [P^{-1}(k+1) + \underline{h}w\underline{h}']\hat{\underline{\theta}}(k)\} \quad , \tag{2.7-18}$$

and finally,

$$\hat{\underline{\theta}}(k+1) = \hat{\underline{\theta}}(k) + P(k+1)\underline{h}(k+1)w(k+1)[z(k+1) - \underline{h}'(k+1)\hat{\underline{\theta}}(k)] \tag{2.7-19}$$

for $k > n - 1$ (expanding-memory identifier). ▲

Equation (2.7-19) is a sequential algorithm for computing $\hat{\underline{\theta}}(k+1)$ from the preceding estimate $\hat{\underline{\theta}}(k)$ and from a linear transformation of the equation error associated with the new measurement made at t_{k+1}. Let us denote this linear transformation as $K^*(k+1)$, where

$$K^*(k+1) = P(k+1)\underline{h}(k+1)w(k+1) \quad . \tag{2.7-20}$$

This $K^*(k+1)$ will be referred to as the *generalized least-squares gain matrix*, or *weighting matrix*. Within the general class of identification algorithms having the structure

$$\hat{\underline{\theta}}(k+1) = \hat{\underline{\theta}}(k) + K(k+1)\tilde{z}(k+1) \tag{2.7-21}$$

where

$$\tilde{z}(k+1) = z(k+1) - \underline{h}'(k+1)\hat{\underline{\theta}}(k) \quad , \tag{2.7-22}$$

we have just determined the weighting matrix $K(k+1) = K^*(k+1)$ that minimizes $\sum_{j=0}^{L+1} w(k+1-j)\tilde{z}^2(k+1-j)$ [see Eq. (2.2-34)]. This provides another meaning to the optimization problem discussed in Section 2.2 and solved in Section 2.3.

Let us summarize the equations necessary for incorporating one additional measurement into our estimate.

Sequential generalized least-squares estimation algorithm: One additional measurement

$$\hat{\underline{\theta}}(k+1) = \hat{\underline{\theta}}(k) + K^*(k+1)[z(k+1) - \underline{h}'(k+1)\hat{\underline{\theta}}(k)] \tag{2.7-23}$$

$$K^*(k+1) = P(k+1)\underline{h}(k+1)w(k+1) \tag{2.7-24}$$

$$P^{-1}(k+1) = P^{-1}(k) + \underline{h}(k+1)w(k+1)\underline{h}'(k+1) \tag{2.7-25}$$

for $k > n - 1$.

The order in which these computations must be performed is: $P^{-1}(k+1) \rightarrow P(k+1) \rightarrow K^*(k+1) \rightarrow \hat{\underline{\theta}}(k+1)$. If k_0 denotes the starting value of k, where $k_0 > n - 1$, then we must know $P^{-1}(k_0)$ and $\hat{\underline{\theta}}(k_0)$ a priori in order to initialize Eqs. (2.7-25) and (2.7-23), respectively. We will have more to say about the initialization problem in Section 2.8.

2.7.2 Generalization of the Sequential Algorithm: More than One Additional Measurement†

Here, we generalize the results of the preceding subsection to the case when p additional measurements, $z(k+1)$, $z(k+2)$, ..., $z(k+p)$, become available. Results are stated without proofs, since they can be obtained by following the derivations in Section 2.7.1 (Problem 2-11).

Let

$$\underline{z}^{o}(k+p) = [z(k+p), z(k+p-1), \ldots, z(k+1)]' \quad , \tag{2.7-26}$$

$$\underline{v}^{o}(k+p) = [v(k+p), v(k+p-1), \ldots, v(k+1)]' \quad , \tag{2.7-27}$$

and

$$H^{o}(k+p) = \begin{pmatrix} \underline{h}'(k+p) \\ \underline{h}'(k+p-1) \\ \vdots \\ \underline{h}'(k+1) \end{pmatrix} \quad . \tag{2.7-28}$$

Then the concatenated measurement equation for all $(L+1)+p$ measurements,

$$\underline{\mathcal{Z}}(k+p) = \mathcal{H}(k+p)\underline{\theta} + \underline{\mathcal{V}}(k+p) \quad , \tag{2.7-29}$$

can also be written as

$$\begin{pmatrix} \underline{z}^{o}(k+p) \\ \hline \underline{\mathcal{Z}}(k) \end{pmatrix} = \begin{pmatrix} H^{o}(k+p) \\ \hline \mathcal{H}(k) \end{pmatrix} \underline{\theta} + \begin{pmatrix} \underline{v}^{o}(k+p) \\ \hline \underline{\mathcal{V}}(k) \end{pmatrix} \quad . \tag{2.7-30}$$

Letting $W(k+p)$ denote the matrix of weights used in the generalized least-squares cost function [Eq. (2.2-35)], where

$$W(k+p) = \left(\begin{array}{c|c} W^{o}(k+p) & 0_1 \\ \hline 0_2 & W(k) \end{array}\right) \quad , \tag{2.7-31}$$

† This section can be omitted on first reading.

0_1 is the $p \times (L + 1)$ zero matrix and 0_2 is the $(L + 1) \times p$ zero matrix. We obtain the following set of equations for incorporating p additional measurements into our estimate.

Sequential generalized least-squares estimation algorithm: p *additional measurements*

$$\hat{\underline{\theta}}(k + p) = \hat{\underline{\theta}}(k) + K^*(k + p)[\underline{z}^o(k + p) - H^o(k + p)\hat{\underline{\theta}}(k)] \quad (2.7\text{-}32)$$

$$K^*(k + p) = P(k + p)H^{o'}(k + p)W^o(k + p) \quad (2.7\text{-}33)$$

$$P^{-1}(k + p) = P^{-1}(k) + H^{o'}(k + p)W^o(k + p)H^o(k + p) \quad (2.7\text{-}34)$$

for $k > n - 1$ and $p \geq 1$.

Observe that when $p = 1$ these equations reduce to their counterparts in Eqs. (2.7-23)-(2.7-25). In the rest of this chapter, our attention will be directed at the "one additional measurement" algorithm. Comparable results can always, and easily, be derived for the "p additional measurements" algorithm.

2.7.3 Matrix Inversion Lemma

Equation (2.7-25) requires the inversion of an $n \times n$ matrix. If there are many unknown parameters, so that n is large, it will be computationally costly to perform this matrix inversion. Fortunately, an alternative is available. This alternative is based upon the so-called "matrix inversion lemma," which we shall now state and prove (*2-6* and *2-7*).

Matrix Inversion Lemma. *If the matrices* P_1, P_2, H, *and* R *satisfy the equation*

$$P_2^{-1} = P_1^{-1} + H'R^{-1}H \quad (2.7\text{-}35)$$

where P_1 *and* R *are nonsingular and* H *is of maximum rank, then*

$$P_2 = P_1 - P_1H'(HP_1H' + R)^{-1}HP_1 \quad . \quad (2.7\text{-}36)$$

Proof.† Premultiply Eq. (2.7-35) by P_2, obtaining

† One way to prove the correctness of the lemma is to show that $P_2P_2^{-1} = I$; however, we prefer to give a constructive proof.

$$I = P_2P_1^{-1} + P_2H'R^{-1}H \quad . \qquad (2.7\text{-}37)$$

Postmultiply this equation by P_1 and then postmultiply the new result by H', obtaining

$$P_1 = P_2 + P_2H'R^{-1}HP_1 \qquad (2.7\text{-}38)$$

and

$$P_1H' = P_2H' + P_2H'R^{-1}HP_1H'$$

$$= P_2H'R^{-1}(R + HP_1H') \quad . \qquad (2.7\text{-}39)$$

Now postmultiply this equation by $(R + HP_1H')^{-1}$ and the resulting equation by HP_1, obtaining

$$P_1H'(R + HP_1H')^{-1} = P_2H'R^{-1} \qquad (2.7\text{-}40)$$

and

$$P_1H'(R + HP_1H')^{-1}HP_1 = P_2H'R^{-1}HP_1 \quad . \qquad (2.7\text{-}41)$$

Subtract Eq. (2.7-41) from P_1,

$$P_1 - P_1H'(R + HP_1H')^{-1}HP_1 = P_1 - P_2H'R^{-1}HP_1 \quad , \qquad (2.7\text{-}42)$$

and substitute Eq. (2.7-38) for the first term on the right-hand side of this equation, obtaining

$$P_1 - P_1H'(R + HP_1H')^{-1}HP_1 = P_2 \qquad (2.7\text{-}43)$$

which is the desired result. ▲

What is the advantage of using what appears to be a more complicated expression for P_2, Eq. (2.7-36), over what appears to be a simpler expression for P_2, Eq. (2.7-35)? Equation (2.7-36) requires the computation of the inverse of only one matrix, whereas at least two inverses are required to obtain P_2 from Eq. (2.7-35). In addition, if P_1 and P_2 are $n \times n$ matrices, H is $n \times m$, and R is $m \times m$, where $m < n$, then Eq. (2.7-35) requires that we invert an $m \times m$ *and* an $n \times n$ matrix, whereas Eq. (2.7-36) requires the inversion of an

$m \times m$ matrix. In the special case when $m = 1$, matrix inversion in Eq. (2.7-36) is replaced by division; however, if we choose to use Eq. (2.7-35) to compute P_2, we must still invert an $n \times n$ matrix, even when $m = 1$.

2.7.4 An Alternative Formulation of the Sequential Generalized Least-Squares Estimation Algorithm

Let us compare Eqs. (2.7-25) and (2.7-35) and establish the following associations: $P_2 \triangleq P(k+1)$, $P_1 \triangleq P(k)$, $H \triangleq \underline{h}'(k+1)$, and $R \triangleq 1/w(k+1)$. From Eq. (2.7-36), we obtain the following alternative expression for $P(k+1)$:

$$P(k+1) = P(k) - P(k)\underline{h}(k+1)\left(\underline{h}'(k+1)P(k)\underline{h}(k+1) + \frac{1}{w(k+1)}\right)^{-1} \times \underline{h}'(k+1)P(k). \tag{2.7-44}$$

The advantage of computing $P(k+1)$ from this expression over computing it from Eq. (2.7-25) is obvious. No matrix inversion is required at all in Eq. (2.7-44), whereas we must invert an $n \times n$ matrix to obtain $P(k+1)$ from Eq. (2.7-25).

Let us now obtain an alternative expression for $K^*(k+1)$. From Eqs. (2.7-24) and (2.7-44),

$$\begin{aligned} K^*(k+1) &= \left[P(k) - P(k)\underline{h}(k+1)\left(\underline{h}'(k+1)P(k)\underline{h}(k+1) + \frac{1}{w(k+1)}\right)^{-1} \times \underline{h}'(k+1)P(k)\right]\underline{h}(k+1)w(k+1) \\ &= P(k)\underline{h}(k+1)\left[I - \left(\underline{h}'(k+1)P(k)\underline{h}(k+1) + \frac{1}{w(k+1)}\right)^{-1} \times \underline{h}'(k+1)P(k)\underline{h}(k+1)\right]w(k+1) \\ &= P(k)\underline{h}(k+1)\left(\underline{h}'(k+1)P(k)\underline{h}(k+1) + \frac{1}{w(k+1)}\right)^{-1} \\ &\quad \times \left(\underline{h}'(k+1)P(k)\underline{h}(k+1) + \frac{1}{w(k+1)} - \underline{h}'(k+1)P(k)\underline{h}(k+1)\right)w(k+1) \\ &= P(k)\underline{h}(k+1)\left(\underline{h}'(k+1)P(k)\underline{h}(k+1) + \frac{1}{w(k+1)}\right)^{-1}. \end{aligned} \tag{2.7-45}$$

Equations (2.7-23), (2.7-45), and (2.7-44) constitute an alternative formulation of our sequential algorithm. Observe that $P(k + 1)$ can be expressed in terms of $K^*(k + 1)$.

Alternative formulation of the sequential generalized least-squares estimation algorithm: One additional measurement

$$\underline{\hat{\theta}}(k + 1) = \underline{\hat{\theta}}(k) + K^*(k + 1)[z(k + 1) - \underline{h}'(k + 1)\underline{\hat{\theta}}(k)] \tag{2.7-46}$$

$$K^*(k + 1) = P(k)\underline{h}(k + 1)\left[\underline{h}'(k + 1)P(k)\underline{h}(k + 1) + \frac{1}{w(k + 1)}\right]^{-1} \tag{2.7-47}$$

$$P(k + 1) = [I - K^*(k + 1)\underline{h}'(k + 1)]P(k) \tag{2.7-48}$$

for $k > n - 1$.

The order in which these computations must be performed differs from the order in which the computations had to be performed for Eqs. (2.7-23)-(2.7-25). Here, we proceed as follows: $P(k) \to K^*(k + 1) \to \underline{\hat{\theta}}(k + 1) \to P(k + 1)$. If k_0 denotes the starting value of k, where $k_0 > n - 1$, then we must know $P(k_0)$ and $\underline{\hat{\theta}}(k_0)$ a priori, in order to initialize Eqs. (2.7-47), (2.7-48), and (2.7-46).

We have demonstrated a significant advantage for using Eqs. (2.7-46)-(2.7-48) over Eqs. (2.7-23)-(2.7-25), when these equations are compared on the basis of matrix inversions; however, other computational factors enter into the comparisons of these formulations, such as accuracy and initialization difficulties. We shall return to a comparison of these formulations in Section 2.12.

2.7.5 Example 6 (Continued)

Let us illustrate some of the results that have been derived in this section by obtaining a sequential algorithm for the least-squares estimate of the scalar θ in the instrument calibration example, Example 6 in Section 2.5. For an expanding memory estimator, we set $L = k$ in Eq. (2.5-44); hence,

$$\hat{\theta}(k) = \frac{1}{k + 1}\sum_{i=0}^{k} z(k - i) \quad . \tag{2.7-49}$$

Before obtaining the sequential algorithm for $\hat{\theta}(k + 1)$ from Eqs. (2.7-23)-(2.7-25), let us obtain it directly from Eq. (2.7-49) by replacing k with k + 1 and performing some algebraic manipulations, as follows:

$$\hat{\theta}(k + 1) = \frac{1}{k + 2} \sum_{i=0}^{k+1} z(k + 1 - i)$$

$$= \frac{1}{k + 2} \left[z(k + 1) + \sum_{i=1}^{k+1} z(k + 1 - i) \right]$$

$$= \frac{1}{k + 2} z(k + 1) + \left(\frac{k + 1}{k + 2} \right) \frac{1}{k + 1} \sum_{j=0}^{k} z(k - j)$$

$$= \frac{1}{k + 2} z(k + 1) + \frac{k + 1}{k + 2} \hat{\theta}(k) \tag{2.7-50}$$

which is the desired result.

Now, we shall obtain the same result from Eqs. (2.7-23) and (2.7-24), making use of the facts that

$$h(k + 1) = 1 \tag{2.7-51}$$

and [see Eqs. (2.7-13) and (2.5-43)]

$$P(k + 1) = [\mathcal{H}'(k + 1)\mathcal{H}(k + 1)]^{-1} = \frac{1}{k + 2} \quad ; \tag{2.7-52}$$

hence,

$$K^*(k + 1) = P(k + 1) = \frac{1}{k + 2} \tag{2.7-53}$$

and

$$\hat{\theta}(k + 1) = \hat{\theta}(k) + \frac{1}{k + 2} [z(k + 1) - \hat{\theta}(k)]$$

$$= \frac{k + 1}{k + 2} \hat{\theta}(k) + \frac{1}{k + 2} z(k + 1) \tag{2.7-54}$$

which is Eq. (2.7-50), exactly. ▲

Usually it is in only the simplest of cases that we can obtain the sequential estimation algorithm by algebraic manipulations; however, we can always obtain it, regardless of the application's difficulty, by means of the results that have been obtained in this section.

2.8 A TECHNIQUE FOR SEQUENTIAL STARTUP OF THE SEQUENTIAL GENERALIZED LEAST-SQUARES ALGORITHM

2.8.1 Introduction

We have seen that the sequential generalized least-squares parameter estimation algorithms, given in either Eqs. (2.7-23)-(2.7-25) or Eqs. (2.7-46)-(2.7-48), iterate for $k > n - 1$. Let us assume that ℓ is the starting value for k, and that $k = \ell, \ell + 1, \ell + 2, \ldots$. Our estimation algorithms require $P(\ell)$ and $\hat{\underline{\theta}}(\ell)$ to start them. Here we show how these algorithms can be used for $k = 0, 1, \ldots, \ell, \ell + 1, \ldots$ such that we get the correct values of $\hat{\underline{\theta}}(\ell + 1)$ and $P(\ell + 1)$, etc. Instead of having to compute $P(\ell)$ and $\hat{\underline{\theta}}(\ell)$, we will show that the entire estimation procedure can be started by choosing

$$P(0) = \left(\frac{1}{a^2} I_n + \underline{h}(0)w(0)\underline{h}'(0)\right)^{-1} \tag{2.8-1}$$

and

$$\hat{\underline{\theta}}(0) = P(0)\left(\frac{1}{a}\underline{\varepsilon} + \underline{h}(0)w(0)z(0)\right) \tag{2.8-2}$$

where a is a very large number, $\underline{\varepsilon}$ is an $n \times 1$ vector of very small numbers ε, and I_n is the $n \times n$ identity matrix.

2.8.2 A Useful Artifice (*2-18*)

Figure 2.8-1 depicts a time scale on which we have shown the measurements $z(0), z(1), \ldots, z(\ell + 1)$, which are used to estimate $\hat{\underline{\theta}}(\ell + 1)$, and a collection of n artificial measurements (recall that there are n unknown parameters in $\underline{\theta}$), $z^a(-n), z^a(-n + 1), \ldots, z^a(-2), z^a(-1)$. These measurements are all set equal to a very small number ε,

$$\begin{pmatrix} z^a(-1) \\ z^a(-2) \\ \vdots \\ z^a(-n) \end{pmatrix} = \begin{pmatrix} \varepsilon \\ \varepsilon \\ \vdots \\ \varepsilon \end{pmatrix} \triangleq \underline{\varepsilon} , \tag{2.8-3}$$

and are *assumed* to be related to $\underline{\theta}$ by the equation

$$\begin{pmatrix} z^a(-1) \\ z^a(-2) \\ \vdots \\ z^a(-n) \end{pmatrix} = \begin{pmatrix} 1/a & 0 & \cdots & 0 \\ 0 & 1/a & \cdots & 0 \\ \vdots & \vdots & \ddots & \vdots \\ 0 & 0 & \cdots & 1/a \end{pmatrix} \begin{pmatrix} \theta_1 \\ \theta_2 \\ \vdots \\ \theta_n \end{pmatrix} = \frac{1}{a} I_n \underline{\theta} \tag{2.8-4}$$

where a is a very large number. Observe that this equation is a concatenation of the n artificial measurements and, as such, it can also be written as

$$\underline{Z}^a(-1) = \mathcal{H}^a(-1)\underline{\theta} \tag{2.8-5}$$

where, in the spirit of Section 2.2, we have let

$$\underline{Z}^a(-1) = \begin{pmatrix} z^a(-1) \\ z^a(-2) \\ \vdots \\ z^a(-n) \end{pmatrix} = \underline{\varepsilon} \tag{2.8-6}$$

and

$$\mathcal{H}^a(-1) = \frac{1}{a} I_n \quad . \tag{2.8-7}$$

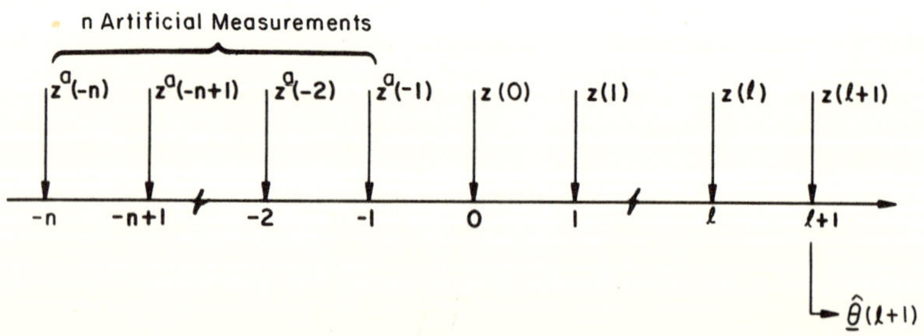

Fig. 2.8-1. Relationship between artificial measurements (superscripted "a") and real measurements.

2.8.3 Objective

First, we shall obtain expressions for $\hat{\underline{\theta}}^a(\ell + 1)$ and $P^a(\ell + 1)$. These quantities utilize the n artificial measurements (hence, the superscript "a" on them) *and* the $\ell + 1$ real measurements. Next, we shall obtain $\hat{\underline{\theta}}(\ell + 1)$ and $P(\ell + 1)$, which make use only of the $\ell + 1$ real measurements. Finally, we shall compare $\hat{\underline{\theta}}(\ell + 1)$ and $\hat{\underline{\theta}}^a(\ell + 1)$, and $P(\ell + 1)$ and $P^a(\ell + 1)$ and show that for ε very small, and a very large, as has been assumed above, $\hat{\underline{\theta}}^a(\ell + 1) \to \hat{\underline{\theta}}(\ell + 1)$ and $P^a(\ell + 1) \to P(\ell + 1)$. Along the way, we shall also obtain expressions for $\hat{\underline{\theta}}^a(0)$ and $P^a(0)$, which can then be used as starting values for $\hat{\underline{\theta}}(0)$ and $P(0)$.

2.8.4 Computation of $\hat{\underline{\theta}}^a(\ell + 1)$ and $P^a(\ell + 1)$

$\hat{\underline{\theta}}^a(-1)$ and $P^a(-1)$

We begin by batch processing the n artificial measurements, using a least-squares algorithm. There is no reason to weight any of these measurements more or less heavily than others, since they are all numerically equal to ε; thus, our choice of the least-squares algorithm. Values of $\hat{\underline{\theta}}^a(-1)$ and $P^a(-1)$ are given by

$$\hat{\underline{\theta}}^a(-1) = [\mathcal{H}^{a'}(-1)\mathcal{H}^a(-1)]^{-1}\mathcal{H}^{a'}(-1)\underline{\mathcal{Z}}^a(-1) \tag{2.8-8}$$

and

$$P^a(-1) = [\mathcal{H}^{a'}(-1)\mathcal{H}^a(-1)]^{-1} \ . \tag{2.8-9}$$

Substituting Eqs. (2.8-6) and (2.8-7) into these equations, we find that

$$\hat{\underline{\theta}}^a(-1) = a\underline{\varepsilon} \tag{2.8-10}$$

and

$$P^a(-1) = a^2 I_n \ . \tag{2.8-11}$$

Incorporation of z(0)

When z(0) is added to the n artificial measurements, our concatenated measurement equation becomes

$$\underline{\mathcal{Z}}^a(0) = \mathcal{H}^a(0)\underline{\theta} + \underline{\mathcal{V}}^a(0) \tag{2.8-12}$$

where

$$\underline{Z}^a(0) = \begin{pmatrix} z(0) \\ \text{------} \\ \underline{Z}^a(-1) \end{pmatrix} , \tag{2.8-13}$$

$$\underline{V}^a(0) = \begin{pmatrix} v(0) \\ \text{----} \\ \underline{0} \end{pmatrix} , \tag{2.8-14}$$

and

$$H^a(0) = \begin{pmatrix} \underline{h}'(0) \\ \text{------} \\ H^a(-1) \end{pmatrix} . \tag{2.8-15}$$

We shall now compute $\hat{\underline{\theta}}^a(0)$ and $P^a(0)$, using the sequential generalized least-squares algorithm in Eqs. (2.7-17) and (2.7-15). Hence,

$$\hat{\underline{\theta}}^a(0) = P^a(0)\{[P^a(-1)]^{-1}\hat{\underline{\theta}}^a(-1) + \underline{h}(0)w(0)z(0)\} \tag{2.8-16}$$

and

$$P^a(0) = \{[P^a(-1)]^{-1} + \underline{h}(0)w(0)\underline{h}(0)\} . \tag{2.8-17}$$

Substituting Eqs. (2.8-10) and (2.8-11) into these expressions, we find that

$$\hat{\underline{\theta}}^a(0) = P^a(0)\left(\frac{1}{a}\,\underline{\varepsilon} + \underline{h}(0)w(0)z(0)\right) \tag{2.8-18}$$

and

$$P^a(0) = \left(\frac{1}{a^2}\,I_n + \underline{h}(0)w(0)\underline{h}'(0)\right)^{-1} . \tag{2.8-19}$$

Incorporation of z(1)

When z(1) is added to the n artificial measurements and z(0), our concatenated measurement equation becomes

$$\underline{Z}^a(1) = H^a(1)\underline{\theta} + \underline{V}^a(1) \tag{2.8-20}$$

where

$$\underline{Z}^a(1) = \begin{pmatrix} z(1) \\ \text{-----} \\ \underline{Z}^a(0) \end{pmatrix} , \tag{2.8-21}$$

$$\underline{V}^a(1) = \begin{pmatrix} v(1) \\ \text{-----} \\ \underline{V}^a(0) \end{pmatrix} , \tag{2.8-22}$$

and

$$H^a(1) = \begin{pmatrix} \underline{h}'(1) \\ \text{-----} \\ H^a(0) \end{pmatrix} . \tag{2.8-23}$$

Values of $\hat{\underline{\theta}}^a(1)$ and $P^a(1)$ are computed in a manner similar to the way in which $\hat{\underline{\theta}}^a(0)$ and $P^a(0)$ were computed, using Eqs. (2.7-17) and (2.7-15):

$$\hat{\underline{\theta}}^a(1) = P^a(1)\{[P^a(0)]^{-1}\hat{\underline{\theta}}^a(0) + \underline{h}(1)w(1)z(1)\} \tag{2.8-24}$$

and

$$P^a(1) = \{[P^a(0)]^{-1} + \underline{h}(1)w(1)\underline{h}'(1)\}^{-1} \quad . \tag{2.8-25}$$

Substituting Eqs. (2.8-18) and (2.8-19) into these expressions, we find, after some algebra, that

$$\hat{\underline{\theta}}^a(1) = P^a(1)\left(\frac{1}{a}\,\underline{\varepsilon} + \sum_{j=0}^{1} \underline{h}(j)w(j)z(j)\right) \tag{2.8-26}$$

and

$$P^a(1) = \left(\frac{1}{a^2}\,I_n + \sum_{j=0}^{1} \underline{h}(j)w(j)\underline{h}'(j)\right)^{-1} \quad . \tag{2.8-27}$$

Generalization: $\hat{\underline{\theta}}^a(\ell + 1)$ and $P^a(\ell + 1)$

Comparing Eqs. (2.8-18) and (2.8-26), and (2.8-19) and (2.8-27), we claim (by analytical extension, or proof by induction) that these results generalize to

$$\hat{\underline{\theta}}^a(\ell + 1) = P^a(\ell + 1)\left(\frac{1}{a}\,\underline{\varepsilon} + \sum_{j=0}^{\ell+1} \underline{h}(j)w(j)z(j)\right) \tag{2.8-28}$$

and

$$P^a(\ell + 1) = \left(\frac{1}{a^2}\,I_n + \sum_{j=0}^{\ell+1} \underline{h}(j)w(j)\underline{h}'(j)\right)^{-1} \quad . \tag{2.8-29}$$

Computation of $\hat{\underline{\theta}}(\ell + 1)$ and $P(\ell + 1)$

We shall obtain expressions for $\hat{\underline{\theta}}(\ell + 1)$ and $P(\ell + 1)$ by batch processing the $\ell + 2$ measurements $z(0), z(1), \ldots, z(\ell + 1)$. From Eqs. (2.7-14), (2.2-15), (2.2-13), and (2.2-36), it is straightforward to show that

$$\begin{aligned}\hat{\underline{\theta}}(\ell + 1) &= P(\ell + 1)[H'(\ell + 1)W(\ell + 1)\underline{Z}(\ell + 1)] \\ &= P(\ell + 1)\left(\sum_{j=0}^{\ell+1} \underline{h}(j)w(j)z(j)\right) \quad ,\end{aligned} \tag{2.8-30}$$

and from Eqs. (2.7-13), (2.2-15), and (2.2-36), one determines, similarly, that

$$P(\ell + 1) = \left(\sum_{j=0}^{\ell+1} \underline{h}(j) w(j) \underline{h}'(j) \right)^{-1} . \tag{2.8-31}$$

2.8.6 Comparisons of Equations

Comparing Eqs. (2.8-29) and (2.8-31), and then Eqs. (2.8-28) and (2.8-30), under our original assumtions that a is very large and ε is very small, we conclude that

$$P^a(\ell + \,) \to P(\ell + 1)$$

and for $a\gg>$ and $\varepsilon\ll<$. (2.8-32)

$$\underline{\hat{\theta}}^a(\ell + 1) \to \underline{\hat{\theta}}(\ell + 1)$$

2.8.7 Conclusions

Rather than specify $\underline{\hat{\theta}}(\ell)$ and $P(\ell)$, and begin the sequential least-squares algorithm for $k = \ell$, we have shown that the algorithm can be started for $k = 0$, as long as $P(0)$ and $\underline{\hat{\theta}}(0)$ are chosen in accordance with Eqs. (2.8-1) and (2.8-2), where a is a very large number and ε is a very small number. All of the information that was assumed about the artificial measurement sequence is contained in $P(0)$ and $\underline{\hat{\theta}}(0)$.

Note also, from Eqs. (2.8-1) and (2.8-2), that *if* $z(0) = 0$, *in which case we can set* $w(0) = 0$, *then the sequential algorithm can be initialized by setting* $\underline{\hat{\theta}}(0) = \underline{0}$, *and* $P(0)$ *to a matrix of large positive numbers*. This is often done in practice, since there are many situations for which we have no measurement available at t_0. Observe, also, that in this case $\underline{\hat{\theta}}(0) \simeq a\underline{\varepsilon}$; and it is clear that, for $\underline{\hat{\theta}}(0) \to \underline{0}$, each element of $\underline{\varepsilon}$ should behave at least like $1/a^2$. For example, if we set $a = 10^5$ we should set $\varepsilon = 1/10^{10}$.

We have shown how to iterate the sequential generalized least-squares estimation algorithm starting with $k = 0$; but what meaning can we give to the estimates obtained by using less than n data? For $k < n - 1$, we know that

$$\underline{Z}(k) = \mathcal{H}(k)\underline{\theta} + \underline{\mathcal{V}}(k) \tag{2.8-33}$$

is an underdetermined system, and that there are many possible solutions (nonunique) to such a system. The solution that we have obtained is related to the generalized (pseudo-) inverse of $H(k)$, $H^{\dagger}(k)$ (*2-8-2-10*). The $H^{\dagger}$ is essentially the inverse of a rectangular matrix where, when $k < n - 1$, the matrix has more columns than rows.

The reader is cautioned to be careful about the meaning and value placed on all estimates for $k < n - 1$.

2.9 GENERALIZATIONS TO VECTOR MEASUREMENTS

Here we shall consider the situation when L + 1 measurements of a vector signal $\underline{y}(t)$ are made at times $t_{k-L}, t_{k-L-1}, \ldots, t_k$, where $k \geq L$. Using our earlier convention to replace t_k by k, the $\underline{z}(k)$ denotes the measured value of $\underline{y}(k)$. We shall demonstrate how all the preceding results are easily extended to the vector measurement situation by means of simple transformations of symbols. Before proceeding, let us discuss some applications where a vector of measurements can occur.

A vector of measurements can occur in any application where it is possible to use more than one sensor; however, it is also possible to obtain a vector of measurements from certain types of individual sensors. In spacecraft applications, it is not unusual to be able to measure attitude (e. g., attitude reference unit), rate (e. g., rate gyro), and acceleration (e. g., accelerometer). In electrical systems applications, it is not uncommon to be able to measure voltages (e. g., voltmeter), currents (e. g., ammeter), and power (e. g., wattmeter). Radar measurements often provide information about range, azimuth, and elevation. Radar is an example of a single (active) sensor that provides a vector of measurements. We could continue this discussion indefinitely, citing examples from many other fields, but this would take us too far afield from our main objective; hence, we invite the reader to pursue this matter further for his own area of interest, as well as others, in Problem 2-18.

We shall assume that $\underline{z}(k)$ is an $m \times 1$ vector, where

$$\underline{z}(k) = \underline{y}(k) + \underline{v}(k) \tag{2.9-1}$$

and

$$E\{\underline{v}(k)\} = \underline{0} \quad . \tag{2.9-2}$$

Each component in $\underline{y}(k)$ is related to an unknown $n \times 1$ parameter vector $\underline{\theta}$ as in Eq. (2.2-5); thus,

$$\underline{y}(k) = H(k)\underline{\theta} \tag{2.9-3}$$

where $H(k)$ is an $m \times n$ matrix given by

$$H(k) = \begin{pmatrix} h_{11}(k) & h_{12}(k) & \cdots & h_{1n}(k) \\ h_{21}(k) & h_{22}(k) & \cdots & h_{2n}(k) \\ \vdots & \vdots & \ddots & \vdots \\ h_{m1}(k) & h_{m2}(k) & \cdots & h_{mn}(k) \end{pmatrix} \quad . \tag{2.9-4}$$

From Eqs. (2.9-1) and (2.9-3), we see that

$$\underline{z}(k) = H(k)\underline{\theta} + \underline{v}(k) \tag{2.9-5}$$

which is the vector analog to Eq. (2.2-9). Proceeding as in Section 2.2, we concatenate the $L + 1$ vector measurements $\underline{z}(k)$, $\underline{z}(k - 1)$, ..., $\underline{z}(k - L)$ to obtain the concatenated measurement equation

$$\underline{Z}(k) = \mathcal{H}(k)\underline{\theta} + \underline{V}(k) \tag{2.9-6}$$

where

$$\underline{Z}(k) = \begin{pmatrix} \underline{z}(k) \\ \hline \underline{z}(k - 1) \\ \hline \vdots \\ \hline \underline{z}(k - L) \end{pmatrix} = \begin{pmatrix} \underline{z}(k) \\ \hline \underline{Z}(k - 1) \end{pmatrix} \quad , \tag{2.9-7}$$

$$\underline{V}(k) = \begin{pmatrix} \underline{v}(k) \\ \hline \underline{v}(k - 1) \\ \hline \vdots \\ \hline \underline{v}(k - L) \end{pmatrix} = \begin{pmatrix} \underline{v}(k) \\ \hline \underline{V}(k - 1) \end{pmatrix} \quad , \tag{2.9-8}$$

and

$$\mathcal{H}(k) = \begin{pmatrix} H(k) \\ \hline H(k-1) \\ \hline \vdots \\ \hline H(k-L) \end{pmatrix} = \begin{pmatrix} H(k) \\ \hline \mathcal{H}(k-1) \end{pmatrix} . \tag{2.9-9}$$

Observe that Eqs. (2.9-6) and (2.2-16) are structurally the same; however, $\underline{\mathcal{Z}}(k)$, $\underline{\mathcal{V}}(k)$, and $\mathcal{H}(k)$ are now $(L + 1)m \times 1$, $(L + 1)m \times 1$, $(L + 1)m \times n$ quantities, respectively.

Symbol $\hat{\underline{\theta}}(k)$ is now the estimate of $\underline{\theta}$ based on $(L + 1)m$ data, or, equivalently, $(L + 1)$ data vectors $\underline{z}(k)$, $\underline{z}(k - 1)$, ..., $\underline{z}(k - L)$. It is obtained by minimizing $J[\hat{\underline{\theta}}(k)]$ in Eq. (2.2-35), where

$$W(k) = \left(\begin{array}{c|c|c|c} w(k) & 0 & \cdots & 0 \\ \hline 0 & w(k-1) & \cdots & 0 \\ \hline \vdots & \vdots & \ddots & \vdots \\ \hline 0 & 0 & \cdots & w(k-L) \end{array}\right) = \left(\begin{array}{c|c} w(k) & 0_1 \\ \hline 0_2 & W(k-1) \end{array}\right) , \tag{2.9-10}$$

$w(k - j)$ is an $m \times m$ matrix of weights that weight the individual components of $\underline{z}(k - j)$, 0_1 is an $m \times Lm$ zero matrix, and 0_2 is an $Lm \times m$ zero matrix.

It is clear from the structural similarity between Eqs. (2.9-6) and (2.2-16) that we can obtain the generalized least-squares estimate of $\underline{\theta}$ from our batch algorithm in Section 2.3, or our sequential algorithm in Section 2.7, by means of the transformations listed in Table 2-3. These results are then summarized.

Generalized least-squares batch algorithm

$$\hat{\underline{\theta}}(k) = [\mathcal{H}'(k)W(k)\mathcal{H}(k)]^{-1}\mathcal{H}'(k)W(k)\underline{\mathcal{Z}}(k) \tag{2.9-11}$$

for $k \geq L$.

Sequential generalized least-squares algorithm: One additional measurement vector

$$\hat{\underline{\theta}}(k + 1) = \hat{\underline{\theta}}(k) + K^*(k + 1)[\underline{z}(k + 1) - H(k + 1)\hat{\underline{\theta}}(k)] \tag{2.9-12}$$

TABLE 2-3
Transformations from Scalar to Vector Measurement Situations, and Vice Versa

Scalar measurement	Vector of measurements
$z(k+1)$	$\underline{z}(k+1)$, an $m \times 1$ vector
$v(k+1)$	$\underline{v}(k+1)$, an $m \times 1$ vector
$w(k+1)$	$w(k+1)$, an $m \times m$ matrix
$\underline{h}'(k+1)$, a $1 \times n$ matrix	$H(k+1)$, an $m \times n$ matrix
$\underline{Z}(k)$, an $(L+1) \times 1$ vector	$\underline{Z}(k)$, an $(L+1)m \times 1$ vector
$\underline{V}(k)$, an $(L+1) \times 1$ vector	$\underline{V}(k)$, an $(L+1)m \times 1$ vector
$W(k)$, an $(L+1) \times (L+1)$ matrix	$W(k)$, an $(L+1)m \times (L+1)m$ matrix
$H(k)$, an $(L+1) \times n$ matrix	$H(k)$, an $(L+1)m \times n$ matrix

$$K^*(k+1) = P(k+1)H'(k+1)w(k+1) \qquad (2.9\text{-}13)$$

$$P^{-1}(k+1) = P^{-1}(k) + H'(k+1)w(k+1)H(k+1) \qquad (2.9\text{-}14)$$

for $k \geq L$; OR

$$\hat{\underline{\theta}}(k+1) = \hat{\underline{\theta}}(k) + K^*(k+1)[\underline{z}(k+1) - H(k+1)\hat{\underline{\theta}}(k)] \qquad (2.9\text{-}15)$$

$$K^*(k+1) = P(k)H'(k+1)[H(k+1)P(k)H'(k+1)$$
$$+ w^{-1}(k+1)]^{-1} \qquad (2.9\text{-}16)$$

$$P(k+1) = [I - K^*(k+1)H(k+1)]P(k) \qquad (2.9\text{-}17)$$

for $k \geq L$.

Observe that $P(k)$ is still an $n \times n$ matrix; but $[H(k+1)P(k)H'(k+1) + w^{-1}(k+1)]$ is now an $m \times m$ matrix, rather than the scalar quantity in Eq. (2.7-47). If $m < n$, as is usually the case, there is still the advantage to using the alternative formulation in Eqs. (2.9-15)-(2.9-17) discussed in Section (2.7-4).

We conclude this section with a discussion on choices for L (the reader is referred back to Section 2.6, where we discussed choices for L, for the case of scalar measurements — m = 1).

Our interest is in the case when there are more measurements than unknown parameters; thus, L, m, and n are constrained by the following generalization of Eq. (2.6-1):

$$(L + 1)m > n \quad . \tag{2.9-18}$$

The solution for L is

$$\begin{aligned} &\text{If } \frac{n}{m} \text{ is an integer,} && L > \frac{n}{m} - 1 \quad ; \\ &\text{If } \frac{n}{m} \text{ is not an integer,} && L \geq \,]\frac{n}{m} - 1[\end{aligned} \tag{2.9-19}$$

where]a[denotes the *greatest integer just greater than* a. For example, if n = 8 and m = 2, then L > 3; but if n = 9 and m = 2, $L \geq 4$. For *expanding-memory estimation*, L is set equal to k, and k must be chosen so that

$$\begin{aligned} &\text{If } \frac{n}{m} \text{ is an integer,} && k > \frac{n}{m} - 1 \quad ; \\ &\text{If } \frac{n}{m} \text{ is not an integer,} && k \geq \,]\frac{n}{m} - 1[\quad . \end{aligned} \tag{2.9-20}$$

We would normally choose the starting value of k as small as possible.

2.10 SEQUENTIAL AND SIMULTANEOUS DATA PROCESSING†

2.10.1 Introduction

Suppose that at each sampling time t_k, there are q statistically independent sources that provide measurement data. The m-dimensional measurement vector‡ $\underline{z}(k)$ can be represented as

---

† Our discussions in this section are quite similar to those in Section III, E of Sorenson (*2-11*). Copyright 1966 by Academic Press, Inc. Used with the permission of Academic Press, Inc.

‡ For the material in this section to be useful, $m \geq 2$.

$$\underline{z}(k) = \begin{pmatrix} \underline{z}_1(k) \\ \text{-----} \\ \underline{z}_2(k) \\ \text{-----} \\ \vdots \\ \text{-----} \\ \underline{z}_q(k) \end{pmatrix} = \begin{pmatrix} H_1(k) \\ \text{-----} \\ H_2(k) \\ \text{-----} \\ \vdots \\ \text{-----} \\ H_q(k) \end{pmatrix} \underline{\theta} + \begin{pmatrix} \underline{v}_1(k) \\ \text{-----} \\ \underline{v}_2(k) \\ \text{-----} \\ \vdots \\ \text{-----} \\ \underline{v}_q(k) \end{pmatrix} . \quad (2.10\text{-}1)$$

Since the sources are statistically independent,

$$E\{\underline{v}_i(k)\underline{v}_j'(k)\} = R_i(k)\delta_{ij} \quad (2.10\text{-}2)$$

where δ_{ij} is the Kroneker delta ($\delta_{ij} = 0$ for all $i \neq j$, and $\delta_{ij} = 1$ for $i = j$), and $i, j = 1, 2, \ldots, q$. The $R_i(k)$ have the dimension $m_i \times m_i$, where

$$\sum_{i=1}^{q} m_i = m \quad . \quad (2.10\text{-}3)$$

Let us assume that $\underline{\theta}$ is estimated by means of the sequential generalized least-squares algorithm in Eqs. (2.9-15)-(2.9-17). The inverse of the matrix $[H(k)P(k - 1)H'(k) + w^{-1}(k)]$ must be computed in order to obtain the estimate $\hat{\underline{\theta}}(k)$. If there are several data sources, the dimension m of this matrix could be quite large. The inversion on a digital computer of a matrix of large dimension is undesirable for several reasons - the amount of storage cells that must be used, the time that is consumed in obtaining the inverse, and the accuracy of the end result. Thus, if the inversion can be circumvented, it is advisable to do so.

An alternative to processing all m measurements in one batch - *simultaneously* - is available, and is one in which the largest matrix that has to be inverted has dimension equal to $\max(m_i)$. This procedure is known as *sequential processing*.

In sequential processing, each set of data is treated separately. Specifically, the data $\underline{z}_1(k)$ are used to obtain an estimate $\hat{\underline{\theta}}_1(k)$, with $\underline{z}(k) \triangleq \underline{z}_1(k)$. When these calculations are completed, $\underline{z}_2(k)$ is processed to obtain the estimate $\hat{\underline{\theta}}_2(k)$. The estimate $\hat{\underline{\theta}}(k - 1)$ is used to initialize the estimate $\hat{\underline{\theta}}_1(k)$; however, $\hat{\underline{\theta}}_1(k)$ is used to initialize $\hat{\underline{\theta}}_2(k)$. Each set of data is processed in this manner until the final set $\underline{z}_q(k)$ has been included. Then time is advanced to t_{k+1} and the cycle is repeated. The sequential processing technique is summarized in Fig. 2.10-1.

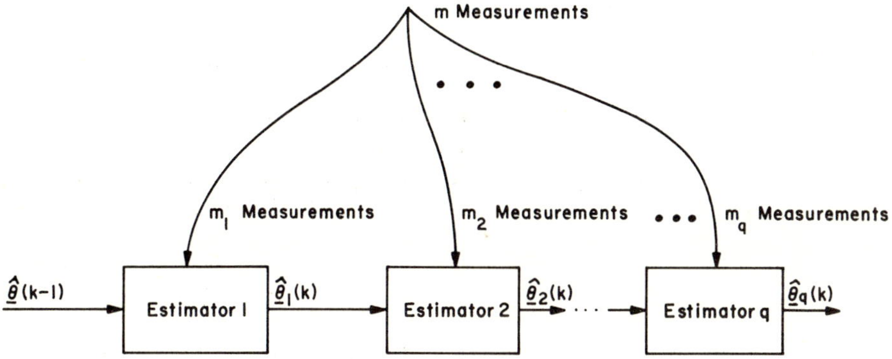

Fig. 2.10-1. Sequential processing of m measurements.

The remarkable property about sequential processing is that

$$\hat{\underline{\theta}}_q(k) = \hat{\underline{\theta}}(k) \quad , \tag{2.10-4}$$

where the estimate on the right-hand side is obtained by simultaneously processing the m measurements. We shall now proceed to prove the truth of Eq. (2.10-4). Our approach will be to show that

$$K_q^*(k) = K^*(k) \tag{2.10-5}$$

and

$$P_q(k) = P(k) \quad . \tag{2.10-6}$$

2.10.2 Simultaneous Processing

In order to *prove* Eqs. (2.10-5) and (2.10-6), we will find it *convenient* to use the expressions for K*(k) and P(k) in Eqs. (2.9-13) and (2.9-14). As we know, *these expressions are analytically equivalent to their counterparts* in Eqs. (2.9-16) and (2.9-17). It must be emphasized, however, that in *implementing* the sequential generalized least-squares algorithm on a digital computer, Eqs. (2.9-16) and

(2.9-17) may be preferred, for the reasons described directly following Eq. (2.9-17).

When the data are processed simultaneously, the gain matrix $K^*(k)$ is given in partitioned form by

$$K^*(k) = P(k)H'(k)W(k) = P(k)[H_1'(k) \mid H_2'(k) \mid \cdots \mid H_q'(k)] \times \begin{pmatrix} w_1(k) & 0 & \cdots & 0 \\ 0 & w_2(k) & \cdots & 0 \\ \vdots & \vdots & \ddots & \vdots \\ 0 & 0 & \cdots & w_q(k) \end{pmatrix}. \tag{2.10-7}$$

Hence

$$K^*(k) = [K_1(k) \mid K_2(k) \mid \cdots \mid K_q(k)] \tag{2.10-8}$$

where

$$K_i(k) = P(k)H_i'(k)w_i(k) \tag{2.10-9}$$

for $i = 1, 2, \ldots, q$. Each of these matrices has the dimension $n \times m_i$.

Representation of $P^{-1}(k)$ is given by

$$P^{-1}(k) = P^{-1}(k-1) + H'(k)W(k)H(k) \tag{2.10-10}$$

Consider $H'(k)W(k)H(k)$ in partitioned form:

$$H'(k)W(k)H(k) = [H_1'(k) \mid H_2'(k) \mid \cdots \mid H_q'(k)] \times \begin{pmatrix} w_1(k) & 0 & \cdots & 0 \\ 0 & w_2(k) & \cdots & 0 \\ \vdots & \vdots & \ddots & \vdots \\ 0 & 0 & \cdots & w_q(k) \end{pmatrix} \begin{pmatrix} H_1(k) \\ H_2(k) \\ \vdots \\ H_q(k) \end{pmatrix}$$

$$= \sum_{i=1}^{q} H_i'(k)w_i(k)H_i(k) \tag{2.10-11}$$

so

$$P^{-1}(k) = P^{-1}(k-1) + \sum_{i=1}^{q} H_i'(k)\omega_i(k)H_i(k) \quad . \qquad (2.10\text{-}12)$$

2.10.3 Sequential Processing

In this policy, the data $\underline{z}_j(k)$ are processed without consideration of the other available data. Denote the estimate and gain and P matrices obtained from this data with a subscript j. Then, for j = 1,

$$\hat{\underline{\theta}}_1(k) = \hat{\underline{\theta}}(k-1) + K_1^*(k)[\underline{z}_1(k) - H_1(k)\hat{\underline{\theta}}(k-1)] \qquad (2.10\text{-}13)$$

where

$$K_1^*(k) = P_1(k)H_1'(k)\omega_1(k) \qquad (2.10\text{-}14)$$

and

$$P_1^{-1}(k) = P^{-1}(k-1) + H_1'(k)\omega_1(k)H_1(k) \quad . \qquad (2.10\text{-}15)$$

Using $\hat{\underline{\theta}}_1(k)$, $K_1^*(k)$, and $P_1(k)$, process $\underline{z}_2(k)$ as follows:

$$\begin{aligned}
\hat{\underline{\theta}}_2(k) &= \hat{\underline{\theta}}_1(k) + K_2^*(k)[\underline{z}_2(k) - H_2(k)\hat{\underline{\theta}}_1(k)] \\
&= \hat{\underline{\theta}}(k-1) + K_1^*(k)[\underline{z}_1(k) - H_1(k)\hat{\underline{\theta}}(k-1)] \\
&\quad + K_2^*(k)\Big\{\underline{z}_2(k) - H_2(k)\{\hat{\underline{\theta}}(k-1) + K_1^*(k)[\underline{z}_1(k) \\
&\quad - H_1(k)\hat{\underline{\theta}}(k-1)]\}\Big\} \\
&= \hat{\underline{\theta}}(k-1) + [I - K_2^*(k)H_2(k)]K_1^*(k)[\underline{z}_1(k) - H_1(k)\hat{\underline{\theta}}(k-1)] \\
&\quad + K_2^*(k)[\underline{z}_2(k) - H_2(k)\hat{\underline{\theta}}(k-1)] \qquad (2.10\text{-}16)
\end{aligned}$$

where

$$K_2^*(k) = P_2(k)H_2'(k)\omega_2(k) \qquad (2.10\text{-}17)$$

and

$$P_2^{-1}(k) = P_1^{-1}(k) + H_2'(k)\omega_2(k)H_2(k) \quad . \qquad (2.10\text{-}18)$$

Now using $\hat{\underline{\theta}}_2(k)$, $K_2^*(k)$, and $P_2(k)$, process $\underline{z}_3(k)$ in a similar manner, obtaining

$$\hat{\underline{\theta}}_3(k) = \hat{\underline{\theta}}_2(k) + K_3^*(k)[\underline{z}_3(k) - H_3(k)\hat{\underline{\theta}}_2(k)] \qquad (2.10\text{-}19)$$

where

$$K_3^*(k) = P_3(k)H_3'(k)w_3(k) \tag{2.10-20}$$

and

$$P_3^{-1}(k) = P_2^{-1}(k) + H_3'(k)w_3(k)H_3(k) \quad . \tag{2.10-21}$$

Introducing Eq. (2.10-16) into Eq. (2.10-19), we obtain

$$\begin{aligned}\hat{\underline{\theta}}_3(k) = \hat{\underline{\theta}}(k-1) &+ \{[I - K_3^*(k)H_3(k)][I \\ &- K_2^*(k)H_2(k)]K_1^*(k)[\underline{z}_1(k) - H_1(k)\hat{\underline{\theta}}(k-1)]\} \\ &+ \{[I - K_3^*(k)H_3(k)]K_2^*(k)[\underline{z}_2(k) - H_2(k)\hat{\underline{\theta}}(k-1)]\} \\ &+ K_3^*(k)[\underline{z}_3(k) - H_3(k)\hat{\underline{\theta}}(k-1)]\end{aligned} \tag{2.10-22}$$

Let us direct our attention at $P_3^{-1}(k)$ in Eq. (2.10-21), making use of Eqs. (2.10-18) and (2.10-15). Clearly,

$$\begin{aligned}P_3^{-1}(k) &= P^{-1}(k-1) + H_1'(k)w_1(k)H_1(k) + H_2'(k)w_2(k)H_2(k) \\ &\quad + H_3'(k)w_3(k)H_3(k) \\ &= P^{-1}(k-1) + \sum_{i=1}^{3} H_i'(k)w_i(k)H_i(k)\end{aligned} \tag{2.10-23}$$

which is identically the same as Eq. (2.10-12), for the case when $q = 3$. The generalization of Eq. (2.10-23) to arbitrary values of q, obviously, is

$$P_q^{-1}(k) = P^{-1}(k-1) + \sum_{i=1}^{q} H_i'(k)w_i(k)H_i(k) \quad . \tag{2.10-24}$$

Comparing the right-hand sides of Eqs. (2.10-24) and (2.10-12), we conclude that

$$P_q^{-1}(k) = P^{-1}(k) \tag{2.10-25}$$

which is equivalent to Eq. (2.10-6).

Let us now direct our attention toward establishing Eq. (2.10-5). Rather than doing this for arbitrary q, we shall do it for $q = 3$ (i. e., for 3 batches of data). Examining the structure of Eq. (2.10-22) reveals that we must show that

$$[I - K_3^*(k)H_3(k)][I - K_2^*(k)H_2(k)]K_1^*(k) = K_1(k) \quad , \tag{2.10-26}$$

$$[I - K_3^*(k)H_3(k)]K_2^*(k) = K_2(k) \quad , \quad (2.10\text{-}27)$$

and

$$K_3^*(k) = K_3(k) \quad (2.10\text{-}28)$$

where $K_i(k)$, i = 1, 2, 3, are given by Eq. (2.10-9), and $K_i^*(k)$, i = 1, 2, 3, are given in Eqs. (2.10-14), (2.10-17), and (2.10-20). For q = 3, we know [Eq. (2.10-25)] that $P_3(k) = P(k)$; hence,

$$K_3^*(k) = P_3(k)H_3'(k)w_3(k) = P(k)H_3'(k)w_3(k) = K_3(k) \quad . \quad (2.10\text{-}29)$$

To prove the correctness of Eqs. (2.10-26) and (2.10-27), we must reexpress $P_3(k)$, $P_2(k)$, and $P_1(k)$. We leave the details of this as an exercise for the reader (Problem 2-21). The generalization of these results from q = 3 to arbitrary values of q is straightforward.

This completes our demonstration of equivalence between sequential and simultaneous data processing.

2.10.4 Procedure for Sequential Processing

We summarize the sequential processing technique that can be used when there are q sets of statistically independent data $\underline{z}_1(k)$, $\underline{z}_2(k)$, ..., $\underline{z}_q(k)$.

(Step 1) Form $K_1^*(k)$, $\hat{\underline{\theta}}_1(k)$, and $P_1(k)$ from Eqs. (2.9-15)-(2.9-17), treating $\underline{z}_1(k)$ as $\underline{z}(k)$:

$$K_1^*(k) = P(k-1)H_1'(k)[H_1(k)P(k-1)H_1'(k) + w_1^{-1}(k)]^{-1} \quad (2.10\text{-}30)$$

$$\hat{\underline{\theta}}_1(k) = \hat{\underline{\theta}}(k-1) + K_1^*(k)[\underline{z}_1(k) - H_1(k)\hat{\underline{\theta}}(k-1)] \quad (2.10\text{-}31)$$

and

$$P_1(k) = [I - K_1^*(k)H_1(k)]P(k-1) \quad . \quad (2.10\text{-}32)$$

(Step 2) Based on $P_1(k)$ and $\hat{\underline{\theta}}_1(k)$, process $\underline{z}_2(k)$, using the relations (for j = 2)

$$K_j^*(k) = P_{j-1}(k)H_j'(k)[H_j(k)P_{j-1}(k)H_j'(k) + w_j^{-1}(k)]^{-1} \quad , \quad (2.10\text{-}33)$$

$$\hat{\underline{\theta}}_j(k) = \hat{\underline{\theta}}_{j-1}(k) + K_j^*(k)[\underline{z}_j(k) - H_j(k)\hat{\underline{\theta}}_{j-1}(k)] \quad , \quad (2.10\text{-}34)$$

and

$$P_j(k) = [I - K_j^*(k)H_j(k)]P_{j-1}(k) \quad . \tag{2.10-35}$$

(Step 3) Repeat Step (2) with the next set of data and with $j = 3$. Continue this policy until $j = q$, when all data will have been processed. Then

$$\underline{\hat{\theta}}_q(k) = \underline{\hat{\theta}}(k) \quad , \tag{2.10-36}$$

$$K_q^*(k) = K^*(k) \quad , \tag{2.10-37}$$

and

$$P_q(k) = P(k) \quad . \tag{2.10-38}$$

2.10.5 Optimal Sequential Processing

It is clear that if we have m statistically independent measurements, matrix inversion can be completely eliminated by means of sequentially processing these measurements one at a time. Of course, we would then have to cycle through the generalized least-squares algorithm n times. The computing time for each cycle will, in this case, depend only upon n, whereas, in general, it will depend upon both n and m.

Let us ask the following question: *Can we determine values for* q, m_1, m_2, ..., m_q *such that total computing time to sequentially process the* q *batches is less than the computing time associated with any other* $q = \bar{q}$ *batches?* The answer to this question is yes, and the interested reader is referred to Ref. (*2-12*), where a similar problem was solved for another optimal sequential estimator, the discrete Kalman filter (see Section 3.8.1). As $m \to n$, very large savings in computing time can be realized by determining the optimal (meaning, minimum with respect to total computing time) decomposition of the m measurements.

2.11 ANALYSES OF THE SEQUENTIAL GENERALIZED LEAST-SQUARES ALGORITHM

2.11.1 Introduction

We have shown that sequential algorithms can be obtained which perform generalized least-squares parameter identification. These algorithms are finite-difference equations which, like differential equations, possess characteristic behaviors and natural modes. Their behavior is unrelated to the data being processed, and unless it is kept under adequate control it can easily make the algorithm's outputs completely useless. What we require is that the natural modes, if ever excited, should of their own accord die out in time. A sequential algorithm whose natural modes eventually die out is then said to be *stable*, and if they fail to die out, or if they build up, the algorithm is said to be *unstable* (*2-2*). Let us illustrate the concepts of stability and instability by means of some examples.

Example 7. We shall investigate the sequence generated by the following first-order finite difference equation:

$$\tilde{\theta}(k + 1) = \mu\tilde{\theta}(k) \tag{2.11-1}$$

for $k = 0, 1, \ldots.$ By direct substitution, it is easily verified that

$$\tilde{\theta}(k) = \mu^k\tilde{\theta}(0) \quad . \tag{2.11-2}$$

Clearly, if $|\mu| < 1$ then $\lim_{k\to\infty}\tilde{\theta}(k) = 0$, and the system in Eq. (2.11-1) is *asymptotically stable*. If $|\mu| = 1$, then $\tilde{\theta}(k) = \tilde{\theta}(0)$ for all k, and $\tilde{\theta}(k)$ is no better than its initial value, which certainly does not seem very desirable when $\tilde{\theta}(k)$ is viewed as parameter identification error. Finally, when $|\mu| > 1$, $\lim_{k\to\infty}\tilde{\theta}(k) \to \infty$, and the system in Eq. (2.11-1) is *unstable*. For this simple system, we see that stability is determined solely by the amplitude of μ (*2-10*). ▲

Example 8. Now let us consider the effects of adding a forcing term to Eq. (2.11-1). To begin, we modify Eq. (2.11-1) to

$$\tilde{\theta}(k + 1) = \mu\tilde{\theta}(k) + c \tag{2.11-3}$$

for $k = 0, 1, \ldots.$ Either by direct substitution, or using Eq. (1.3-58), one finds

$$\tilde{\theta}(k) = \mu^k\tilde{\theta}(0) + \frac{\mu^k - 1}{\mu - 1}c \quad . \tag{2.11-4}$$

Obviously, the effect of $\tilde{\theta}(0)$ will disappear for large values of k only if $|\mu| < 1$, as in Example 7. If c is a bounded constant then, for $|\mu| < 1$,

$$\lim_{k\to\infty} \tilde{\theta}(k) = \frac{c}{1 - \mu} \quad . \tag{2.11-5}$$

We have just demonstrated a concept from stability theory known as *bounded-input/bounded-output stability*. It applies to forced systems and is a statement of the fact that: if an unforced system is stable, then that same system, when excited by a bounded input function, will have a bounded output response (*2-10*).

In order to reinforce these remarks, let us consider the stability of the system

$$\tilde{\theta}(k + 1) = \mu\tilde{\theta}(k) + k \quad . \tag{2.11-6}$$

The solution for $\tilde{\theta}(k)$ is

$$\tilde{\theta}(k) = \mu^k\tilde{\theta}(0) + \mu^{k-2} + 2\mu^{k-3} + 3\mu^{k-4} + \cdots + (k - 1) \quad . \tag{2.11-7}$$

Clearly, even when $|\mu| < 1$, $\lim_{k\to\infty}\tilde{\theta}(k) \to \infty$, because of the term $(k - 1)$ which appears in this solution. ▲

Example 9. Return to Example 8, but now c is a zero-mean random variable. This means that $\tilde{\theta}(k)$ will also be a random variable for each k. Since it is a random variable, we are more interested in $E\{\tilde{\theta}(k)\}$ and its limiting behavior than in $\tilde{\theta}(k)$'s. Taking the expectation on both sides of Eq. (2.11-4), we find

$$E\{\tilde{\theta}(k)\} = \mu^k E\{\tilde{\theta}(0)\} \tag{2.11-8}$$

which is of exactly the same form as Eq. (2.11-2); hence, if $|\mu| < 1$, then

$$\lim_{k\to\infty} E\{\tilde{\theta}(k)\} = 0 \quad . \tag{2.11-9}$$

When $\tilde{\theta}(k)$ *is parameter identification error*

$$\tilde{\theta}(k) = \theta - \hat{\theta}(k) \quad , \tag{2.11-10}$$

Eq. (2.11-9) can then be rewritten as†

---

† In this section, and in most of the book, θ is deterministic.

$$\lim_{k\to\infty} E\{\hat{\theta}(k)\} = \theta \tag{2.11-11}$$

which is a property known as *asymptotic unbiasedness* [compare Eq. (2.11-11) with Eq. (2.5-2)]. This example illustrates the very important concept of *stability in the mean*. It is closely related to stability, but is more apropos when dealing with random variables, or their extensions, random processes. ▲

Determining the stability properties of higher-order difference equations is much more complicated than the preceding examples would have us believe. It is assumed that the reader has some familiarity with stability of differential equations or difference equations. For both types of equations, stability of the unforced system is determined by the locations of the *roots* of the *characteristic equation*. These roots are often called *eigenvalues* (*2-13* and *2-14*). A differential system is asymptotically stable if all of its eigenvalues are negative. A discrete system, on the other hand, is asymptotically stable if all of its eigenvalues are of amplitude less than unity. In Example 7, μ is the system's eigenvalue, and as we have demonstrated, the system in Eq. (2.11-1) is asymptotically stable when $|\mu| < 1$.

When a system is described as a vector first-order discrete-time state equation, its stability properties are determined by the eigenvalues of the state transition matrix.

The eigenvalues of a matrix A are found from the equation (*2-15*)

$$A\underline{x} = \lambda\underline{x} \tag{2.11-12}$$

where $\underline{x}$ is the eigenvector associated with the eigenvalue λ.

With this as a brief introduction to some important stability concepts, let us direct our attention to the stability of the sequential generalized least-squares estimation algorithm.

2.11.2 Identification-Error System

We shall determine the finite difference equation for the parameter identification error $\underline{\tilde{\theta}}(k)$. This equation will be referred to as the *identification-error system*. It is obtained here from Eqs. (2.9-12), (2.9-13), and (2.9-5), as follows:

$$\begin{aligned}\tilde{\underline{\theta}}(k+1) &= \underline{\theta} - \hat{\underline{\theta}}(k+1)\\ &= \underline{\theta} - \hat{\underline{\theta}}(k) - K^*(k+1)[\underline{z}(k+1) - H(k+1)\hat{\underline{\theta}}(k)]\\ &= \tilde{\underline{\theta}}(k) - P(k+1)H'(k+1)\omega(k+1)[H(k+1)\tilde{\underline{\theta}}(k)\\ &\quad + \underline{v}(k+1)]\\ &= [I - P(k+1)H'(k+1)\omega(k+1)H(k+1)]\tilde{\underline{\theta}}(k)\\ &\quad - P(k+1)H'(k+1)\omega(k+1)\underline{v}(k+1) \quad . \qquad (2.11\text{-}13)\end{aligned}$$

We now make use of the analog of Eq. (2.7-15) for a vector of measurements,

$$P(k+1) = [\mathcal{H}'(k)W(k)\mathcal{H}(k) + H'(k+1)\omega(k+1)H(k+1)]^{-1} \quad . \qquad (2.11\text{-}14)$$

Substituting this expression into Eq. (2.11-13), we obtain the desired *identification-error system:*

$$\tilde{\underline{\theta}}(k+1) = A(k+1,\, k)\tilde{\underline{\theta}}(k) + B(k+1,\, k)\underline{v}(k+1) \qquad (2.11\text{-}15)$$

where

$$\begin{aligned}A(k+1,\, k) &= [\mathcal{H}'(k)W(k)\mathcal{H}(k) + H'(k+1)\omega(k+1)H(k+1)]^{-1}\\ &\quad \times \mathcal{H}'(k)W(k)\mathcal{H}(k) \quad , \qquad (2.11\text{-}16)\end{aligned}$$

$$\begin{aligned}B(k+1,\, k) &= [\mathcal{H}'(k)W(k)\mathcal{H}(k) + H'(k+1)\omega(k+1)H(k+1)]^{-1}\\ &\quad \times H'(k+1)\omega(k+1) \quad , \qquad (2.11\text{-}17)\end{aligned}$$

and, $k \geq k_0$. The identification-error system is linear, time varying, forced, and stochastic; hence, we shall be interested in *stability in the mean* for it [see Example 9].

2.11.3 Stability of the Unforced Identification-Error System

Theorem 2-4. *The eigenvalues of the unforced identification-error system*

$$\tilde{\underline{\theta}}(k + 1) = A(k + 1, k)\tilde{\underline{\theta}}(k) \qquad (2.11\text{-}18)$$

are all between zero and unity when $\mathcal{H}(k)$ *and* $H(k + 1)$ *are of maximum rank (2-18).*

Proof. It is well known that the characteristic roots of the system in Eq. (2.11-18) are the eigenvalues of $A(k + 1, k)$, which can be found from the equation

$$A(k + 1, k)\underline{x} = \lambda\underline{x} \quad . \qquad (2.11\text{-}19)$$

Using Eq. (2.11-16), this last expression can also be written as

$$(\mathcal{H}'\mathcal{W}\mathcal{H} + H'wH)^{-1}\mathcal{H}'\mathcal{W}\mathcal{H}\underline{x} = \lambda\underline{x} \qquad (2.11\text{-}20)$$

from which it follows that

$$(1 - \lambda)\mathcal{H}'\mathcal{W}\mathcal{H}\underline{x} = \lambda H'wH\underline{x} \quad . \qquad (2.11\text{-}21)$$

Premultiplying both sides by $\underline{x}'$, we find

$$(1 - \lambda)\underline{x}'\mathcal{H}'\mathcal{W}\mathcal{H}\underline{x} = \lambda\underline{x}'H'wH\underline{x} \quad . \qquad (2.11\text{-}22)$$

Now, $\mathcal{W}(k)$ and $w(k + 1)$ are positive definite [see discussion just prior to Eq. (2.2-38)], and, since $\mathcal{H}(k)$ and $H(k + 1)$ are of maximum rank, both $\mathcal{H}'\mathcal{W}\mathcal{H}$ and $H'wH$ are positive definite (*2-1* and *2-15*). This means that $\underline{x}'\mathcal{H}'\mathcal{W}\mathcal{H}\underline{x} > 0$ and $\underline{x}'H'wH\underline{x} > 0$ for all $\underline{x} \neq 0$; hence, the two sides of this equation can be equal only if

$$(1 - \lambda) > 0 \qquad (2.11\text{-}23)$$

and

$$\lambda > 0 \qquad (2.11\text{-}24)$$

which proves the assertion that

$$0 < \lambda < 1 \quad . \qquad (2.11\text{-}25)$$

▲

2.11.4 Stability in the Mean for the Identification-Error System

Theorem 2-5. *For the identification-error system in Eq. (2.11-15), if*

$$B(k, k - 1) \quad \text{and} \quad \underline{v}(k) \quad ,$$

$$A(k, k-1)B(k-1, k-2) \text{ and } \underline{v}(k-1) \quad ,$$
$$\vdots$$

and

$$A(k, k-1) \cdots A(3, 2)A(2, 1)B(1, 0) \text{ and } \underline{v}(1)$$

are statistically independent, then

$$\lim_{k\to\infty} E\{\hat{\underline{\theta}}(k)\} = \underline{\theta} \quad . \tag{2.11-26}$$

Proof. Writing out Eq. (2.11-15) for k = 0, 1, 2, we see that

$$\tilde{\underline{\theta}}(1) = A(1, 0)\tilde{\underline{\theta}}(0) + B(1, 0)\underline{v}(1) \quad , \tag{2.11-27}$$

$$\begin{aligned}\tilde{\underline{\theta}}(2) &= A(2, 1)\tilde{\underline{\theta}}(1) + B(2, 1)\underline{v}(2) \\ &= A(2, 1)A(1, 0)\tilde{\underline{\theta}}(0) + A(2, 1)B(1, 0)\underline{v}(1) \\ &\quad + B(2, 1)\underline{v}(2) \quad ,\end{aligned} \tag{2.11-28}$$

and

$$\begin{aligned}\tilde{\underline{\theta}}(3) &= A(3, 2)\tilde{\underline{\theta}}(2) + B(3, 2)\underline{v}(3) \\ &= A(3, 2)A(2, 1)A(1, 0)\tilde{\underline{\theta}}(0) \\ &\quad + A(3, 2)A(2, 1)B(1, 0)\underline{v}(1) \\ &\quad + A(3, 2)B(2, 1)\underline{v}(2) \\ &\quad + B(3, 2)\underline{v}(3) \quad .\end{aligned} \tag{2.11-29}$$

From the pattern emerging in these solutions, we deduce the general solution of Eq. (2.11-15) to be

$$\begin{aligned}\tilde{\underline{\theta}}(k) &= \prod_{i=1}^{k} A(i, i-1)\tilde{\underline{\theta}}(0) \\ &\quad + \sum_{i=1}^{k-1} \prod_{j=i+1}^{k} A(j, j-1)B(i, i-1)\underline{v}(i) \\ &\quad + B(k, k-1)\underline{v}(k) \quad .\end{aligned} \tag{2.11-30}$$

Since we are only concerned with the behavior of this equation as $k \to \infty$, we apply the results of Theorem 2-4 to this expression, recognizing that stability of the unforced system means

$$\lim_{k\to\infty} \prod_{i=1}^{k} A(i, i-1)\underline{\tilde{\theta}}(0) \to \underline{0} \quad . \tag{2.11-31}$$

Hence,

$$\lim_{k\to\infty} E\{\underline{\tilde{\theta}}(k)\} = \lim_{k\to\infty} \sum_{i=1}^{k-1} E\{\prod_{j=i+1}^{k} A(j, j-1)B(i, i-1)\underline{v}(i)\}$$

$$+ \lim_{k\to\infty} E\{B(k, k-1)\underline{v}(k)\} = \underline{0} \tag{2.11-32}$$

by the assumptions made in the statement of this theorem, and the fact that $\underline{v}(j)$ is zero mean. ▲

Corollary 2-3. *For the identification error system in Eq. (2.11-15), if* $H(k)$ *is a deterministic matrix, then the conditions in the statement of Theorem 2-5 are satisfied and* $\lim_{k\to\infty} E\{\underline{\hat{\theta}}(k)\} = \underline{\theta}$. ▲

Corollary 2-4. *For the identification error system in Eq. (2.11-15), if* $H(k)$ *and* $\underline{V}(k)$ *are independent then the conditions in the statement of Theorem 2-5 are satisfied and* $\lim_{k\to\infty} E\{\underline{\hat{\theta}}(k)\} = \underline{\theta}$. ▲

The importance of Corollary 2-4 is that it demonstrates that the conditions for unbiasedness also guarantee asymptotic unbiasedness. In retrospect, we should have predicted this result directly from the definition of unbiasedness. Unbiasedness is a small sample property of an estimator, whereas asymptotic unbiasedness is a large sample property; that is to say, unbiasedness does not depend on how many measurements are used to obtain $\underline{\hat{\theta}}(k)$, whereas asymptotic unbiasedness does. Obviously, unbiasedness implies asymptotic unbiasedness; however, asymptotic unbiasedness does not necessarily imply unbiasedness. For example, $\hat{\sigma}^2(k)$ in Eq. (2.5-41) is an unbiased estimate of σ^2. Consider the following estimator of σ^2,

$$\hat{\hat{\sigma}}^2(k) = \frac{\underline{\tilde{Z}}'(k)\underline{\tilde{Z}}(k)}{L+1} \quad . \tag{2.11-33}$$

Since

$$E\{\hat{\sigma}^2(k)\} = \frac{\sigma^2(L + 1 - n)}{L + 1} \neq \sigma^2 \quad , \qquad (2.11\text{-}34)$$

$\hat{\sigma}^2(k)$ is not an unbiased estimate of σ^2; however, since

$$\lim_{L\to\infty} E\{\hat{\sigma}^2(k)\} = \sigma^2 \quad , \qquad (2.11\text{-}35)$$

$\hat{\sigma}^2(k)$ is an asymptotically unbiased estimate of σ^2.

2.12 COMPUTATIONAL ASPECTS OF GENERALIZED LEAST-SQUARES ESTIMATION

2.12.1 Batch-Processing Algorithm

The batch generalized least-squares estimation algorithm in Eq. (2.3-12) requires the computation of the inverse of an $n \times n$ matrix $H'(k)W(k)H(k)$. $W(k)$ has been assumed to be a positive definite matrix; hence, as we have pointed out in Sections 2.2 and 2.3, $H'(k)W(k)H(k)$ will be positive definite if $H(k)$ has full column rank (*2-1*). In this case, $[H'(k)W(k)H(k)]^{-1}$ will be nonsingular, and the inverse will be *computable*.

For vector measurements, $H(k)$ is an $(L + 1)m \times n$ matrix, and estimation does not begin until $H(k)$ has more rows than columns (see Sections 2.6 and 2.9). It is for this reason that $H(k)$ must have full column rank rather than full row rank (Problem 2-28).

If measurements are spaced too closely together, we can expect trouble, since the rows of $H(k)$ will become too similar; hence, measurements must be spaced far enough apart in time so that the signals being measured change by meaningful amounts between the first and last measurements. The reader is referred to Section 8.8 in Morrison (*2-2*) for additional discussions.

2.12.2 Sequential-Processing Algorithms

We have derived two formulations of sequential generalized least-squares estimation algorithms. The *original formulation* is in Eqs. (2.9-12)-(2.9-14), and, an *alternative formulation* is in Eqs. (2.9-15)-(2.9-17).

In on-line applications, where speed of computation is usually the most important consideration, the alternative formulation is preferable to the original formulation. This is because a smaller matrix needs to be inverted in the alternative formulation - an $m \times m$ matrix rather than an $n \times n$ matrix (m usually is much less than n). For related discussions, see Sections 2.7.4, 2.9, and 2.10.1.

The original formulation often is more useful than the alternative formulation in analytical studies. Observe, for example, that we used it in Section 2.10, where we proved the equivalence between sequential and simultaneous data processing.

Not only is the original formulation more useful than the alternative formulation during the analytical study of sequential startup, but, as we shall now demonstrate, it is to be preferred during its actual implementation. In Section 2.8.7, we saw that one often starts up the sequential algorithms at $k = 0$, by setting $\hat{\underline{\theta}}(0) = \underline{0}$ and $P(0)$ to a matrix of large positive numbers. Let us consider the case when

$$P(0) = \uparrow I \tag{2.12-1}$$

where by $\uparrow$ we mean a very large scalar, perhaps the largest that can be carried out by the computer in which the computations are being performed.

Using the original formulation, we find that, for $k = 0$, $P^{-1}(1) = H'(1)\mathcal{W}(1)H(1)$ and, therefore,

$$K^*(1) = [H'(1)\mathcal{W}(1)H(1)]^{-1}H'(1)\mathcal{W}(1) \quad .$$

If $H(1)$ is of maximum rank, $K^*(1)$ is computable.

Using the alternative formulation we find, first, that $K^*(1) = H'(1)[H(1)H'(1)]^{-1}$, and then, that

$$P(1) = \uparrow\{I - H'(1)[H(1)H'(1)]^{-1}H(1)\} \quad ; \tag{2.12-2}$$

however, this matrix is singular. To see this, postmultiply both sides of Eq. (2.12-2) by $H'(1)$, obtaining

$$P(1)H'(1) = \dagger\{H'(1) - H'(1)[H(1)H'(1)]^{-1}H(1)H'(1)\}$$
$$= 0 \quad . \tag{2.12-3}$$

We have shown that $P(1)H'(1) = 0$ without either $P(1)$ or $H'(1)$ being null, which means that $P(1)$ is singular. In fact, once $P(1)$ becomes singular, all other $P(j)$, $j \geq 2$, will be singular (Problem 2-29).

Actually, we do not need to compute the inverse of $P(1)$ using the alternative formulation; thus, the singularity of this matrix does not in itself present a drawback to using the alternative formulation for sequential initialization. In the next chapter, we shall show that when $W^{-1}(k) = E\{\underline{V}(k)\underline{V}'(k)\} = R(k)$, $P(k)$ is the covariance matrix of the identification error $\underline{\tilde{\theta}}(k)$. It will be difficult to maintain the properties of a covariance matrix if $P(k)$ is singular; hence, it is advisable to initialize the sequential algorithms using the original formulation. However, it is also advisable to switch to the alternative formulation as soon after initialization as possible, in order to reduce computing time.

Let us now investigate the effect of a computational error in the gain matrix on the computation of $P(k + 1)$. Denoting a small error in $K^*(k + 1)$ as $\delta K^*(k + 1)$, then from Eq. (2.9-17), we find

$$P(k + 1) + \delta P(k + 1) = [I - K^*(k + 1)H(k + 1)$$
$$- \delta K^*(k + 1)H(k + 1)]P(k) \quad . \tag{2.12-4}$$

Hence,

$$\delta P(k + 1) = -\delta K^*(k + 1)H(k + 1)P(k) \tag{2.12-5}$$

which means that the computational error $\delta K^*(k + 1)$ is propagated into the calculation of $P(k + 1)$, which is undesirable. We shall next derive another equation for $P(k + 1)$ which, as we show, eliminates the propagation of $\delta K^*(k + 1)$ (*2-19* and *2-20*).

Lemma. *Another equation for computing* $P(k + 1)$ *is*

$$P(k + 1) = [I - K^*(k + 1)H(k + 1)]P(k)[I - K^*(k + 1)H(k + 1)]'$$
$$+ K^*(k + 1)\mathcal{W}^{-1}(k + 1)K^{*\prime}(k + 1) \quad . \tag{2.12-6}$$

Proof. Expand Eq. (2.12-6), obtaining

$$\begin{aligned}P(k+1) &= P(k) - K^*(k+1)H(k+1)P(k) \\ &\quad - P(k)H'(k+1)K^{*\prime}(k+1) \\ &\quad + K^*(k+1)H(k+1)P(k)H'(k+1)K^{*\prime}(k+1) \\ &\quad + K^*(k+1)\mathcal{W}^{-1}(k+1)K^{*\prime}(k+1) \\ &= [I - K^*(k+1)H(k+1)]P(k) \\ &\quad - P(k)H'(k+1)K^{*\prime}(k+1) \\ &\quad + K^*(k+1)[H(k+1)P(k)H'(k+1) \\ &\quad + \mathcal{W}^{-1}(k+1)]K^{*\prime}(k+1) \quad . \qquad (2.12\text{-}7)\end{aligned}$$

However, from Eq. (2.9-16) we know that

$$\begin{aligned}K^*(k+1) &= P(k)H'(k+1)[H(k+1)P(k)H'(k+1) \\ &\quad + \mathcal{W}^{-1}(k+1)]^{-1} \quad . \qquad (2.12\text{-}8)\end{aligned}$$

Thus, substituting Eq. (2.12-8) into the third term on the right-hand side of Eq. (2.12-7), we find

$$\begin{aligned}P(k+1) &= [I - K^*(k+1)H(k+1)]P(k) \\ &\quad - P(k)H'(k+1)K^{*\prime}(k+1) \\ &\quad + P(k)H'(k+1)K^{*\prime}(k+1) \\ &= [I - K^*(k+1)H(k+1)]P(k) \qquad (2.12\text{-}9)\end{aligned}$$

which is Eq. (2.9-17). ▲

We remark that Eq. (2.12-6) was not simply pulled out of the air. It appears quite naturally in the derivation of the discrete Kalman filter [*2-17*, Eq. (5.62)] and, as we show in Section 3.8.1 there is a close relationship between Kalman filtering for state-estimation and minimum-variance (and, subsequently, generalized least-squares) parameter identification.

Let us return to our intended objective which is to show that when $P(k+1)$ is computed using Eq. (2.12-6), $\delta P(k+1) \approx 0$. In this case,

$$\begin{aligned} P(k+1) + \delta P(k+1) = & \{I - [K^*(k+1) + \delta K^*(k+1)]H(k+1)\}P(k) \\ & \times \{I - [K^*(k+1) + \delta K^*(k+1)]H(k+1)\}' \\ & + [K^*(k+1) + \delta K^*(k+1)]w^{-1}(k+1)[K^*(k+1) \\ & + \delta K^*(k+1)]' \quad . \end{aligned} \tag{2.12-10}$$

Expanding on the right-hand side of this expression, it is easily shown that

$$\begin{aligned} \delta P(k+1) = & -\delta K^* HP(k)[I - K^*H]' - [I - K^*H]P(k)H'\delta K^{*\prime} \\ & + \delta K^* HP(k)H'\delta K^* + K^* w^{-1}\delta K^{*\prime} + \delta K^* w^{-1} K^{*\prime} \\ & + \delta K^* w^{-1} \delta K^{*\prime} \quad . \end{aligned} \tag{2.12-11}$$

To within second-order terms in δK^*, $\delta P(k + 1)$ becomes

$$\begin{aligned} \delta P(k+1) & \simeq -\delta K^*[HP(k)(I - K^*H)' - w^{-1}K^{*\prime}] \\ & \quad - [(I - K^*H)P(k)H' - K^*w^{-1}]\delta K^{*\prime} \\ & \simeq -\delta K^*[HP(k+1) - w^{-1}K^{*\prime}] \\ & \quad - [P(k+1)H' - K^*w^{-1}]\delta K^{*\prime} \\ & = 0 \quad , \end{aligned} \tag{2.12-12}$$

where we have used the facts that $P(k)$ is a symmetric matrix (Problem 2-27), $P(k + 1) = P(k)(I - K^*H)'$, and $K^* = P(k + 1)H'w$. ▲

Although computing time is increased when Eq. (2.12-6) is used in place of Eq. (2.9-17), it is often worth the increase. We are trading additional computing time for increased accuracy.

2.13 APPLICABILITY OF GENERALIZED LEAST-SQUARES PARAMETER ESTIMATION ALGORITHMS

We have shown that if $\mathcal{H}(k)$ and $\underline{\mathcal{V}}(k)$ are statistically independent, then batch estimates will be *unbiased*, and sequential estimates will be *asymptotically unbiased* [Theorem 2-1 in Section 2.5 and Corollary 2.4 in Section 2.11.4]. In this section, both unbiased and asymptotically unbiased estimates will be referred to as unbiased estimates.

For the applications we are interested in (e. g., identifying coefficients of a finite-difference equation), unbiasedness is a very desirable property. We will judge the success or failure of generalized least-squares estimation, for a specific application, on whether or not the estimated parameters are unbiased. We have already passed such judgment on five applications in Examples 1 through 5 of Section 2.5. Here we shall demonstrate that generalizations are possible.

For convenience, our discussions are for the scalar measurement situation. To begin, let us reconcile the scalar stochastic measurement equation, Eq. (2.2-9)

$$z(k) = \underline{h}'(k)\underline{\theta} + v(k) \quad , \tag{2.13-1}$$

with the measured output of the scalar equation-error identification system in Fig. 1.2-2c. They are identical when $\underline{h}(k)$ is equated to $\underline{x}(k)$, the input to the "identification representation" block; hence, let us write Eq. (2.13-1) as

$$z(k) = \underline{x}'(k)\underline{\theta} + v(k) \quad . \tag{2.13-2}$$

The concatenated measurement equation for L + 1 measurements, Eq. (2.2-12), becomes

$$\begin{pmatrix} z(k) \\ z(k-1) \\ \vdots \\ z(k-L) \end{pmatrix} = \begin{pmatrix} x_1(k) & x_2(k) & \cdots & x_n(k) \\ x_1(k-1) & x_2(k-1) & \cdots & x_n(k-1) \\ \vdots & \vdots & \ddots & \vdots \\ x_1(k-L) & x_2(k-L) & \cdots & x_n(k-L) \end{pmatrix} \begin{pmatrix} \theta_1 \\ \theta_2 \\ \vdots \\ \theta_n \end{pmatrix} + \begin{pmatrix} v(k) \\ v(k-1) \\ \vdots \\ v(k-L) \end{pmatrix} \quad . \tag{2.13-3}$$

We shall consider the following cases: (1) all x_i's are known perfectly at $t_k, t_{k-1}, \ldots, t_{k-L}$; and (2) some or all x_i's are measured with measurement errors at t_k, t_{k-1}, ..., t_{k-L}. We also assume that, in both cases, $\underline{x}(k)$, $\underline{x}(k-1)$, ..., $\underline{x}(k-L)$ are independent of $v(k)$, $v(k-1)$, ..., $v(k-L)$; for if this is not true, then $H(k)$ and $\underline{V}(k)$ are dependent, and our estimates may be biased.

It is clear that when all x_i's are known perfectly, $H(k)$ and $\underline{V}(k)$ are independent; therefore, in this case, generalized least-squares parameter estimates will be unbiased. Examples which satisfy these requirements are Examples 1, 2, and 3 in Section 2.5.

When (some or) all x_i's are measured with measurement errors,

$$x_i(k) = r_i(k) - n_i(k) \tag{2.13-4}$$

for $i = 1, 2, \ldots, n$ (see Fig. 1.2-2c). In this case, Eq. (2.13-3) must be rewritten in terms of the measured quantities $r_i(k)$ as

$$\begin{pmatrix} z(k) \\ z(k-1) \\ \vdots \\ z(k-L) \end{pmatrix} = \begin{pmatrix} r_1(k) & r_2(k) & \cdots & r_n(k) \\ r_1(k-1) & r_2(k-1) & \cdots & r_n(k-1) \\ \vdots & \vdots & \ddots & \vdots \\ r_1(k-L) & r_2(k-L) & \cdots & r_n(k-L) \end{pmatrix} \begin{pmatrix} \theta_1 \\ \theta_2 \\ \vdots \\ \theta_n \end{pmatrix} + \begin{pmatrix} v_1(k) \\ v_1(k-1) \\ \vdots \\ v_1(k-L) \end{pmatrix} \tag{2.13-5}$$

where

$$v_1(k-j) = v(k-j) - \theta_1 n_1(k-j) - \theta_2 n_2(k-j) - \cdots - \theta_n n_n(k-j) \tag{2.13-6}$$

for $j = 0, 1, \ldots, L$. Clearly, $H(k)$ and $\underline{V}(k)$ are no longer independent, since both $r_i(k)$ and $v_1(k)$, for example, depend upon $n_i(k)$. Hence, in this case, generalized least-squares estimates may or may not be biased. Examples 4 and 5 in Section 2.5 serve to illustrate this case.

To summarize, we have shown that *generalized least-squares parameter estimates will be unbiased when all* x_i*'s* (inputs to the equation-error identification system) *are known perfectly at* t_k, t_{k-1}, ..., t_{k-L}; *they may or may not be biased when some or all* x_i*'s are measured with measurement errors at* t_k, t_{k-1}, ..., t_{k-L}. The uncertainty about bias, in the latter situation, usually cannot be resolved ahead of time. Such uncertainty detracts from the usefulness of generalized least-squares estimation when $\mathcal{H}(k)$ and $\underline{\mathcal{V}}(k)$ are dependent.

PROBLEMS

2-1. (a) Using discrete Legendre polynomials, obtain a quadratic least-squares estimating polynomial for the observation vector

$$\underline{z} = (\sin 90°, \sin 67\tfrac{1}{2}°, \sin 45°, \sin 22\tfrac{1}{2}°, \sin 0°)' \quad .$$

Arrange the estimate into a power series

$$\hat{z}(\zeta) = \alpha_0 + \alpha_1\zeta + \alpha_2\zeta^2 \quad .$$

(b) Starting with the power series

$$\hat{z}(\zeta) = \beta_0 + \beta_1\zeta + \beta_2\zeta^2 \quad ,$$

obtain the least-squares estimates for β_0, β_1, and β_2, using the data in (a).

(c) Compare the estimates obtained in (a) and (b). Which estimate gives better results for sin ζ at values of ζ other than those used to obtain the data in (a)?

2-2. Computers often have difficulties processing numbers that are either very large or very small; hence, in many estimation problems variables are transformed prior to estimation to variables whose numerical values are guaranteed, by their a priori choice, not to cause numerical difficulties. Suppose $x \in [b, h]$. Show that

$$x' = \frac{M - m}{h - b}x + \frac{hm - bM}{h - b}$$

transforms all values of x to the range [m, M]. If b and h are known ahead of time, we can work with x' instead of x, since m and M are prespecified by the designer based upon the computer considerations described above.

2-3.[†] The United States production of steel in millions of short tons (1 short ton = 2000 pounds) during the years 1946-1956 is given in Table P2-3.

(a) Find the equation of a least-squares line fitting the data, using the batch-estimation algorithm.

(b) Estimate the production of steel during the years 1957 and 1958 and compare with the true values of 112.7 and 85.3 millions of short tons, respectively.

TABLE P2-3[a]

Year	U. S. production of steel (millions of short tons)
1946	66.6
1947	84.9
1948	88.6
1949	78.0
1950	96.8
1951	105.2
1952	93.2
1953	111.6
1954	88.3
1955	117.0
1956	115.2

[a] *Source:* American Iron and Steel Institute.

2-4. Table P2-4 gives experimental values of the pressure P of a given mass of gas corresponding to various values of the volume V. According to thermodynamic principles, a relationship having the

[†] This problem as well as the next five have been taken for the most part from Ref. *2-4*.

form $PV^{\gamma} = C$, where γ and C are constants, should exist between the variables.

(a) Estimate γ and C using least-squares estimation.

(b) Estimate P when V = 100.0 in.^{3}.

TABLE P2-4

Volume V (in.^{3}):	54.3	61.8	72.4	88.7	118.6	194.0
Pressure P (lb/in.^{2}):	61.2	49.5	37.6	28.4	19.2	10.1

2-5. Table P2-5 shows the final grades in algebra and economics obtained by ten students selected at random from a large group of students.

(a) Find the least-square line fitting the data, using X as the independent variable.

(b) Find the least-square line fitting the data, using Y as the independent variable.

(c) If a student receives a grade of 75 in algebra, what is his expected grade in economics? If a student receives a grade of 95 in economics, what is his expected grade in algebra?

TABLE P2-5

Algebra (X):	75	80	93	65	87	71	98	68	84	77
Economics (Y):	82	78	86	72	91	80	95	72	89	74

2-6. Table P2-6 shows the number of farm workers in the United States (in millions) during the years 1949-1957.

(a) Find a least-square line fitting this time series.

(b) Estimate the number of farm workers in the year 1948 and compare with the actual value (10.36 million).

(c) Predict the number of farm workers in 1958 (true value is 7.53 million). Discuss the possible sources of error in such a prediction.

TABLE P2-6[a]

Year:	1949	1950	1951	1952	1953	1954	1955	1956	1957
Number of farm workers (millions):	9.96	9.93	9.55	9.15	8.86	8.64	8.36	7.82	7.58

[a] *Source:* Department of Agriculture.

2-7. The consumer price index for medical care in the United States is given in Table P2-7 for the years 1950-1957. (The *reference period* or *base period* 1947-1949 is assigned the value 100 which actually means 100%. The index for 1952, for example, is 117.2 and shows that during 1952 the average price of medical care was 117.2% of what it was in the base period, i. e., it increased by 17.2%)

(a) Find a least-square line fitting the data.

(b) Predict the price index for medical care during 1958 and compare with the true value (144.4).

(c) In what year can we expect the index of medical costs to be double that of 1947-1949 assuming that the trends in the data continue?

TABLE P2-7[a]

Year:	1950	1951	1952	1953	1954	1955	1956	1957
Consumer price index for medical care (1947-1949 = 100):	106.0	111.1	117.2	121.3	125.2	128.0	132.6	138.0

[a] *Source:* Bureau of Labor Statistics.

2-8. Table P2-8 shows the birth rate per 1000 population in the United States during the years 1915-1955 in five-year intervals.

(a) Find a least-squares parabola fitting the data.

(b) Explain why the equation obtained in (a) is not useful for extrapolation purposes.

TABLE P2-8[a]

Year:	1915	1920	1925	1930	1935	1940	1945	1950	1955
Birth rate per 1000 population:	25.0	23.7	21.3	18.9	16.9	17.9	19.5	23.6	24.6

[a] *Source:* Department of Health, Education and Welfare.

2-9. Show that if $\underline{\hat{\theta}}$ is an unbiased estimate of $\underline{\theta}$, $a\underline{\hat{\theta}} + \underline{b}$ is an unbiased estimate of $a\underline{\theta} + \underline{b}$. Is $\hat{\theta}^2$ an unbiased estimate of θ^2? [Use the least-squares batch algorithm.]

2-10. Suppose that the coefficients $a_1, a_2, \ldots, a_n$ in the finite-difference equation

$$y(k + n) + a_1 y(k + n - 1) + \cdots + a_n y(k) = \omega(k) + m(k)$$

are identified using the formulation based on the solution to the finite-difference equation described in Section 1.3.3. Equation (1.3-74) is the starting point for our analyses.

(a) Denoting measured values of $y(k)$ as $z(k)$ where $z(k) = y(k) + n(k)$, and assuming that $m(k)$ can be measured perfectly, obtain the concatenated measurement equation for $L + 1$ measurements of y.

(b) Are the generalized least-squares estimates of the a parameters unbiased? Explain.

2-11. Derive the sequential generalized least-squares algorithm for p additional measurements, given by Eqs. (2.7-32)-(2.7-34).

2-12. The following weighting matrix weights past measurements less heavily than the most recent measurements:

$$W(k + 1) = \left(\begin{array}{c|c} \bar{w}(k + 1) & 0_1 \\ \hline 0_2 & \beta^{t_{k+1}-t_k} W(k) \end{array}\right)$$

where $0 < \beta < 1$.

(a) Show that

$$\bar{w}(j) = w(j)\beta^{t_{k+1}-t_j}$$

for $j = 0, 1, \ldots, k + 1$.

(b) How must the equations for the sequential generalized least-squares estimation algorithm [Eqs. (2.7-23)-(2.7-25)] be modified for this weighting matrix? The estimator thus obtained is referred to as a *fading memory estimator* (2-2).

2-13. Owing to large initial errors, it is sometimes useful to weight more recent observations more heavily than older observations. The standard average of the function $z^2(j)$ is $\frac{1}{(k + 1)} \sum_{i=0}^{k} z^2(i)$. The *linear reinforcement average* of $z^2(j)$, denoted LRA(j), is obtained sequentially from the expression

$$\text{LRA}(k + 1) = \gamma\ \text{LRA}(k) + (1 - \gamma)z^2(k + 1)$$

$$\text{LRA}(0) = z^2(0)$$

where $k = 0, 1, \ldots$.

(a) For what values of γ are past values of z^2 weighted less heavily than newer values?

(b) By what factor is $z^2(k_1)/[z^2(k_1 + j)]$ weighted (j is an arbitrary positive integer)?

(c) For what value of γ does the linear reinforcement average reduce to the standard average?

(d) Does the linear reinforcement average lead to a fading-memory estimator (see Problem 2-12)? Be sure to indicate what $W(k + 1)$ is, for the linear reinforcement average.

2-14. The alternative formulation of the sequential generalized least-squares estimation algorithm is given in Eqs. (2.7-46)-(2.7-48) for the case of one additional measurement. Derive the comparable algorithm for the case of p additional measurements.

2-15. Prove that

$$(I_n + XY)^{-1} = I_n - X(I_m + YX)^{-1}Y$$

where X is $n \times m$ and Y is $m \times n$. This is another matrix-inversion lemma.

2-16. Using mathematical induction, prove Eqs. (2.8-28) and (2.8-29).

2-17. For the data in Table P2-17, do the following.

(a) Obtain the least-square line by means of the batch-processing least-squares estimation algorithm.

TABLE P2-17

ζ	$y(\zeta)$
0	1
1	3
2	6
3	8

(b) Obtain the least-square line by means of the sequential least-squares estimation algorithm, using the sequential startup technique (let $a = 10^8$ and $\varepsilon = 10^{-16}$).

2-18. Give some measurements that can be made in the following fields: medicine, biology, economics, sociology, and ecology.

2-19. Derive both the batch and sequential generalized least-squares estimation algorithms for the case when we desire to process p additional vector measurements instead of just one additional vector measurement. Obtain a solution for L that is comparable to Eq. (2.9-19).

2-20. Sketch the surface defined by Eq. (2.9-19) for $n = 1, 2, \ldots, 10$, $m = 1, 2, \ldots, 10$, and $n \geq m$.

2-21. Prove Eqs. (2.10-26) and (2.10-27).

2-22. Explain how sequential data processing would be used to estimate $\underline{x}(0)$ in Eq. (2.2-24) if $n = 10$, $\underline{z}(k) = H\underline{x}(k) + \underline{v}(k)$, and $H = I_n$.

2-23. Which of the following systems are asymptotically stable?

(a) $\tilde{\theta}(k + 1) - \tilde{\theta}(k) = 0$.

(b) $\tilde{\theta}(k + 2) - 2\tilde{\theta}(k + 1) + \tilde{\theta}(k) = 0$.

(c) $\tilde{\theta}(k + 1) - \frac{1}{2}\tilde{\theta}(k) = 0$.

(d) $\tilde{\theta}(k + 2) + \frac{1}{4}\tilde{\theta}(k) = 0$.

(e) $\tilde{\theta}(k + 2) - 3\tilde{\theta}(k + 1) + 2\tilde{\theta}(k) = 0$.

(f) $\tilde{\theta}(k + 2) + 4\tilde{\theta}(k) = 0$.

2-24. Verify Eq. (2.11-30) for k = 1, 2, and 3.

2-25. Prove Corollary 2-3 (*Hint:* Concatenate the conditions in the statement of Theorem 2-5).

2-26. Prove Corollary 2-4.

2-27. Prove that P(k) is a symmetric matrix.

2-28. Verify that if W(k) is positive definite then $\mathcal{H}'(k)W(k)\mathcal{H}(k)$ is positive definite if and only if $\mathcal{H}(k)$ has full column rank.

2-29. Using the alternative formulation for the sequential generalized least-squares estimation algorithm, show that, if P(1) is singular, all P(j), $j > 1$, will be singular as well.

2-30. Let us investigate the question of bias for Problems 2-3, 2-4, and 2-6.

(a) In Problem 2-3, is the least-square line an unbiased estimate of the data if the values given for U. S. production of steel, in Table P2-3, are in error by some random amounts?

(b) In Problem 2-4, are the least-squares estimates of γ and C unbiased if P and V cannot be measured perfectly? Is the least-squares formulation applicable in this situation?

(c) In Problem 2-6, assume that the number of farm workers is normally distributed, with the mean value given in Table P2-6 and variance equal to σ^2. Is the least-square line an unbiased estimate of the data in Table P2-6 under this assumption?

2-31. In many cases it is not known a priori how many terms of the approximation in Eq. (2.4-1) should be used. The least-squares estimation problem usually has to be solved for several values of N. When the least-squares problem is solved for a model with N parameters and we wish to determine a model with N + 1 parameters, it seems to be a waste of computing effort to start from scratch. Instead, it seems preferable to use the results obtained for N parameters to obtain the model with N + 1 parameters. In this problem we will investigate how this should be done (K. J. Åström, *Lectures on the Identification Problem - The Least Squares Method*, Rept. 6806, Lund Institute of Technology, Division of Automatic Control, September 1968).

When introducing extra parameters $\hat{\theta}_j$ where $j > N$ we get additional terms in Eq. (2.4-1). This means that additional columns are added to the $H(q)$ matrix and that additional elements are added to the vector $\underline{\theta}$. Increasing N to $N + \ell$, $H(q)$ and $\underline{\theta}$ can be written in partitioned form as

$$\mathcal{H}(q) = \left(\begin{array}{ccc|ccc} \phi_1(\zeta_q) & \cdots & \phi_N(\zeta_q) & \phi_{N+1}(\zeta_q) & \cdots & \phi_{N+\ell}(\zeta_q) \\ \phi_1(\zeta_{q-1}) & \cdots & \phi_N(\zeta_{q-1}) & \phi_{N+1}(\zeta_{q-1}) & \cdots & \phi_{N+\ell}(\zeta_{q-1}) \\ \vdots & \ddots & \vdots & \vdots & \ddots & \vdots \\ \phi_1(\zeta_1) & \cdots & \phi_N(\zeta_1) & \phi_{N+1}(\zeta_1) & \cdots & \phi_{N+\ell}(\zeta_1) \end{array}\right)$$

$$= [\mathcal{H}_1(q) \mid \mathcal{H}_2(q)]$$

and

$$\underline{\theta}' = (\theta_1, \ldots, \theta_N \mid \theta_{N+1}, \ldots, \theta_{N+\ell})$$

$$= (\underline{\theta}_1' \mid \underline{\theta}_2') \quad .$$

(a) Show that $\hat{\underline{\theta}}_2(k)$ can be obtained from $\hat{\underline{\theta}}_1(k)$ by means of the algorithm

$$\hat{\underline{\theta}}_2(k) = [\mathcal{H}_2'(k)\mathcal{H}_2(k)]^{-1}\mathcal{H}_2'(k)[\underline{Z}(k) - \mathcal{H}_1(k)\hat{\underline{\theta}}_1(k)] \quad .$$

(b) Discuss the computational advantages associated with evaluating $\hat{\underline{\theta}}_2(k)$ from the expression in (a) as compared with computing the entire $\hat{\underline{\theta}}$ vector when the dimension of $\hat{\underline{\theta}}$ is increased from N to N + ℓ.

(c) Devise a test from the expression for $\hat{\underline{\theta}}_2(k)$ in (a), to determine the optimum value for ℓ.

REFERENCES

2-1. F. A. Graybill, *An Introduction to Linear Statistical Models*, Vol. I, McGraw-Hill, New York, 1961.

2-2. N. Morrison, *Introduction to Sequential Smoothing and Prediction*, McGraw-Hill, New York, 1969.

2-3. M. Abromowitz and I. A. Stegun, *Handbook of Mathematical Functions*, National Bureau of Standards, Appl. Math. Ser. No. 55, June 1964.

2-4. M. R. Siegel, *Schaum's Outline of Theory and Problems of Statistics*, Schaum Publishing, New York, 1961.

2-5. A. Papoulis, *Probability, Random Variables, and Stochastic Processes*, McGraw-Hill, New York, 1965.

2-6. Y. C. Ho, On the stochastic approximation method and optimal filtering theory, *J. Math. Anal. Appl.*, 6, 152-154 (1962).

2-7. I. A. Gura, An algebraic solution of the state estimation problem, *AIAA J.*, 7, 1242-1247 (1969).

2-8. R. Penrose, A generalized inverse for matrices, *Proc. Cambridge Phil. Soc.*, 51, part 3, 406-413 (1955).

2-9. T. N. E. Greville, Some applications of the pseudo inverse of a matrix, *Soc. Ind. Appl. Math. Rev.*, 2, 15-22 (1960).

2-10. L. A. Zadeh and C. A. Desoer, *Linear System Theory, The State Space Approach*, McGraw-Hill, New York, 1963.

2-11. H. W. Sorenson, Kalman filtering techniques, in *Advances in Control Systems*, Vol. 3 (C. T.Leondes, ed.), Academic Press, New York, 1966.

2-12. J. M. Mendel, Computational requirements for a discrete Kalman filter, *IEEE Trans. Autom. Contr.*, AC-16, 748-758, December (1971).

2-13. L. S. Pontryagin, *Ordinary Differential Equations*, Addison-Wesley, Reading, Mass., 1962.

2-14. F. B. Hildebrand, *Methods of Applied Mathematics*, Prentice-Hall, New Jersey, 1952.

2-15. R. Bellman, *Introduction to Matrix Analysis*, 2nd Ed., McGraw-Hill, New York, 1970.

2-16. S. Perlis, *Theory of Matrices*, Addison-Wesley, Cambridge, Mass., 1952.

2-17. J. S. Meditch, *Stochastic Optimal Linear Estimation and Control*, McGraw-Hill, New York, 1969.

2-18. R. C. K. Lee, *Optimal Estimation, Identification, and Control*, Research Monograph No. 28, The M.I.T. Press, Cambridge, Mass., 1964.

2-19. M. Aoki, *Optimization of Stochastic Systems: Topics in Discrete-Time Systems*, Academic Press, New York, 1967.

2-20. A. H. Jazwinski, *Stochastic Processes and Filtering Theory*, Academic Press, New York, 1970.

2-21. J. Kmenta, *Elements of Econometrics*, Macmillan, New York, 1971.

2-22. P. C. Young, An instrumental variable method for real-time identification of a noisy process, *Automatica*, 6, 271-287 (1970).

3

Minimum – Variance Parameter Estimation

3.1 INTRODUCTION

The generalized least-squares estimate, obtained in Chapter 2, does not make use of the statistics of the measurement-error process, in any way. In many applications, it is not uncommon to know the mean and variance of the measurement error. If one believes that it is good policy to make use of all a priori information, then he must seek an estimate that utilizes the statistics of the measurement errors. Minimum-variance parameter estimates achieve this objective.

3.2 FORMULATION AND STATEMENT OF THE ESTIMATION PROBLEM

We shall consider the situation[†] when L + 1 measurements of a vector signal $\underline{y}(t)$ are made at times t_{k-L}, t_{k-L+1}, ..., and t_k, where $k \geq L$. The $\underline{z}(j)$, the measured value of $\underline{y}(t)$ at t_j, is related to $\underline{y}(j)$ by the expression

---

[†] For more detailed discussions on formulations of the estimation problem, see Sections 2.2 (scalar measurements) and 2.9 (vector measurements).

$$\underline{z}(j) = \underline{y}(j) + \underline{v}(j) \tag{3.2-1}$$

where $\underline{v}(j)$ is an $m \times 1$ vector random measurement-error process, with the following *known statistics:*

$$E\{\underline{v}(j)\} = \underline{0} \tag{3.2-2}$$

and

$$E\{\underline{v}(j)\underline{v}'(j)\} = R(j) \tag{3.2-3}$$

for $j = k - L, k - L + 1, \ldots, k$, and $k \geq L$. $R(j)$ *is an* $m \times m$ *positive definite covariance matrix that is assumed to be known a priori.* Such information is often available from manufacturers of the sensors that are used to obtain the data.

Each component in $\underline{y}(k)$ is assumed to be linearly related to the n components of an unknown parameter vector $\underline{\theta}$; that is to say,

$$\underline{y}(k) = H(k)\underline{\theta} \tag{3.2-4}$$

where

$$H(k) = \begin{pmatrix} h_{11}(k) & h_{12}(k) & \cdots & h_{1n}(k) \\ h_{21}(k) & h_{22}(k) & \cdots & h_{2n}(k) \\ \vdots & \vdots & \ddots & \vdots \\ h_{m1}(k) & h_{m2}(k) & \cdots & h_{mn}(k) \end{pmatrix} . \tag{3.2-5}$$

Hence

$$\underline{z}(j) = H(j)\underline{\theta} + \underline{v}(j) \quad . \tag{3.2-6}$$

Proceeding as in Sections 2.2 and 2.9, we concatenate the L + 1 vector measurements, $\underline{z}(k)$, $\underline{z}(k - 1)$, ..., and $\underline{z}(k - L)$, obtaining the *concatenated measurement equation*

$$\underline{\mathcal{Z}}(k) = \mathcal{H}(k)\underline{\theta} + \underline{\mathcal{V}}(k) \quad , \tag{3.2-7}$$

where $\underline{\mathcal{Z}}(k)$ is the $(L + 1)m \times 1$ *concatenated measurement vector,*

$$\underline{\mathcal{Z}}(k) = \begin{pmatrix} \underline{z}(k) \\ \hline \underline{z}(k-1) \\ \hline \vdots \\ \hline \underline{z}(k-L) \end{pmatrix} , \tag{3.2-8}$$

$\underline{\mathcal{V}}(k)$ is the $(L + 1)m \times 1$ *concatenated measurement error vector,*

$$\underline{\mathcal{V}}(k) = \begin{pmatrix} \underline{v}(k) \\ \hline \underline{v}(k-1) \\ \hline \vdots \\ \hline \underline{v}(k-L) \end{pmatrix} , \tag{3.2-9}$$

and $\mathcal{H}(k)$ is the $(L + 1)m \times n$ *concatenated observation matrix,*

$$\mathcal{H}(k) = \begin{pmatrix} H(k) \\ \hline H(k-1) \\ \hline \vdots \\ \hline H(k-L) \end{pmatrix} . \tag{3.2-10}$$

Observe that the statistics of $\underline{\mathcal{V}}(k)$ are

$$E\{\underline{\mathcal{V}}(k)\} = \underline{0} \tag{3.2-11}$$

and

$$E\{\underline{\mathcal{V}}(k)\underline{\mathcal{V}}'(k)\} = \mathcal{R}(k) = \left(\begin{array}{c|c|c|c} R(k) & 0 & \cdots & 0 \\ \hline 0 & R(k-1) & \cdots & 0 \\ \hline \vdots & \vdots & \ddots & \vdots \\ \hline 0 & 0 & \cdots & R(k-L) \end{array}\right) . \tag{3.2-12}$$

The symbol $\hat{\underline{\theta}}(k)$ denotes the estimate of $\underline{\theta}$ based on $(L + 1)m$ data or, equivalently, $(L + 1)$ data vectors, $\underline{z}(k)$, $\underline{z}(k - 1)$, ..., and $\underline{z}(k - L)$. We shall assume that $\hat{\underline{\theta}}(k)$ is related to $\underline{\mathcal{Z}}(k)$ by means of a linear transformation $F(k)$; that is,

$$\hat{\underline{\theta}}(k) = F(k)\underline{\mathcal{Z}}(k) . \tag{3.2-13}$$

An optimum choice of $F(k)$ is desired, one that makes use of the statistics of the measurement-error process.

We shall determine F(k) *by minimizing the error variance of each parameter* $\theta_1, \theta_2, \ldots, \theta_n$, *subject to the constraint that the estimate must be unbiased.* The resulting estimate shall be referred to as the *unbiased minimum-variance estimate.*[†]

Before proceeding to the solution of this optimization problem, let us investigate how unbiasedness constrains the admissible choices of F(k).

3.3 IMPLICATION OF THE UNBIASEDNESS CONSTRAINT

Theorem 3-1. *Let* F(k) *and* $H(k)$ *be non-random. Then* $\hat{\underline{\theta}}(k)$ *is an unbiased estimate of* $\underline{\theta}$ *if, and only if,*

$$F(k)H(k) = I \quad . \tag{3.3-1}$$

Proof. (A) *(Necessity)* From Eqs. (3.2-13) and (3.2-7), we find

$$\hat{\underline{\theta}}(k) = F(k)H(k)\underline{\theta} + F(k)\underline{V}(k) \quad . \tag{3.3-2}$$

Hence, under the assumptions that $\hat{\underline{\theta}}(k)$ is an unbiased estimate of $\underline{\theta}$ [see Eq. (2.5-2)], and F(k) and $H(k)$ are nonrandom, we obtain

$$E\{\hat{\underline{\theta}}(k)\} = \underline{\theta} = F(k)H(k)\underline{\theta} \quad , \tag{3.3-3}$$

or

$$[I - F(k)H(k)]\underline{\theta} = \underline{0} \tag{3.3-4}$$

for all $k \geq L$. Obviously, for $\underline{\theta} \neq \underline{0}$, Eq. (3.3-1) is the solution to this equation.

(B) *(Sufficiency)* From Eq. (3.3-2) and the nonrandomness of F(k) and $H(k)$, we have

$$E\{\hat{\underline{\theta}}(k)\} = F(k)H(k)\underline{\theta} \quad . \tag{3.3-5}$$

Assuming the truth of Eq. (3.3-1), it must be, that

$$E\{\hat{\underline{\theta}}(k)\} = \underline{\theta} \tag{3.3-6}$$

which means, of course, that $\hat{\underline{\theta}}(k)$ is an unbiased estimate of $\underline{\theta}$. ▲

---

† This estimate is also known as the best linear unbiased estimate (BLUE) of $\underline{\theta}$.

In the special case when no measurement errors are present, so that $\underline{V}(k) = \underline{0}$, *and* $(L + 1)m = n$, so that $H(k)$ is a square matrix, Eq. (3.3-1) provides us with the following solution for $F(k)$:

$$F(k) = H^{-1}(k) \qquad (3.3\text{-}7)$$

which means that

$$\hat{\underline{\theta}}(k) = F(k)\underline{Z}(k) = H^{-1}(k)[H(k)\underline{\theta}] = \underline{\theta} \quad . \qquad (3.3\text{-}8)$$

$\hat{\underline{\theta}}(k)$ exactly equals $\underline{\theta}$; for this reason, Eq. (3.3-1) is sometimes called the *exactness constraint* (*3-1*). Since $\underline{V}(k) \neq \underline{0}$ in our applications, and $(L + 1)m$ is usually not equal to n, we prefer to call Eq. (3.3-1) the *unbiasedness constraint.*

In solving for the unbiased minimum variance $F(k)$, it will be convenient to partition this $n \times (L + 1)m$ matrix as follows:

$$F(k) = \begin{pmatrix} \underline{f}_1'(k) \\ \text{-----} \\ \underline{f}_2'(k) \\ \text{-----} \\ \vdots \\ \text{-----} \\ \underline{f}_n'(k) \end{pmatrix} \qquad (3.3\text{-}9)$$

where

$$\underline{f}_i'(k) = [f_{i1}(k), f_{i2}(k), \ldots, f_{i\alpha}(k)] \qquad (3.3\text{-}10)$$

and, for notational convenience,

$$\alpha \triangleq (L + 1)m \quad . \qquad (3.3\text{-}11)$$

Let us demonstrate that Eq. (3.3-1) can also be written as

$$H'(k)\underline{f}_i(k) = \underline{e}_i \qquad (3.3\text{-}12)$$

for $i = 1, 2, \ldots, n$, where $\underline{e}_i$ is the ith unit vector

$$\underline{e}_i = (0 \;\; 0 \;\; \cdots \;\; 0 \;\; 1 \;\; 0 \;\; \cdots \;\; 0)' \quad , \qquad (3.3\text{-}13)$$

in which the only nonzero element occurs in the ith position. This is obtained by taking the transpose of Eq. (3.3-1) and using the decomposition of $F(k)$, in Eq. (3.3-9), as follows:

$$I = \mathcal{H}'(k)F'(k) = \mathcal{H}'(k)[\underline{f}_1(k) \vdots \cdots \vdots \underline{f}_n(k)]$$
$$= [\mathcal{H}'(k)\underline{f}_1(k) \vdots \cdots \vdots \mathcal{H}'(k)\underline{f}_n(k)] \quad . \quad (3.3\text{-}14)$$

The $n \times n$ identity matrix can also be written, in terms of the unit vectors in Eq. (3.3-13), as

$$I = (\underline{e}_1 \vdots \cdots \vdots \underline{e}_n) \quad . \quad (3.3\text{-}15)$$

Equation (3.3-12) is obtained by equating respective column vectors of I in Eqs. (3.3-14) and (3.3-15).

3.4 COMPUTATION OF UNBIASED, MINIMUM-VARIANCE F(k)

Recall our proclamation at the end of Section 3.2 that F(k) will be determined by minimizing the error variance of each parameter, θ_1, θ_2, ..., and θ_n, subject to the unbiasedness constraint. We shall use the method of Lagrange multipliers to accomplish this task (*3-2* and *3-3*).

For notational convenience, let $\sigma^2_{\hat{\theta}_i}(k)$ denote the error variance of the ith parameter; that is to say,

$$\sigma^2_{\hat{\theta}_i}(k) = E\{[\theta_i - \hat{\theta}_i(k)]^2\} \quad . \quad (3.4\text{-}1)$$

We shall determine $\underline{f}_i$ such that the performance function

$$J_i = \sigma^2_{\hat{\theta}_i}(k) + \underline{\lambda}_i'(k)[\mathcal{H}'(k)\underline{f}_i(k) - \underline{e}_i] \quad (3.4\text{-}2)$$

is minimum for $i = 1, 2, \ldots, n$. $\underline{\lambda}_i(k)$ is an $n \times 1$ vector of Lagrange multipliers associated with the ith unbiasedness constraint. Below, we show that $\sigma^2_{\hat{\theta}_i}(k)$ is a function only of $\underline{f}_i(k)$; hence, it is only necessary to include the ith unbiasedness constraint in J_i, as we have done.

<u>Theorem 3-2</u>. *The unbiased minimum-variance estimate is*[†]

[†] Some authors prefer to use different symbols for generalized least-squares, unbiased minimum-variance, etc., estimates of $\underline{\theta}$. We have chosen to use the generic symbol $\hat{\underline{\theta}}$ for estimate and to let the right-hand side of the equation for $\hat{\underline{\theta}}(k)$ denote the type of estimate.

$$\hat{\underline{\theta}}(k) = [H'(k)R^{-1}(k)H(k)]^{-1}H'(k)R^{-1}(k)\underline{Z}(k) \tag{3.4-3}$$

for $k \geq L$.

According to Plackett (*3-4*), Gauss (*3-5*) was the first to derive this result; however, he restricted $R(k)$ to be diagonal. In 1934, Aitken (*3-6*) generalized the results to nondiagonal $R(k)$.

Proof. (Ref. *3-3*) To begin, let us show that

$$J_i = \underline{f}_i'(k)R(k)\underline{f}_i(k) + \underline{\lambda}_i'(k)[H'(k)\underline{f}_i(k) - \underline{e}_i] \tag{3.4-4}$$

for i = 1, 2, ..., n. We shall make use of Eqs. (3.4-2), (3.4-1), (3.3-12), and (3.2-7), and the following equivalent representation of Eq. (3.2-13) [Problem 3-3]:

$$\hat{\theta}_i(k) = \underline{f}_i'(k)\underline{Z}(k) \quad . \tag{3.4-5}$$

Proceeding, we find

$$\begin{aligned} J_i &= E\{[\theta_i - \hat{\theta}_i(k)]^2\} + \underline{\lambda}_i'(H'\underline{f}_i - \underline{e}_i) \\ &= E\{(\theta_i - \underline{f}_i'\underline{Z})^2\} + \underline{\lambda}_i'(H'\underline{f}_i - \underline{e}_i) \\ &= E\{\theta_i^2 - 2\theta_i\underline{Z}'\underline{f}_i + (\underline{Z}'\underline{f}_i)^2\} + \underline{\lambda}_i'(H'\underline{f}_i - \underline{e}_i) \\ &= E\{\theta_i^2 - 2\theta_i(H\underline{\theta} + \underline{V})'\underline{f}_i + [(H\underline{\theta} + \underline{V})'\underline{f}_i]^2\} + \underline{\lambda}_i'(H'\underline{f}_i - \underline{e}_i) \\ &= E\{\theta_i^2 - 2\theta_i\underline{\theta}'\underline{e}_i - 2\theta_i\underline{V}'\underline{f}_i + (\underline{\theta}'\underline{e}_i + \underline{V}'\underline{f}_i)^2\} + \underline{\lambda}_i'(H'\underline{f}_i - \underline{e}_i) \quad . \end{aligned} \tag{3.4-6}$$

Now

$$\theta_i = \underline{\theta}'\underline{e}_i \quad . \tag{3.4-7}$$

Hence, continuing the development of the expectation in Eq. (3.4-6), we find

$$\begin{aligned} J_i &= E\{(\underline{V}'\underline{f}_i)^2\} + \underline{\lambda}_i'(H'\underline{f}_i - \underline{e}_i) \\ &= E\{(\underline{f}_i'\underline{V})(\underline{V}'\underline{f}_i)\} + \underline{\lambda}_i'(H'\underline{f}_i - \underline{e}_i) \\ &= \underline{f}_i'E\{\underline{V}\underline{V}'\}\underline{f}_i + \underline{\lambda}_i'(H'\underline{f}_i - \underline{e}_i) \\ &= \underline{f}_i'R\underline{f}_i + \underline{\lambda}_i'(H'\underline{f}_i - \underline{e}_i) \end{aligned} \tag{3.4-8}$$

which is the desired result.

We shall obtain $\underline{f}_i$ that minimizes Eq. (3.4-8) using vector calculus (see Sections 2.3.1 and 2.3.2). A necessary condition for minimizing J_i is $(d/d\underline{f}_i)J_i = \underline{0}$ (i = 1, 2, ..., n); hence,

$$2R\underline{f}_i + H\underline{\lambda}_i = \underline{0} \quad , \tag{3.4-9}$$

from which we determine $\underline{f}_i$, as

$$\underline{f}_i = -\frac{1}{2}R^{-1}H\underline{\lambda}_i \quad . \tag{3.4-10}$$

A second necessary condition for minimizing J_i is $(d/d\underline{\lambda}_i)J_i = 0$, which gives us the unbiasedness constraints

$$H'\underline{f}_i = \underline{e}_i \quad , \tag{3.4-11}$$

for i = 1, 2, ..., n.

The unknown vector of Lagrange multipliers $\underline{\lambda}_i$ is found, from Eqs. (3.4-10) and (3.4-11), to be

$$\underline{\lambda}_i = -2(H'R^{-1}H)^{-1}\underline{e}_i \tag{3.4-12}$$

whereupon, we find

$$\underline{f}_i = R^{-1}H(H'R^{-1}H)^{-1}\underline{e}_i \tag{3.4-13}$$

for i = 1, 2, ..., n. Then F(k) is reconstructed from $\underline{f}_i$ as follows:

$$\begin{aligned} F'(k) &= (\underline{f}_1 \,\vdots\, \underline{f}_2 \,\vdots\, \cdots \,\vdots\, \underline{f}_n) \\ &= R^{-1}H(H'R^{-1}H)^{-1}(\underline{e}_1 \,\vdots\, \underline{e}_2 \,\vdots\, \cdots \,\vdots\, \underline{e}_n) \\ &= R^{-1}H(H'R^{-1}H)^{-1} \quad . \end{aligned} \tag{3.4-14}$$

Hence

$$F(k) = [H'(k)R^{-1}(k)H(k)]^{-1}H'(k)R^{-1}(k) \quad . \tag{3.4-15}$$

The unbiased minimum-variance estimate in Eq. (3.4-3) is obtained by substituting Eq. (3.4-15) into Eq. (3.2-13). ▲

For the optimal estimate in Eq. (3.4-3), let us compute the error variance $\sigma^2_{\hat{\theta}_i}(k)$ of the ith parameter. Observe from Eqs. (3.4-2) and (3.3-12) that $\sigma^2_{\hat{\theta}_i}(k) = J_i$; hence, it follows from Eqs. (3.4-4) and (3.3-12) that

$$\sigma^2_{\hat{\theta}_i}(k) = \underline{f}_i'(k)R(k)\underline{f}_i(k) \quad . \tag{3.4-16}$$

Substituting the optimal value of $\underline{f}_i(k)$ from Eq. (3.4-13) into $\sigma^2_{\hat{\theta}_i}(k)$, we find

$$\sigma^2_{\hat{\theta}_i}(k) = \underline{e}_i'[H'(k)R(k)H(k)]^{-1}\underline{e}_i \quad . \qquad (3.4\text{-}17)$$

3.5 COMPARISON OF UNBIASED MINIMUM-VARIANCE AND GENERALIZED LEAST-SQUARES ESTIMATES

We are struck by the close similarity between the unbiased minimum-variance estimate in Eq. (3.4-3) and the generalized least-squares estimate in Eq. (2.3-12).

Theorem 3-3. *The unbiased minimum-variance estimate of* $\underline{\theta}$ *is the special case of the generalized least-squares estimate of* $\underline{\theta}$, *when*

$$W(k) = R^{-1}(k) \quad . \qquad (3.5\text{-}1)$$

The proof of this important link between two seemingly dissimilar approaches to parameter identification is obvious. ▲

The expression $R^{-1}(k)$ *is a weighting matrix that stresses the contributions of precise measurements and plays down the contributions of less precise measurements.* The minimum-variance technique leads to a weighting matrix that is quite sensible.

Corollary 3-1. *All results obtained in Chapter 2 for generalized least-squares estimates of* $\underline{\theta}$ *apply as well to minimum-variance estimates of* $\underline{\theta}$, *when* $W(k) = R^{-1}(k)$. ▲

The importance of this result cannot be underestimated. We shall have more to say about it, and shall make use of it shortly, in Sections 3.6 and 3.7.

Corollary 3-2. *If*

$$R(k) = \sigma^2 I \qquad (3.5\text{-}2)$$

then the unbiased minimum-variance *and* least-squares *estimates of* $\underline{\theta}$ *are identical.*

The proof of this result, often known as the *Gauss-Markoff Theorem,* is left as an exercise to the reader (Problem 3-4) (*3-7*).

Observe that the Gauss-Markoff Theorem is applicable only to homoskedastic measurement-error processes. ▲

3.6 SOME PROPERTIES OF UNBIASED MINIMUM-VARIANCE PARAMETER ESTIMATES

In this section, we present some results that are comparable to those presented for generalized least-squares estimates in Section 2.5. We do not have to study the question of bias, since our minimum-variance estimate has been synthesized subject to the a priori constraint of unbiasedness. Let us, therefore, direct our attention to the covariance of parameter estimation error.

Theorem 3-4. *For the unbiased minimum-variance estimate of* $\underline{\theta}$ *given in Eq. (3.4-3), if* $\underline{V}(k)$ *has zero mean, then*

$$\text{Cov}\,[\underline{\tilde{\theta}}(k)] = [H'(k)R^{-1}(k)H(k)]^{-1} \quad . \tag{3.6-1}$$

Proof. The proof follows directly from an application of Corollary 3-1 to Eq. (2.5-27), which appears in Corollary 2-1.† ▲

Observe the great simplification of the expression for Cov $[\underline{\tilde{\theta}}(k)]$ in Eq. (2.5-27), when $W(k) = R^{-1}(k)$. Note, also, that Eq. (3.6-1) includes $\sigma^2_{\hat{\theta}_i}(k)$ given in Eq. (3.4-17), since the ith diagonal element of Cov $[\underline{\tilde{\theta}}(k)]$ is $\sigma^2_{\hat{\theta}_i}(k)$. For homoskedastic measurement error processes, $R(k) = \sigma^2 I$, and Cov $[\underline{\tilde{\theta}}(k)] = \sigma^2[H'(k)H(k)]^{-1}$.

Let us next examine Cov $[\underline{\tilde{\theta}}(k)]$ in more detail. The ith diagonal element of Cov $[\underline{\tilde{\theta}}(k)]$ is $\sigma^2_{\hat{\theta}_i}(k)$, and it was the minimization of these diagonal elements that led to the solution for $\underline{\hat{\theta}}(k)$, obtained in Theorem 3-2. A natural question to ask, is: *does* $\underline{\hat{\theta}}(k)$ *in Eq. (3.4-3) give the minimum error-covariance estimate of all unbiased estimates?*

Theorem 3-5. $\underline{\hat{\theta}}(k)$ *in Eq. (3.4-3) gives the minimum error-covariance estimate of all unbiased estimates that are linearly related to* $\underline{Z}(k)$ *(3-3).*

---

† $H(k)$ was assumed to be deterministic in Theorem 3-1.

Proof. We shall show that

$$\Sigma = \mathrm{Cov}\,[\tilde{\underline{\theta}}(k)]\Big|_{\substack{\text{Arbitrary}\\ \text{unbiased}\\ \text{estimate}}} - \mathrm{Cov}\,[\tilde{\underline{\theta}}(k)]\Big|_{\substack{\text{Unbiased}\\ \text{minimum-variance}\\ \text{estimate}}} \tag{3.6-2}$$

is positive definite. For convenience, we write Σ as

$$\Sigma = \Sigma_2 - \Sigma_1 \quad . \tag{3.6-3}$$

Let F_a denote the transformation matrix in Eq. (3.2-13) for the arbitrary unbiased estimate. The symbol F will be reserved for the unbiased minimum-variance transformation matrix.

Clearly,

$$\begin{aligned}\Sigma_2 &= E\{(\underline{\theta} - F_a\underline{Z})(\underline{\theta} - F_a\underline{Z})'\}\\ &= E\{(\underline{\theta} - F_aH\underline{\theta} - F_a\underline{V})(\underline{\theta} - F_aH\underline{\theta} - F_a\underline{V})'\}\\ &= E\{(F_a\underline{V})(F_a\underline{V})'\}\\ &= F_aRF_a' \quad ,\end{aligned} \tag{3.6-4}$$

and

$$\Sigma_1 = FRF' \tag{3.6-5}$$

as well; hence,

$$\begin{aligned}\Sigma &= F_aRF_a' - FRF'\\ &= F_aRF_a' - FRF' + 2(H'R^{-1}H)^{-1} - (H'R^{-1}H)^{-1} - (H'R^{-1}H)^{-1}\\ &= F_aRF_a' - FRF' + 2(H'R^{-1}H)^{-1}(H'R^{-1}RF')\\ &\quad - (H'R^{-1}H)^{-1}(H'R^{-1}RF_a') - (F_aRR^{-1}H)(H'R^{-1}H)^{-1}\end{aligned} \tag{3.6-6}$$

where we have made repeated use of the unbiasedness constraints

$$H'F' = H'F_a' = F_aH_a = I \quad . \tag{3.6-7}$$

Making use of the structure of F(k) in Eq. (3.4-15), we see that Σ can also be written as

$$\Sigma = F_a R F_a' - FRF' + 2FRF' - FRF_a' - F_a RF'$$

$$= (F_a - F)R(F_a - F)' \quad . \qquad (3.6\text{-}8)$$

In order to investigate the definiteness of Σ, consider the definiteness of $\underline{a}'\Sigma\underline{a}$, where $\underline{a}$ is an arbitrary nonzero vector (*3-8* and *3-9*).

$$\underline{a}'\Sigma\underline{a} = \underline{a}'(F_a - F)R(F_a - F)'\underline{a}$$

$$= [(F_a - F)'\underline{a}]'R[(F_a - F)'\underline{a}] \quad . \qquad (3.6\text{-}9)$$

By virtue of the fact that F is unique (Problem 3-5), $(F_a - F)'\underline{a}$ is a nonzero $(L + 1)m \times 1$ vector; hence, the definiteness of Σ is determined by the definiteness of R. The matrix R is positive definite; thus, Σ is positive definite, and the proof is complete. ▲

These results serve as further confirmation that choosing F(k) as we did in Section 3.4, by minimizing only the diagonal elements of Cov $[\tilde{\underline{\theta}}(k)]$, is sound.

We conclude this section with a brief discussion on the invariance of unbiased minimum-variance estimates to scale changes.

Theorem 3-6. *The unbiased minimum-variance estimate of* $\underline{\theta}$, *given in Eq. (3.4-3), is invariant under scale changes.*

Proof (3-1). Assume that observers A and B are observing a process, simultaneously using the same set of sensors. Observer A reads the measurements in one set of units and B in another. Let M be the *diagonal matrix* of scale factors relating A to B (e. g., 16 oz to the pound); then the total observation vectors of A and B, $\underline{Z}_A(k)$ and $\underline{Z}_B(k)$, will be numerically related by

$$\underline{Z}_B(k) = M\underline{Z}_A(k)$$

$$= MH_A(k)\underline{\theta} + M\underline{V}_A(k) \qquad (3.6\text{-}10)$$

which means that

$$H_B(k) = MH_A(k) \quad , \qquad (3.6\text{-}11)$$

$$\underline{V}_B(k) = M\underline{V}_A(k) \quad , \qquad (3.6\text{-}12)$$

and, subsequently, that

$$\mathcal{R}_B(k) = M\mathcal{R}_A(k)M' = M\mathcal{R}_A(k)M \quad . \tag{3.6-13}$$

Letting $\hat{\underline{\theta}}_A(k)$ and $\hat{\underline{\theta}}_B(k)$ denote the unbiased minimum-variance estimates associated with the data from observers A and B, respectively, we find that

$$\begin{aligned}
\hat{\underline{\theta}}_B(k) &= [\mathcal{H}_B'(k)\mathcal{R}_B^{-1}(k)\mathcal{H}_B(k)]^{-1}\mathcal{H}_B'(k)\mathcal{R}_B^{-1}(k)\underline{Z}_B(k) \\
&= [\mathcal{H}_A'M(M\mathcal{R}_AM)^{-1}M\mathcal{H}_A]^{-1}\mathcal{H}_A'M(M\mathcal{R}_AM)^{-1}M\underline{Z}_A \\
&= [\mathcal{H}_A'MM^{-1}\mathcal{R}_A^{-1}M^{-1}M\mathcal{H}_A]^{-1}\mathcal{H}_A'MM^{-1}\mathcal{R}_A^{-1}M^{-1}M\underline{Z}_A \\
&= (\mathcal{H}_A'\mathcal{R}_A^{-1}\mathcal{H}_A)^{-1}\mathcal{H}_A'\mathcal{R}_A^{-1}\underline{Z}_A \\
&= \hat{\underline{\theta}}_A(k) \quad .
\end{aligned} \tag{3.6-14}$$

▲

We hasten to point out that generalized least-squares estimates are not invariant under scale changes; thus, one must be very careful about using generalized least-squares estimation on measurements that are of mixed dimensions (e. g., distance, in feet, and velocity, in miles/hour). For additional discussions, see Problem 3-6.

3.7 SEQUENTIAL UNBIASED MINIMUM-VARIANCE ESTIMATION

We have demonstrated in Corollary 3-1 that all results obtained in Chapter 2 for generalized least-squares estimation of $\underline{\theta}$ apply as well to unbiased minimum-variance estimation of $\underline{\theta}$ when W(k) is set equal to $\mathcal{R}^{-1}(k)$; hence, sequential unbiased minimum-variance algorithms comparable to those derived in Sections 2.7 and 2.9 are obtained merely by setting

$$w(k + 1) = R^{-1}(k + 1) \tag{3.7-1}$$

in Eqs. (2.9-12)-(2.9-17). These algorithms are summarized next.[†]

[†] It may be helpful at this point for the reader to review the material in Sections 2.7 and 2.9.

Sequential unbiased minimum-variance algorithm: One additional measurement vector[†]

$$\underline{\hat{\theta}}(k + 1) = \underline{\hat{\theta}}(k) + K^{o}(k + 1)[\underline{z}(k + 1) - H(k + 1)\underline{\hat{\theta}}(k)] \qquad (3.7\text{-}2)$$

$$K^{o}(k + 1) = P(k + 1)H'(k + 1)R^{-1}(k + 1) \qquad (3.7\text{-}3)$$

$$P^{-1}(k + 1) = P^{-1}(k) + H'(k + 1)R^{-1}(k + 1)H(k + 1) \qquad (3.7\text{-}4)$$

for $k \geq L$; *or*

$$\underline{\hat{\theta}}(k + 1) = \underline{\hat{\theta}}(k) + K^{o}(k + 1)[\underline{z}(k + 1) - H(k + 1)\underline{\hat{\theta}}(k)] \qquad (3.7\text{-}5)$$

$$K^{o}(k + 1) = P(k)H'(k + 1)[H(k + 1)P(k)H'(k + 1) + R(k + 1)]^{-1} \qquad (3.7\text{-}6)$$

$$P(k + 1) = [I - K^{o}(k + 1)H(k + 1)]P(k) \qquad (3.7\text{-}7)$$

for $k \geq L$.

By means of Corollary 3-1, we also conclude that: (1) these sequential algorithms may be applied starting with $k = 0$, using the *sequential startup technique* described in Section 2.8; (2) *sequential data processing techniques*, described in Section 2.10, may be used in order to minimize computing time associated with the digital computer implementations of these algorithms; and, (3) if we use these algorithms starting with $k = 0$, initialization should be made using the "original" formulation, in Eqs. (3.7-2)-(3.7-4); but it is advisable to switch to the "alternate" formulation in Eqs. (3.7-5)-(3.7-7), as soon after initialization as possible, in order to reduce computing time (see Section 2.12).

The matrix P(k) has no physical meaning attached to it in sequential generalized least-squares estimation; however, it has a very important physical meaning in sequential unbiased minimum-variance estimation, namely

---

† To distinguish between the optimal weighting matrices for unbiased minimum-variance and generalized least-squares estimators, we use K^{o} for the former weighting matrix and K^{*} for the latter.

$$P(k) = \text{Cov}\,[\tilde{\underline{\theta}}(k)] \quad . \tag{3.7-8}$$

Recall that P(k) was defined in Eq. (2.7-13) as

$$P(k) = [H'(k)W(k)H(k)]^{-1} \quad . \tag{3.7-9}$$

Setting $W(k) = R^{-1}(k)$, P(k) becomes

$$P(k) = [H'(k)R^{-1}(k)H(k)]^{-1} \quad . \tag{3.7-10}$$

We have already demonstrated, in Theorem 3-4, that

$$\text{Cov}\,[\tilde{\underline{\theta}}(k)] = [H'(k)R^{-1}(k)H(k)]^{-1} \quad ; \tag{3.7-11}$$

hence, the relationship between P(k) and Cov $[\tilde{\underline{\theta}}(k)]$ in Eq. (3.7-8).

This is a very interesting and important result; for it means that a measure of confidence in $\hat{\underline{\theta}}(k)$ - Cov $[\tilde{\underline{\theta}}(k)]$ - is computed every iteration, as part of the computation of $\hat{\underline{\theta}}(k)$. The matrix P(k) is often referred to in the literature as the *estimation-error covariance matrix*, for obvious reasons.

3.8 RELATIONSHIP OF SEQUENTIAL UNBIASED MINIMUM-VARIANCE ESTIMATION TO DISCRETE KALMAN FILTERING†

3.8.1 Discrete Kalman Filter

Kalman filtering is a popular method for obtaining minimum-variance estimates *of signals* from noisy measurements (*3-10* - *3-14*). The discrete Kalman filter, which is easily programmed on a digital computer, provides state estimates for the following dynamic system:

$$\underline{x}(k+1) = \Phi(k+1, k)\underline{x}(k) + \Psi(k+1, k)\underline{u}(k) + \Gamma(k+1, k)\underline{\omega}(k) \tag{3.8-1}$$

$$\underline{z}(k+1) = H(k+1)\underline{x}(k+1) + \underline{v}(k+1) \tag{3.8-2}$$

where

† The reader unfamiliar with Kalman filtering may want to skip this section. It is doubtful that he will find the material in it as illuminating (or even illuminating at all) as will the reader who is already familiar with Kalman filtering.

$\underline{x} \equiv n \times 1$ state vector

$\underline{\omega} \equiv r \times 1$ disturbance vector

$\underline{z} \equiv m \times 1$ measurement vector

$\underline{v} \equiv m \times 1$ measurement noise vector

$\underline{u} \equiv p \times 1$ control or test signal vector

$\Phi \equiv n \times n$ state transition matrix

$\Psi \equiv n \times p$ control distribution matrix

$\Gamma \equiv n \times r$ disturbance transition matrix

$H \equiv m \times n$ measurement matrix

$k \equiv$ discrete-time index $(k = 0, 1, \ldots)$

The process† $\{\underline{\omega}(k), k = 0, 1, \ldots\}$ is an r-dimensional gaussian white sequence for which

$$E\{\underline{\omega}(k)\} = \underline{0} \tag{3.8-3}$$

for all $k = 0, 1, \ldots,$ and

$$E\{\underline{\omega}(j)\underline{\omega}'(k)\} = Q(k)\delta_{jk} \tag{3.8-4}$$

for all $j, k = 0, 1, \ldots,$ where $Q(k)$ is a positive semidefinite $r \times r$ matrix. The process $\{\underline{v}(k + 1), k = 0, 1, \ldots\}$ is an m-dimensional gaussian white sequence for which

$$E\{\underline{v}(k + 1)\} = \underline{0} \tag{3.8-5}$$

for all $k = 0, 1, \ldots,$ and

$$E\{\underline{v}(j + 1)\underline{v}'(k + 1)\} = R(k + 1)\delta_{jk} \tag{3.8-6}$$

for all $j, k = 0, 1, \ldots,$ where $R(k + 1)$ is a positive semidefinite $m \times m$ matrix. The two stochastic processes are assumed to be independent of each other, so that

$$E\{\underline{v}(j)\underline{\omega}'(k)\} = 0 \tag{3.8-7}$$

for all $j = 1, 2, \ldots,$ and $k = 0, 1, \ldots.$

---

† The rest of this paragraph is taken, essentially, from pages 168-169, and 176 of Ref. *3-12*. Copyright 1969 by McGraw-Hill, Inc. Used with permission of McGraw-Hill Book Company.

The initial state $\underline{x}(0)$ is assumed to be a gaussian random vector, with mean

$$E\{\underline{x}(0)\} = \underline{0} \tag{3.8-8}$$

and $n \times n$ positive semidefinite covariance matrix

$$E\{\underline{x}(0)\underline{x}'(0)\} = P(0) \quad . \tag{3.8-9}$$

It is further assumed that $\underline{x}(0)$ is independent of $\{\underline{\omega}(k),\ k = 0, 1, \ldots\}$ and $\{\underline{v}(k),\ k = 0, 1, \ldots\}$.

The *optimal filtered estimate* of $\underline{x}(k + 1)$, denoted $\hat{\underline{x}}(k + 1|k + 1)$, is given† by the recursive relations:

$$\hat{\underline{x}}(k + 1|k) = \Phi(k + 1, k)\hat{\underline{x}}(k|k) + \Psi(k + 1, k)\underline{u}(k) \tag{3.8-10}$$

and

$$\hat{\underline{x}}(k + 1|k + 1) = \hat{\underline{x}}(k + 1|k) + K(k + 1)[\underline{z}(k + 1) - H(k + 1)\hat{\underline{x}}(k + 1|k)] \tag{3.8-11}$$

for $k = 0, 1, \ldots$, where $\hat{\underline{x}}(0|0) = \underline{0}$. The $K(k + 1)$ is an $n \times m$ matrix which is specified by the set of relations

$$K(k + 1) = P(k + 1|k)H'(k + 1)[H(k + 1)P(k + 1|k)H'(k + 1) + R(k + 1)]^{-1} \tag{3.8-12}$$

$$P(k + 1|k) = \Phi(k + 1, k)P(k|k)\Phi'(k + 1, k) + \Gamma(k + 1, k)Q(k)\Gamma'(k + 1, k) \tag{3.8-13}$$

$$P(k + 1|k + 1) = [I - K(k + 1)H(k + 1)]P(k + 1|k) \tag{3.8-14}$$

for $k = 0, 1, \ldots$, where I is the $n \times n$ identity matrix and $P(0|0) = P(0)$ is the initial condition for Eq. (3.8-13).

---

† The expression $\hat{\underline{x}}(k|j)$ is shorthand notation for "the estimate of $\underline{x}(k)$ obtained at t_k, and based on the measurements $\underline{z}(1)$, $\underline{z}(2)$, ..., and $\underline{z}(j)$." The expression $\hat{\underline{x}}(k + 1|k)$ is known as the single-stage prediction of $\underline{x}(k)$, whereas $\hat{\underline{x}}(k + 1|k + 1)$ is known as the filtered estimate of $\underline{x}(k)$.

Equations (3.8-10)-(3.8-14) constitute the discrete Kalman filter. Equation (3.8-10) is often referred to as the *prediction equation*, since it provides the optimal estimate of $\underline{x}(k+1)$ using only the measurements $\underline{z}(1)$, $\underline{z}(2)$, ..., and $\underline{z}(k)$. The new measurement $\underline{z}(k+1)$ at t_{k+1} is blended together with $\hat{\underline{x}}(k+1|k)$ in Eq. (3.8-11) - the *correction equation*. The matrix $K(k+1)$ is often referred to as the *Kalman gain matrix* or *weighting matrix*. The $P(k+1|k)$ and $P(k+1|k+1)$ have definite physical meanings. Each is a conditional state estimation-error covariance matrix; that is to say,

$$P(k+1|k) = E\left\{\tilde{\underline{x}}(k+1|k)\tilde{\underline{x}}'(k+1|k) \middle| \underline{z}(1), \underline{z}(2), \ldots, \underline{z}(k)\right\} \tag{3.8-15}$$

and

$$P(k+1|k+1) = E\left\{\tilde{\underline{x}}(k+1|k+1)\tilde{\underline{x}}'(k+1|k+1) \middle| \underline{z}(1), \underline{z}(2), \ldots, \underline{z}(k+1)\right\} \tag{3.8-16}$$

where

$$\tilde{\underline{x}}(k+1|k) = \underline{x}(k+1) - \hat{\underline{x}}(k+1|k) \tag{3.8-17}$$

and

$$\tilde{\underline{x}}(k+1|k+1) = \underline{x}(k+1) - \hat{\underline{x}}(k+1|k+1) \quad . \tag{3.8-18}$$

3.8.2 Comparison of Sequential Unbiased Minimum-Variance Algorithm and Discrete Kalman Filter

The relationship between the discrete Kalman filter and the sequential unbiased minimum-variance parameter estimator is provided in the following theorem.

Theorem 3-7. *The sequential unbiased minimum-variance estimator of a constant parameter* $\underline{\theta}$ *can be obtained from the discrete Kalman filter equations, by setting*

$$\underline{x}(k) = \underline{\theta} \quad , \tag{3.8-19}$$

$$\Phi(k+1, k) = I \quad , \tag{3.8-20}$$

$$\Psi(k+1, k) = 0 \quad , \tag{3.8-21}$$

and

$$Q(k) = 0 \quad . \tag{3.8-22}$$

Proof. By direct substitution of Eqs. (3.8-19)-(3.8-22) into Eqs. (3.8-10)-(3.8-14), one obtains:

$$\hat{\underline{\theta}}(k + 1|k) = \hat{\underline{\theta}}(k|k) \quad , \tag{3.8-23}$$

$$\hat{\underline{\theta}}(k + 1|k + 1) = \hat{\underline{\theta}}(k + 1|k) + K(k + 1)[\underline{z}(k + 1) - H(k + 1)\hat{\underline{\theta}}(k + 1|k)] \quad , \tag{3.8-24}$$

$$P(k + 1|k) = P(k|k) \quad , \tag{3.8-25}$$

$$K(k + 1) = P(k + 1|k)H'(k + 1)[H(k + 1)P(k + 1|k)H'(k + 1) + R(k + 1)]^{-1} \quad , \tag{3.8-26}$$

and

$$P(k + 1|k + 1) = [I - K(k + 1)H(k + 1)]P(k + 1|k) \quad . \tag{3.8-27}$$

Combining Eqs. (3.8-23) and (3.8-25) with Eqs. (3.8-24), (3.8-26), and (3.8-27), we find that

$$\hat{\underline{\theta}}(k + 1|k + 1) = \hat{\underline{\theta}}(k|k) + K(k + 1)[\underline{z}(k + 1) - H(k + 1)\hat{\underline{\theta}}(k|k)] \quad , \tag{3.8-28}$$

$$K(k + 1) = P(k|k)H'(k + 1)[H(k + 1)P(k|k)H'(k + 1) + R(k + 1)]^{-1} \quad , \tag{3.8-29}$$

and

$$P(k + 1|k + 1) = [I - K(k + 1)H(k + 1)]P(k|k) \quad . \tag{3.8-30}$$

These equations are, obviously, exactly the same as Eqs. (3.7-5), (3.7-6), and (3.7-3), except for some minor differences in notation.

▲

Let us take a closer look at the conditions of the preceding theorem, and some of their consequences. Under the condition that $Q(k) = 0$ we see that $\underline{\omega}(k) = \underline{0}$ for all k; and under the conditions that $\Phi(k + 1, k) = I$ and $\Psi(k + 1, k) = 0$, we see that

$$\underline{x}(k + 1) = \underline{x}(k) \tag{3.8-31}$$

which means, of course, that $\underline{x}(k) = \text{constant} - \underline{\theta}$.

Equations (3.8-23) and (3.8-25) explain why we have not chosen to adopt the more cumbersome notation $\hat{\underline{\theta}}(k+1|k+1)$ for $\hat{\underline{\theta}}(k+1)$. Because we are identifying a vector of constant parameters, there are no dynamics associated with $\underline{\theta}$; hence, the predicted estimate of $\underline{\theta}$, $\hat{\underline{\theta}}(k+1|k)$, cannot be different from the corrected $\hat{\underline{\theta}}(k+1|k+1)$. When we treat the problem of tracking time-varying parameters, in Chapter 6, we will find it helpful to distinguish between the two estimates, $\hat{\underline{\theta}}(k+1|k)$ and $\hat{\underline{\theta}}(k+1|k+1)$. The reason for this is that we may think of a time-varying parameter as described by some underlying dynamics. These dynamics will have some associated state transition matrix that is different from the identity matrix; hence, $\hat{\underline{\theta}}(k+1|k) \neq \hat{\underline{\theta}}(k|k)$, as is the case when $\underline{\theta}$ is constant [Eq. (3.8-23)].

3.9 RELATIONSHIP OF UNBIASED MINIMUM-VARIANCE ESTIMATION TO MAXIMUM-LIKELIHOOD ESTIMATION

Here we shall take a brief look at the so-called *maximum-likelihood estimate* of the unknown parameter vector $\underline{\theta}$. Once again, our starting point is the concatenated measurement equation

$$\underline{Z}(k) = H(k)\underline{\theta} + \underline{V}(k) \tag{3.9-1}$$

where $\underline{Z}(k)$, $\underline{V}(k)$, and $H(k)$ are defined in Eqs. (3.2-8), (3.2-9), and (3.2-10), respectively.

Recall that in Chapter 2 the only assumption made regarding $\underline{V}(k)$ was that $E\{\underline{V}(k)\} = \underline{0}$. In the present chapter, not only did we assume that $E\{\underline{V}(k)\} = \underline{0}$, but we also assumed that $E\{\underline{V}(k)\underline{V}'(k)\} = R(k)$ is known to us a priori.

In this section, we shall assume that $V(k)$ is a zero-mean, multivariate gaussian distributed random vector. As in preceding sections of this chapter

$$E\{\underline{V}(k)\underline{V}'(k)\} = R(k) \tag{3.9-2}$$

is assumed known. The multivariate gaussian density function of $\underline{V}(k)$ is (see Ref. *3-7*, for example)

$$f(\underline{V}) = \frac{1}{\sqrt{(2\Pi)^{\alpha}|R(k)|}} \exp\left[-\frac{1}{2}\underline{V}'(k)R^{-1}(k)\underline{V}(k)\right] \qquad (3.9\text{-}3)$$

where

$$\alpha = \dim \underline{V}(k) = (L + 1)m \quad . \qquad (3.9\text{-}4)$$

It is well known (*3-8* and *3-15*), that linear transformations on, and linear combinations of, gaussian random vectors are gaussian random vectors. For this reason, it is clear that when $\underline{V}(k)$ is gaussian, $\underline{Z}(k)$ is as well. The multivariate gaussian density function of $\underline{Z}(k)$, derived from $f(\underline{V})$, is

$$f[\underline{Z}(k)] = \frac{1}{\sqrt{(2\Pi)^{\alpha}|R(k)|}} \exp\left\{-\frac{1}{2}[\underline{Z}(k) - H(k)\underline{\theta}]'R^{-1}(k)[\underline{Z}(k) - H(k)\underline{\theta}]\right\} \quad . \qquad (3.9\text{-}5)$$

The *principle of maximum likelihood* states that the estimate of $\underline{\theta}$, $\hat{\underline{\theta}}(k)$, shall be chosen so that the value of $f[\underline{Z}(k)]$ is maximized for a particular realization of $\underline{Z}(k)$; that is to say, the maximum likelihood estimator of $\underline{\theta}$ is that value of $\underline{\theta}$, denoted $\hat{\underline{\theta}}(k)$, that would generate the observed sample $\underline{Z}(k)$ most often. We hasten to point out that this principle is in no way limited to multivariate gaussian density functions, or to applications in which measurements are related linearly to the unknown parameters (see *3-16* and *3-17*).

Theorem 3-8. *When* $f[\underline{Z}(k)]$ *is multivariate gaussian, the principle of maximum likelihood leads to the unbiased minimum-variance estimate of* $\underline{\theta}$.

Proof. To find the estimate that maximizes $f[\underline{Z}(k)]$, we must set $(d/d\underline{\theta})f[\underline{Z}(k)] = \underline{0}$; however, notice that $f[\underline{Z}(k)]$ is maximized when the argument of the exponential function is minimized. Let $\underline{\theta}_{ML}$ denote the maximum likelihood estimate of $\underline{\theta}$. It is found as the solution to

$$\frac{d}{d\underline{\theta}_{ML}}[\underline{Z}(k) - H(k)\underline{\theta}_{ML}]'R^{-1}(k)[\underline{Z}(k) - H(k)\underline{\theta}_{ML}] = \underline{0} \quad . \qquad (3.9\text{-}6)$$

This equation should be compared with Eq. (2.2-35) and the discussion immediately preceding it. We conclude that $\underline{\theta}_{ML} = \hat{\underline{\theta}}(k)$ given in Eq. (2.3-12), where $W(k) = R^{-1}(k)$; but by Theorem 3-3, this estimate is the unbiased minimum-variance estimate of $\underline{\theta}$. ▲

The principle of maximum likelihood is, generally speaking, quite broad; however, it does require considerably more a priori knowledge about the statistics of $\underline{V}(k)$ - or $\underline{Z}(k)$, when $\underline{Z}(k)$ is not related to $\underline{\theta}$ linearly - than we have had to assume in our preceding analyses. Of course, when $\underline{V}(k)$ is multivariate gaussian, its statistics (e. g., higher-order moments) can all be determined from $E\{\underline{V}(k)\}$ and $R(k)$ (*3-15*). It is this well-known fact, plus the fact that the concatenated measurement equation - the starting equation for all of our preceding analyses, both in this chapter and Chapter 2 - relates the measurements linearly to $\underline{\theta}$, which cause the maximum likelihood estimate of $\underline{\theta}$ to equal the unbiased minimum-variance estimate.

Corollary 3-3. *If* $f[\underline{Z}(k)]$ *is multivariate gaussian and*

$$R(k) = \sigma^2 I \quad , \tag{3.9-7}$$

then the unbiased minimum-variance, least-squares, *and* maximum-likelihood *estimates of* $\underline{\theta}$ *are identical.*

The proof of this very important result follows directly from Theorem 3-8 and Corollary 3-2. We leave the derivation of the maximum-likelihood estimator of σ^2 as an exercise (Problem 3-15).

3.10 SENSITI 'ITY CONSIDERATIONS[†]

The estimation of parameters by means of unbiased minimum-variance estimation algorithms presupposes knowledge of the underlying statistical and dynamical parameters which characterize the data generation model. When an estimation algorithm processes data which are generated by a mechanism whose actual statistical and dynamical

[†] The discussions in this section can also be extended to sequential generalized least-squares algorithms.

parameters differ from those used in the estimation algorithm, *suboptimal estimation* of the parameters occurs.

In this section, we shall consider a sensitivity analysis of a suboptimal sequential unbiased minimum-variance parameter estimation algorithm. The measure of actual performance will be the parameter estimation-error covariance matrix. Our approach is patterned after the approach taken by Jazwinski (*3-13*) and Lainiotis and Sims (*3-18*), who presented similar results for the discrete Kalman filter.

To begin our analysis, let us distinguish between the *actual measurement system* and the *modeled measurement system*. The actual measurement system is described by the $m \times 1$ vector measurement equation

$$\underline{z}(k) = H(k)\underline{\theta} + \underline{v}(k) \tag{3.10-1}$$

and the statistics of $\underline{v}(k)$,

$$E\{\underline{v}(k)\} = \underline{0} \tag{3.10-2}$$

$$E\{\underline{v}(k)\underline{v}'(k)\} = R(k) \quad . \tag{3.10-3}$$

The modeled measurement system is described by a similar set of equations:

$$\underline{z}_m(k) = H_m(k)\underline{\theta} + \underline{v}_m(k) \tag{3.10-4}$$

where

$$E\{\underline{v}_m(k)\} = \underline{0} \tag{3.10-5}$$

and

$$E\{\underline{v}_m(k)\underline{v}_m'(k)\} = R_m(k) \quad . \tag{3.10-6}$$

In these equations, $\underline{z}_m(k)$ is the modeled measurement vector, $\underline{v}_m(k)$ is the modeled measurement-error process, $H_m(k)$ is the modeled observation matrix, and $R_m(k)$ is the modeled measurement-error covariance matrix. The vector $\underline{z}_m(k)$ differs from $\underline{z}(k)$ in two ways: $H_m(k)$ differs from $H(k)$, and $R_m(k)$ differs from $R(k)$.

The suboptimal sequential unbiased minimum-variance parameter estimation algorithm is [Eqs. (3.7-5)-(3.7-7)]:

$$\hat{\underline{\theta}}_m(k+1) = \hat{\underline{\theta}}_m(k) + K_m^o(k+1)[\underline{z}(k+1) - H_m(k+1)\hat{\underline{\theta}}_m(k)] \quad , \tag{3.10-7}$$

$$K_m^o(k+1) = P_m(k)H_m'(k+1)[H_m(k+1)P_m(k)H_m'(k+1) + R_m(k+1)]^{-1} \quad , \tag{3.10-8}$$

and

$$P_m(k+1) = [I - K_m^o(k+1)H_m(k+1)]P_m(k) \quad . \tag{3.10-9}$$

For computational reasons discussed in Section 2.12.2 (also see Section 3.11), we prefer to use the following equivalent expression for $P_m(k+1)$ [Eq. (2.12-6) with $w(k+1) = R^{-1}(k+1)$]:

$$P_m(k+1) = [I - K_m^o(k+1)H_m(k+1)]P_m(k)[I - K_m^o(k+1)H_m(k+1)]' + K_m^o(k+1)R_m(k+1)K_m^{o'}(k+1) \quad . \tag{3.10-10}$$

Observe in Eq. (3.10-7) that the actual measurement $\underline{z}(k+1)$, and not $\underline{z}_m(k+1)$, is used as the basis for updating $\hat{\underline{\theta}}_m(k)$ to $\hat{\underline{\theta}}_m(k+1)$. Also, $P_m(k) \neq \text{Cov}\,[\tilde{\underline{\theta}}(k)]$; whereas, for the actual measurement system, we saw that $P(k) = \text{Cov}\,[\tilde{\underline{\theta}}(k)]$ [Eq. (3.7-8)].

Let us proceed to compute the parameter estimation-error covariance matrix for the modeled measurement system. This matrix can then be compared with P(k) in order to determine the effects of modeling errors.

Quite possibly, only some, rather than all elements in H(k) and R(k), have modeling uncertainty associated with their numerical values. Let us group the total collection of such elements from H(k) and R(k) into the vector $\underline{\alpha}$ (Problem 3-7). We proceed to compute the conditional parameter estimation-error covariance matrix $P(k|\underline{\alpha})$ where

$$P(k|\underline{\alpha}) = E\{[\underline{\theta} - \hat{\underline{\theta}}_m(k)][\underline{\theta} - \hat{\underline{\theta}}_m(k)]'|\underline{\alpha}\} \quad . \tag{3.10-11}$$

Expanding the right-hand side of this expression, we find

$$P(k|\underline{\alpha}) = \underline{\theta}\underline{\theta}' - \underline{\theta}E\{\hat{\underline{\theta}}_m'(k)|\underline{\alpha}\} - E\{\hat{\underline{\theta}}_m(k)|\underline{\alpha}\}\underline{\theta}' + E\{\hat{\underline{\theta}}_m(k)\hat{\underline{\theta}}_m'(k)|\underline{\alpha}\} \quad . \tag{3.10-12}$$

We shall derive sequential equations for obtaining $E\{\hat{\underline{\theta}}_m(k)|\underline{\alpha}\}$ and $E\{\hat{\underline{\theta}}_m(k)\hat{\underline{\theta}}_m'(k)|\underline{\alpha}\}$.

Consider $E\{\underline{\hat{\theta}}_m(k)|\underline{\alpha}\}$ first. Take the conditional expectation of both sides of Eq. (3.10-7), to obtain

$$E\{\underline{\hat{\theta}}_m(k+1)|\underline{\alpha}\} = E\{\underline{\hat{\theta}}_m(k)|\underline{\alpha}\} + K_m^o(k+1)\Big[E\{\underline{z}(k+1)|\underline{\alpha}\} - H_m(k+1)E\{\underline{\hat{\theta}}_m(k)|\underline{\alpha}\}\Big]$$
$$= \Big[I - K_m^o(k+1)H_m(k+1)\Big]E\{\underline{\hat{\theta}}_m(k)|\underline{\alpha}\} + K_m^o(k+1)H(k+1)\underline{\theta} \qquad (3.10\text{-}13)$$

for $k \geq L$, or $k \geq 0$, assuming use of the sequential startup technique described in Section 2.8. In the latter case, we frequently choose $\underline{\hat{\theta}}_m(0) = \underline{0}$; hence, the sequential algorithm in Eq. (3.10-13) would be initialized by setting $E\{\underline{\hat{\theta}}_m(0)|\underline{\alpha}\} = \underline{0}$. In the former case, we would initialize Eq. (3.10-13) by setting $E\{\underline{\hat{\theta}}_m(L)|\underline{\alpha}\} = \underline{\hat{\theta}}_m(L)$.

Next, let us obtain a sequential algorithm for $E\{\underline{\hat{\theta}}_m(k)\underline{\hat{\theta}}_m'(k)|\underline{\alpha}\}$. From Eqs. (3.10-7) and (3.10-1),

$$\underline{\hat{\theta}}_m(k+1) = [I - K_m^o(k+1)H_m(k+1)]\underline{\hat{\theta}}_m(k) + K_m^o(k+1)H(k+1)\underline{\theta} + K_m^o(k+1)\underline{v}(k+1) \quad . \qquad (3.10\text{-}14)$$

It is left as an exercise for the reader (Problem 3-8) to demonstrate that

$$\begin{aligned} E\{\underline{\hat{\theta}}_m(k+1)\underline{\hat{\theta}}_m'(k+1)|\underline{\alpha}\} &= \Lambda(k+1)E\{\underline{\hat{\theta}}_m(k)\underline{\hat{\theta}}_m'(k)|\underline{\alpha}\}\Lambda'(k+1) \\ &+ K_m^o(k+1)H(k+1)\underline{\theta}\underline{\theta}'H'(k+1)K_m^{o'}(k+1) \\ &+ K_m^o(k+1)R(k+1)K_m^{o'}(k+1) \\ &+ \Lambda(k+1)E\{\underline{\hat{\theta}}_m(k)|\underline{\alpha}\}\underline{\theta}'H'(k+1)K_m^{o'}(k+1) \\ &+ K_m^o(k+1)H(k+1)\underline{\theta}E\{\underline{\hat{\theta}}_m'(k)|\underline{\alpha}\}\Lambda'(k+1) \end{aligned} \qquad (3.10\text{-}15)$$

for $k \geq L$, or $k \geq 0$, assuming use of the sequential startup technique. For notational convenience,

$$[I - K_m^o(k+1)H_m(k+1)] = \Lambda(k+1) \quad . \tag{3.10-16}$$

Equation (3.10-15) is initialized by setting

$$E\{\hat{\underline{\theta}}_m(L)\hat{\underline{\theta}}_m'(L)|\underline{\alpha}\} = \hat{\underline{\theta}}_m(L)\hat{\underline{\theta}}_m'(L) \tag{3.10-17}$$

or, if $k \geq 0$,

$$E\{\hat{\underline{\theta}}_m(0)\hat{\underline{\theta}}_m'(0)|\underline{\alpha}\} = \hat{\underline{\theta}}_m(0)\hat{\underline{\theta}}_m'(0) \quad . \tag{3.10-18}$$

In order to obtain $P(k|\underline{\alpha})$, computations should be performed in the following order: $P_m(k) \to K_m^o(k+1) \to \Lambda(k+1) \to E\{\hat{\underline{\theta}}_m(k)|\underline{\alpha}\} \to E\{\hat{\underline{\theta}}_m(k)\hat{\underline{\theta}}_m'(k)|\underline{\alpha}\} \to P(k+1|\underline{\alpha}) \to P_m(k+1)$.

Finally, the effects of differences between the actual and modeled measurement systems can be obtained by computing the *sensitivity matrix* $S(k+1|\underline{\alpha})$, where

$$S(k+1|\underline{\alpha}) = P(k+1) - P(k+1|\underline{\alpha}) \quad . \tag{3.10-19}$$

In addition to the computations required to obtain $P(k+1|\underline{\alpha})$, we must also compute $P(k+1)$, using Eqs. (3.7-6) and (3.7-7) [or (2.12-6)] (Problem 3-9).

Computing $S(k+1|\underline{\alpha})$ is difficult, even for the simplest of measurement systems [e. g., $z(k) = h\theta + v(k)$, where θ is scalar (Problem 3-10)]; however, that is often the price one must pay when he asks difficult questions - and the question, "How sensitive is the sequential unbiased minimum-variance estimator to modeling errors?" is a difficult one. Fortunately for us, all of the equations which must be evaluated in order to obtain $S(k+1|\alpha)$ are easily programmed for digital simulations.

3.11 COMPUTATIONAL ASPECTS OF UNBIASED MINIMUM-VARIANCE ESTIMATION

3.11.1 Batch Processing Algorithm

The batch unbiased minimum-variance estimation algorithm, in Eq. (3.4-3), requires the computation of two matrix inverses, $\mathcal{R}^{-1}(k)$ and $[\mathcal{H}'(k)\mathcal{R}^{-1}(k)\mathcal{H}(k)]^{-1}$. Often, $\mathcal{R}$ is diagonal, in which case it is a trivial matter to obtain $R^{-1}(k)$. For a discussion on inverting $[H'(k)R^{-1}(k)H(k)]$, see Section 2.12.1.

3.11.2 Sequential Processing Algorithms

All of the discussions in Section 2.12.2, on computational aspects of sequential generalized least-squares estimation algorithms, apply to sequential unbiased minimum-variance estimation algorithms as well (Corollary 3-1). Let us summarize the important results.

(1) In on-line applications, where speed of computation is usually the most important consideration, the alternate formulation in Eqs. (3.7-5)-(3.7-7) is preferable to the original formulation in Eqs. (3.7-2)-(3.7-4). This is because a smaller matrix needs to be inverted in the alternate formulation - an $m \times m$ matrix rather than an $n \times n$ matrix.

(2) The original formulation should be used to initiate the sequential startup technique, when the sequential estimation algorithms are started at $k = 0$. This is because, when $P(0)$ is very very large, $P(1)$ becomes singular when it is computed by means of the alternate formulation; however, it does not become singular when it is computed by means of the original formulation. It is also advisable to switch to the alternate formulation as soon after initialization as possible, in order to reduce computing time.

(3) When $P(k + 1)$ is computed by means of Eq. (3.7-7), errors in the computation of $K^o(k + 1)$ are propagated into $P(k + 1)$; hence, it is advisable to use the following equivalent expression to compute $P(k + 1)$ (also, see Problem 3-11).

$$P(k + 1) = [I - K^o(k + 1)H(k + 1)]P(k)[I - K^o(k + 1)H(k + 1)]' + K^o(k + 1)R(k + 1)K^o(k + 1) \quad . \tag{3.11-1}$$

Although computing time is increased when Eq. (3.11-1) is used in place of Eq. (3.7-7), the increased accuracy is usually well worth it.

A computational problem that is solely related to unbiased minimum-variance estimation occurs when measurements become "too good."

For example, consider the situation when all measurements are made with very little measurement error; that is to say, let (3-1)

$$R(k + 1) = \varepsilon I \tag{3.11-2}$$

where $\varepsilon <<<$. From Eqs. (3.7-7) and (3.7-6) [or (3.11-1) and (3.7-6)],

$$P(k + 1) = P(k) - P(k)H'(k + 1)[H(k + 1)P(k)H'(k + 1) + \varepsilon I]^{-1} \times H(k + 1)\, P(k)$$

$$\approx P(k) - P(k)H'(k + 1)[H(k + 1)P(k)H'(k + 1)]^{-1} \times H(k + 1)P(k) \quad ; \tag{3.11-3}$$

but, as we demonstrate next, $P(k + 1)$ is singular. To see this, postmultiply both sides of Eq. (3.11-3) by $H'(k + 1)$, obtaining

$$P(k + 1)H'(k + 1) = P(k)H'(k + 1) - P(k)H'(k + 1) = 0 \quad . \tag{3.11-4}$$

We have demonstrated that $P(k + 1)H'(k + 1) = 0$ without either $P(k + 1)$ or $H'(k + 1)$ being null; hence, $P(k + 1)$ must be singular. Observe, also, that all $P(k + 1)$, $k \geq 0$, are singular when Eq. (3.11-2) is true for $k \geq 0$.

About the only remedy to this unfortunate situation is to model $R(k)$ different from εI. In that case, the resulting estimator is suboptimal, and sensitivity considerations of the suboptimal parameter estimator may have to be studied (see Section 3.10).

When we stop to think about why difficulty occurs when measurements are too good, an explanation emerges. Everything we have done in this chapter has assumed the existence of some randomness, so that $E\{[\tilde{\theta}_i(k)]^2\}$ is a meaningful measure of parameter estimation error. The only source of randomness in our measurement system is the measurement-error process, $\underline{v}(k)$; hence, when $\underline{v}(k)$ is not random, which occurs, for example, when $\underline{v}(k) = \underline{0}$, then $E\{[\tilde{\theta}_i(k)]^2\}$ is no longer meaningful. In short, when $\underline{v}(k) = \underline{0}$, we have a solution to a problem formulation which no longer is valid.

One technique for estimating $\underline{\theta}$ from perfect measurements is described in Chapter 4. A nice feature of that technique is that it is also applicable to the situations of almost perfect measurements and noisy measurements (Chapter 5).

3.12 APPLICABILITY OF UNBIASED MINIMUM-VARIANCE PARAMETER ESTIMATION ALGORITHMS

All of the results in this chapter are limited to the situation when $\mathcal{H}(k)$ is nonrandom.[†] This was made very clear in the statement of Theorem 3-1, in Section 3.3.

Just as we did in Section 2.13, let us limit our discussions here to the scalar measurement situation, in which case

$$z(k) = \underline{x}'(k)\underline{\theta} + v(k) \quad , \qquad (3.12\text{-}1)$$

where we have reconciled the scalar measurement equation, (3.2-6) with m = 1, and the measured output of the scalar equation-error identification system in Fig. 1.2-2c, by setting $\underline{x}(k) = \underline{h}(k)$. The concatenated measurement equation for L + 1 measurements, Eq. (3.2-7), becomes

$$\begin{pmatrix} z(k) \\ z(k-1) \\ \vdots \\ z(k-L) \end{pmatrix} = \begin{pmatrix} x_1(k) & x_2(k) & \cdots & x_n(k) \\ x_1(k-1) & x_2(k-1) & \cdots & x_n(k-1) \\ \vdots & \vdots & \ddots & \vdots \\ x_1(k-L) & x_2(k-L) & \cdots & x_n(k-L) \end{pmatrix} \begin{pmatrix} \theta_1 \\ \theta_2 \\ \vdots \\ \theta_n \end{pmatrix} + \begin{pmatrix} v(k) \\ v(k-1) \\ \vdots \\ v(k-L) \end{pmatrix} \qquad (3.12\text{-}2)$$

and, therefore,

[†] Nonrandomness of F(k) is assured by the nonrandomness of $\mathcal{H}(k)$ [see Eq. (3.4-15)].

$$\mathcal{H}(k) = \begin{pmatrix} x_1(k) & x_2(k) & \cdots & x_n(k) \\ x_1(k-1) & x_2(k-1) & \cdots & x_n(k-1) \\ \vdots & \vdots & \ddots & \vdots \\ x_1(k-L) & x_2(k-L) & \cdots & x_n(k-L) \end{pmatrix}. \quad (3.12\text{-}3)$$

Obviously, $\mathcal{H}(k)$ is nonrandom if and only if $\underline{x}(j)$ is nonrandom for $j = k - L, \ldots, k - 1, k$. This means that the techniques of this chapter are limited to those applications for which the input vector to the scalar equation-error identification system in Fig. 1.2-2c is deterministic. Applications that we have studied in Chapter 2 for which the techniques of the present chapter can be applied, are: (1) estimating the coefficients in a polynomial model (Example 1, Sections 2.2 and 2.5); (2) estimating the initial condition vector $\underline{x}(0)$ of a linear, time-varying, deterministic, discrete system (Example 2, Sections 2.2 and 2.5); (3) estimating the weights in a superposition summation, for nonrandom inputs (Example 3, Sections 2.2 and 2.5); (4) curve fitting (Sections 2.4.1, 2.4.2, and 2.4.3); (5) identifying the coefficients of a deterministic finite-difference equation, assuming that all signals needed for identification can be measured perfectly (usually not possible, though); and (6) identifying aerodynamic parameters when angle of attack and fin deflection can be measured perfectly (usually not possible, though) (Example 5, Section 2.5).

Applications that we have studied in Chapter 2 for which the techniques of the present chapter cannot be applied are: (1) identifying the coefficients of a deterministic finite-difference equation, assuming that some or all signals needed for identification cannot be measured perfectly (Example 4, Section 2.5); (2) identifying the coefficients of a randomly excited finite-difference equation (Example 4, Section 2.5); and (3) identifying aerodynamic parameters when angle of attack and/or fin deflection cannot be measured without measurement errors (Example 5, Section 2.5).

Interestingly enough, we see that unbiased minimum-variance estimation is inapplicable to exactly the same applications for which generalized least-squares estimation was inapplicable. The reasons are not the same, however. We have also demonstrated in Section 3.11 that unbiased minimum-variance estimation is inapplicable when measurements become "too good."

PROBLEMS

3-1. Starting with the assumption that $\hat{\underline{\theta}}(k) = F(k)\underline{Z}(k)$, rederive the generalized least-squares batch estimation algorithm, Eq. (2.3-12).

3-2. Assume that $\mathcal{H}(k)$ is random, and that $\hat{\underline{\theta}}(k) = F(k)\underline{Z}(k)$.

(a) Show that unbiasedness of the estimate is attained when $E\{\mathcal{H}(k)F(k)\} = I$.

(b) At what point in the derivation of $\hat{\underline{\theta}}(k)$ do the computations break down because $\mathcal{H}(k)$ is random?

3-3. Show that Eq. (3.4-5) is equivalent to Eq. (3.2-13).

3-4. Prove the so-called Gauss-Markoff Theorem, stated in Corollary 3-2. Under what conditions will $\mathcal{R}(k) = \sigma^2 I$?

3-5. Prove that the unbiased minimum-variance estimate of $\underline{\theta}$ is unique.

3-6. Show that the generalized least-squares estimate of $\underline{\theta}$ is not invariant under scale changes. One way to circumvent this difficulty is to normalize the data. This can be accomplished by redefining the weighting matrix W to be $W_N'WW_N$. The matrix W_N is a diagonal normalization matrix whose ith diagonal element is $1/(\tilde{z}_i)_{nominal}$, where $(\tilde{z}_i)_{nominal}$ is chosen a priori. Show that by means of this normalization z_i and $\hat{z}_i$ are being compared on a *percentage basis*, which is invariant to scale changes. Explain one way that $(\tilde{z}_i)_{nominal}$ could be chosen.

3-7. What is the maximum dimension of $\underline{\alpha}$ in Section 3-10?

3-8. Derive Eq. (3.10-15).

3-9. Make a flowchart showing the computations required to simultaneously obtain $\hat{\underline{\theta}}_m(k+1)$, $P(k+1|\underline{\alpha})$, and $S(k+1|\underline{\alpha})$.

3-10. Obtain the sensitivity equations for the scalar system $z(k) = h\theta + v(k)$ when: (a) $h_m = h$ but $R_m \neq R$; and (b) $R_m = R$ but $h_m \neq h$. Can the sensitivity equations be solved analytically for this simplest of all systems?

3-11. Show that all linear estimators,

$$\hat{\underline{\theta}}(k+1) = \hat{\underline{\theta}}(k) + K(k+1)[\underline{z}(k+1) - H(k+1)\hat{\underline{\theta}}(k)] \quad ,$$

have parameter estimation-error covariance matrices $P(k)$ that satisfy Eq. (3.11-1), where K^o is replaced by K.

3-12.† A series of measurements $\underline{z}(k)$ are made, where $\underline{z}(k) = H\underline{\theta} + \underline{v}(k)$, H is an $m \times n$ constant matrix, $E\{\underline{v}(k)\} = \underline{0}$, and $\text{Cov}[\underline{v}(k)] = R$ is a constant matrix.

(a) Using the two formulations of the sequential unbiased minimum-variance estimator, show that: (i) $P(k+1)H' = P(k)H'[HP(k)H' + R]^{-1}R$, and (ii) $HP(k) = R[HP(k-1)H' + R]^{-1}HP(k-1)$.

(b) Next, show that (i) $P(k)H' = P(k-2)H'[2HP(k-2)H' + R]^{-1}R$; (ii) $P(k)H' = P(k-3)H'[3HP(k-3)H' + R]^{-1}R$; and (iii) $P(k)H' = P(0)H'[kHP(0)H' + R]^{-1}R$.

(c) Finally, show that the asymptotic form $(k \to \infty)$ for the minimum-variance estimate of $\underline{\theta}$ is

$$\hat{\underline{\theta}}(k+1) = \hat{\underline{\theta}}(k) + \frac{1}{(k+1)} P(0)H'[HP(0)H']^{-1}[\underline{z}(k+1) - H(k+1)\hat{\underline{\theta}}(k)] \quad .$$

This equation, with its $1/(k+1)$ weighting function, represents a form of multidimensional stochastic approximation; hence, this problem

---

† This problem is based on material in the paper On the stochastic approximation method and optimal filtering, by Y. C. Ho, *J. of Math. Anal. and Appl.*, 6, 152-154 (1962).

serves to provide a link between minimum-variance estimation and stochastic-gradient estimation (Chapter 5).

3-13. Throughout this chapter we have assumed that $E\{\underline{V}(k)\} = \underline{0}$. Let us investigate the situation when this assumption is not true.

(a) Assume that $E\{\underline{V}(k)\} = \underline{V}_0$ where $\underline{V}_0$ is known a priori. How are the estimation equations modified in this situation?

(b) Assume that $E\{\underline{V}(k)\} = \underline{V}_1$ where $\underline{V}_1$ is constant but is unknown. How are the estimation equations modified so that estimates of $\underline{V}_1$ can be obtained as well as estimates of $\underline{\theta}$?

(c) Assume that $E\{\underline{V}(k)\} = \underline{V}_1$ where $\underline{V}_1$ is constant but unknown. Now, however, we do not wish to estimate $\underline{V}_1$. How is the estimation procedure modified so that only estimates of $\underline{\theta}$ are obtained? (*Hint:* Construct a different concatenated measurement vector, one in which the effects of $\underline{V}_1$ are not present.)

3-14. Refer to Problem 2-3. Assume that the figures given in Table P2-3 for U. S. production of steel are in error by some zero mean random amount. The variance of this error is 2% of the yearly U. S. production of steel listed in the table. Determine the unbiased minimum-variance line fitting the data. What is $R(k)$? (The solution to this problem requires a calculator or computer.)

3-15. Consider the multivariate gaussian situation when $R(k) = \sigma^2 I$, that is to say when $\underline{V}(k)$ is homoskedastic.

(a) Show that the maximum likelihood estimator of σ^2, denoted $\hat{\sigma}^2_{ML}$, is

$$\hat{\sigma}^2_{ML} = \frac{(\underline{Z} - H\underline{\theta})'(\underline{Z} - H\underline{\theta})}{L + 1} .$$

(*Hint:* Maximizing $\log f[\underline{Z}(k)]$ is equivalent to maximizing $f[\underline{Z}(k)]$.)

(b) Show that $\hat{\sigma}^2_{ML}$ is biased, but that it is asymptotically unbiased.

3-16. Equation (3.2-12) assumes that $\underline{V}(k)$ is independent from sample to sample (off-diagonal terms of $R(k)$ are all zero). Let us consider the situation below when this is not so. Now

$$\underline{z}(k+1) = H(k+1)\underline{\theta} + \underline{v}(k+1)$$

where

$$\underline{v}(k+1) = A_V \underline{v}(k) + \underline{\mu}(k) \quad .$$

For our purposes, we shall assume A_V is known, and that $\underline{\mu}(k)$ is a discrete zero-mean white sequence with

$$E\{\underline{\mu}(k)\underline{\mu}'(k)\} = R_\mu(k) \quad .$$

Show that by working with the measurement difference $\underline{z}^*(k+1) = \underline{z}(k+1) - A_V\underline{z}(k)$, the unbiased minimum-variance estimate of $\underline{\theta}$ can be obtained as

$$\hat{\underline{\theta}}(k) = [\mathcal{H}^{*\prime}\mathcal{R}_\mu^{-1}(k)\mathcal{H}^*(k)]^{-1}\mathcal{H}^{*\prime}(k)\mathcal{R}_\mu^{-1}(k)\underline{\mathcal{Z}}^*(k)$$

where $H^*(k+1) = H(k+1) - A_V H(k)$, and, $\underline{\mathcal{Z}}^*(k)$ and $\mathcal{H}^*(k)$ are concatenated vectors of $\underline{z}^*$ and H^*, respectively.

REFERENCES

3-1. N. Morrison, *Introduction to Sequential Smoothing and Prediction*, McGraw-Hill, New York, 1969.

3-2. A. E. Taylor, *Advanced Calculus*, Ginn, New York, 1955.

3-3. I. A. Gura, *An Algebraic Approach to Optimal State Estimation*, Rept. No. SSD 70072R, Hughes Aircraft Co., Space Systems Division, March 1967.

3-4. R. L. Plackett, A historical note on the method of least-squares, *Biometrika*, 36, 458-460 (1949).

3-5. K. F. Gauss, *Theory of the Motion of the Heavenly Bodies Moving About the Sun in Conic Sections*, 1809. English translation, Dover Publications, New York, 1963.

3-6. A. C. Aitken, On least-squares and linear combinations of observations, *Proc. Roy. Soc. Edinburgh*, A55, 42-47 (1934).

3-7. F. A. Graybill, *An Introduction to Linear Statistical Models*, Vol. I, McGraw-Hill, New York, 1961.

3-8. K. Ogata, *State Space Analysis of Control Systems*, Prentice-Hall, Englewood Cliffs, New Jersey, 1967.

3-9. R. Bellman, *Introduction to Matrix Analysis*, (2nd ed.), McGraw-Hill, New York, 1970.

3-10. R. E. Kalman, A new approach to linear filtering and prediction problems, *Trans. ASME, Series D: Journal of Basic Engineering*, 82, 35-45 (1960).

3-11. R. E. Kalman and R. S. Bucy, New results in linear filtering and prediction theory, *Trans. ASME, Series D: Journal of Basic Engineering*, 83, 95-108 (1961).

3-12. J. S. Meditch, *Stochastic Optimal Linear Estimation and Control*, McGraw-Hill, New York, 1969.

3-13. A. H. Jazwinski, *Stochastic Processes and Filtering Theory*, Academic Press, New York, 1970.

3-14. H. W. Sorenson, Kalman filtering techniques, in *Advances in Control Systems*, Vol. 3 (C. T. Leondes, ed.), Academic Press, New York, 1966.

3-15. A. Papoulis, *Probability, Random Variables, and Stochastic Processes*, McGraw-Hill, New York, 1965.

3-16. R. C. K. Lee, *Optimal Estimation, Identification, and Control*, Research Monograph No. 28, The M.I.T. Press, Cambridge, Mass., 1964.

3-17. H. L. Van Trees, *Detection, Estimation, and Modulation Theory, Part I*, Wiley, New York, 1968.

3-18. D. G. Lainiotis and F. L. Sims, Sensitivity analysis of discrete Kalman filters, *Int. J. of Control*, 12, 657-669 (1970).

4

Deterministic—Gradient Parameter Estimation

4.1 INTRODUCTION

The starting point for the generalized least-squares and unbiased minimum-variance parameter estimation techniques is the concatenated measurement equation, $\underline{Z}(k) = H(k)\underline{\theta} + \underline{V}(k)$. This measurement equation brings together the L + 1 measurements[†] z(k - L), z(k - L + 1), ..., and z(k). After deriving the classical batch processing algorithms for both of these techniques, we derived sequential processing algorithms for them, in Chapters 2 and 3. These sequential algorithms have the following common structure:

New estimate = Old estimate + Linear transformation of equation error (4.1-1)

or, mathematically,

$$\hat{\underline{\theta}}(k + 1) = \hat{\underline{\theta}}(k) + K(k + 1)[z(k + 1) - \underline{h}'(k + 1)\hat{\underline{\theta}}(k)] \quad . \tag{4.1-2}$$

[†] In the vector measurement situation, $\underline{Z}(k)$ is an $(L + 1)m \times 1$ vector.

The linear transformation matrix for the sequential generalized least-squares algorithm $K^*(k+1)$ and the linear transformation matrix for the sequential unbiased minimum-variance algorithm $K^o(k+1)$ are both computed from additional recursive matrix equations that are, computationally speaking, rather complicated.

It is, therefore, quite logical to ask the question, "Are there other optimal estimation algorithms for estimating $\underline{\theta}$ that have the structure of Eq. (4.1-2), but that have easy-to-compute transformation matrices?" We shall discuss one such algorithm in this chapter. It is one in which estimates of $\underline{\theta}$ are obtained at t_{k+1} from the preceding estimate at t_k and from the gradient of a performance function that provides a measure of identification error.

4.2 FORMULATION, STATEMENT, AND SOLUTION OF THE ESTIMATION PROBLEM

Suppose that a measurement of a scalar signal $y(t)$ is made at time t_k, where $k \geq 0$. The t_k, t_{k+1}, ..., etc., do not have to be uniformly spaced. *A basic assumption in this chapter is that the measurement of* $y(t)$ *is made without measurement error.* The situation, when it is not possible to satisfy this assumption, is discussed in great detail, in Chapter 5.

The signal $y(t)$ is assumed to be a linear combination of n parameters θ_1, θ_2, ..., and θ_n; that is to say,

$$y(t) = x_1(t)\theta_1 + x_2(t)\theta_2 + \cdots + x_n(t)\theta_n \tag{4.2-1}$$

where *it is also assumed that* $x_1(t)$, $x_2(t)$, ..., *and* $x_n(t)$ *can be measured without measurement errors.* Chapter 5 treats the more general situation, when measurements of $x_1(t)$, $x_2(t)$, ..., and $x_n(t)$ are corrupted by measurement noise. Letting

$$\underline{x}(t) = [x_1(t), x_2(t), \ldots, x_n(t)]' \tag{4.2-2}$$

and

$$\underline{\theta} = (\theta_1, \theta_2, \ldots, \theta_n)' \quad , \tag{4.2-3}$$

the measured value of $y(t)$ at t_k can be written more compactly as

$$y(k) = \underline{x}'(k)\underline{\theta} \tag{4.2-4}$$

for $k \geq 0$. Equation (4.2-4) is the starting point for all of the analyses in this chapter. It is a scalar deterministic measurement equation.

The following approximation (model) to y(k) is assumed:

$$\hat{y}(k) = \underline{x}'(k)\hat{\underline{\theta}}(k) \tag{4.2-5}$$

where

$$\hat{\underline{\theta}}(k) = [\hat{\theta}_1(k), \hat{\theta}_2(k), \ldots, \hat{\theta}_n(k)]'$$

is the kth approximation of $\underline{\theta}$.

The error between $\hat{y}(k)$ and y(k) is $\tilde{y}(k)$:

$$\tilde{y}(k) = y(k) - \hat{y}(k) \quad . \tag{4.2-6}$$

A direct approach to obtain $\hat{\underline{\theta}}(k)$ is to minimize some measure of the error between $\hat{\underline{\theta}}(k)$ and $\underline{\theta}$, such as $||\underline{\theta} - \hat{\underline{\theta}}(k)||^2$, where† $||\underline{\theta} - \hat{\underline{\theta}}(k)||$ is the Euclidean norm. This approach is inapplicable in practice because $\underline{\theta}$ is not known ahead of time; that is to say, the solution to the problem of minimizing $||\underline{\theta} - \hat{\underline{\theta}}(k)||^2$ is $\hat{\underline{\theta}}(k) = \underline{\theta}$, and since $\underline{\theta}$ is not known a priori, this solution is of no value.

A less direct approach for obtaining $\hat{\underline{\theta}}(k)$ is to choose it in a manner that minimizes some measure of the error, $\tilde{y}(k)$, and to then demonstrate that, for that choice, $\hat{\underline{\theta}}(k)$ converges to $\underline{\theta}$. Here we define an instantaneous quadratic measure of the error $\tilde{y}(k)$ as

$$J[\hat{\underline{\theta}}(k)] = \frac{1}{2}\tilde{y}^2(k) \tag{4.2-7}$$

for $k \geq 0$, and obtain $\hat{\underline{\theta}}(k)$ by minimizing $J[\hat{\underline{\theta}}(k)]$ with respect to each of the n components of $\hat{\underline{\theta}}(k)$.

The minimum value of $J[\hat{\underline{\theta}}(k)]$ is obviously zero which occurs when $\hat{\underline{\theta}}(k) = \underline{\theta}$ [unless $\tilde{\underline{\theta}}(k)$ is orthogonal to $\underline{x}(k)$, in which case $J[\hat{\underline{\theta}}(k)] = 0$, but $\hat{\underline{\theta}}(k) \neq \underline{\theta}$ (we shall discuss this situation in more

---

† $$||\underline{a}|| = \sqrt{\sum_{i=1}^{n} a_i^2} \quad .$$

detail in Section 4.5)]. The specific method for minimizing $J[\hat{\underline{\theta}}(k)]$ is the *gradient descent procedure*. Recall, from differential vector calculus (*4-1*), that the gradient of $J[\hat{\underline{\theta}}(k)]$ is in the direction of maximum increase of $J[\hat{\underline{\theta}}(k)]$; hence, in the gradient descent procedure[†] the estimate of $\underline{\theta}$ is changed in the direction of the negative gradient (*4-2* - *4-4*), as follows:

$$\hat{\underline{\theta}}(k+1) = \hat{\underline{\theta}}(k) - R(k) \operatorname*{grad}_{\hat{\underline{\theta}}(k)} J[\hat{\underline{\theta}}(k)] \qquad (4.2\text{-}8)$$

for $k \geq 0$, where $\hat{\underline{\theta}}(0)$ is specified a priori, and $R(k)$ is an $n \times n$ symmetric weighting matrix with elements remaining to be specified.

Proceeding to the computation of $\text{grad}_{\hat{\underline{\theta}}(k)} J[\hat{\underline{\theta}}(k)]$, it is easily shown from Eqs. (4.2-4)-(4.2-7) that

$$J[\hat{\underline{\theta}}(k)] = \frac{1}{2}\tilde{y}^2(k) = \frac{1}{2}[y(k) - \hat{y}(k)]^2$$
$$= \frac{1}{2}\{\underline{x}'(k)[\underline{\theta} - \hat{\underline{\theta}}(k)]\}^2 \quad ; \qquad (4.2\text{-}9)$$

hence [see Eq. (2.3-1)],

$$\operatorname*{grad}_{\hat{\underline{\theta}}(k)} J[\hat{\underline{\theta}}(k)] = \frac{d}{d\hat{\underline{\theta}}(k)} J[\hat{\underline{\theta}}(k)] = -\tilde{y}(k)\underline{x}(k) \quad . \qquad (4.2\text{-}10)$$

The *deterministic-gradient parameter identification algorithm* in Eq. (4.2-8) becomes

$$\hat{\underline{\theta}}(k+1) = \hat{\underline{\theta}}(k) + R(k)\underline{x}(k)\tilde{y}(k) \qquad (4.2\text{-}11)$$

for $k \geq 0$. Observe that this algorithm is of the form in Eq. (4.1-1); however, observe also that $\hat{\underline{\theta}}(k+1)$ in Eq. (4.2-11) is based upon the measurement at t_k, whereas the algorithms that were obtained in Chapters 2 and 3 updated $\hat{\underline{\theta}}(k)$ using the measurement at t_{k+1}. The reason for this apparent difference can be traced back to what we mean by the notation $\hat{\underline{\theta}}(k+1)$.

In Chapters 2 and 3, $\hat{\underline{\theta}}(k+1)$ was used to mean the estimate of $\underline{\theta}$ based on the $L + 2$ data samples, $z(k+1)$, $z(k)$, ..., and $z(k-L)$. The argument of $\hat{\underline{\theta}}$ was chosen to correspond with the most recent

---

[†] Our objective is to minimize $J[\hat{\underline{\theta}}(k)]$, not to maximize it.

measurement, $z(k + 1)$. Making the association in this manner is quite common in estimation theory.

The gradient descent procedure originated as a procedure for sequentially locating the minimum of a multivariable nonlinear function. The notation $\hat{\underline{\theta}}(k + 1)$ does not mean the same thing in optimization theory as it does in estimation theory. In optimization theory, $\hat{\underline{\theta}}(k + 1)$ is the $(k + 1)$th *trial value* of $\underline{\theta}$. The gradient of $J[\hat{\underline{\theta}}(k)]$ utilizes information only about $J[\hat{\underline{\theta}}(k)]$ at the kth trial value of $\underline{\theta}$, as implied by the dependence of J on $\hat{\underline{\theta}}(k)$; hence, in this chapter, as well as the next chapter, $\hat{\underline{\theta}}(k + 1)$ should be thought of as the $(k + 1)$th trial value of $\underline{\theta}$.

Important observation. The $\tilde{y}(k)$ is a scalar deterministic equation error of exactly the type that was discussed in Chapter 1, Section 1.2.3. The $y(k)$ in Eq. (4.2-4) represents the perfectly measured output of the identification representation in Fig. 1.2-3, and $\hat{y}(k)$ represents the output of the model. We depict these relationships in Fig. 4.2-1. ▲

The gradient identification algorithm in Eq. (4.2-11) has been applied to a variety of identification problems. Among the applications are: (1) determination of a plant's differential equation during its normal operation (*4-5*); (2) determination of the sampled impulse response of a system (*4-6*); (3) determination of the kernels of a Volterra representation for nonlinear systems (*4-7*); and (4) identification of aerodynamic parameters for adaptive control of a high-performance aerospace vehicle (*4-8* and *4-9*).

We conclude this section with five examples that illustrate how a variety of identification problems can be cast into the scalar equation-error formulation.

Example 1. *Identification of Coefficients in a Polynomial Model.* As in Example 1 of Section 2.2, we consider a signal y which is known a priori to be related to another parameter ζ by an $(n - 1)$th degree polynomial in ζ,

$$y(\zeta) = \theta_1 + \theta_2\zeta + \theta_3\zeta^2 + \cdots + \theta_n\zeta^{n-1} \tag{4.2-12}$$

where θ_1, θ_2, ..., and θ_n are unknown coefficients. For specific values of ζ, we measure $y(\zeta)$ without error; that is to say, for $\zeta = \zeta_i$

$$y(\zeta_i) = \theta_1 + \theta_2\zeta_i + \theta_3\zeta_i^2 + \cdots + \theta_n\zeta_i^{n-1} \tag{4.2-13}$$

is measured. Letting

$$\begin{aligned} x_1(i) &= 1 \\ x_2(i) &= \zeta_i \\ x_3(i) &= \zeta_i^2 \\ &\vdots \\ x_n(i) &= \zeta_i^{n-1} \end{aligned} \tag{4.2-14}$$

and

$$y(i) = y(\zeta_i) \quad , \tag{4.2-15}$$

Eq. (4.2-13) can be written as

$$y(i) = x_1(i)\theta_1 + x_2(i)\theta_2 + \cdots + x_n(i)\theta_n(i) \tag{4.2-16}$$

which is of exactly the same form as the deterministic measurement equation, (4.2-4). The $y(i)$ is modeled by $\hat{y}(i)$ in Eq. (4.2-5). ▲

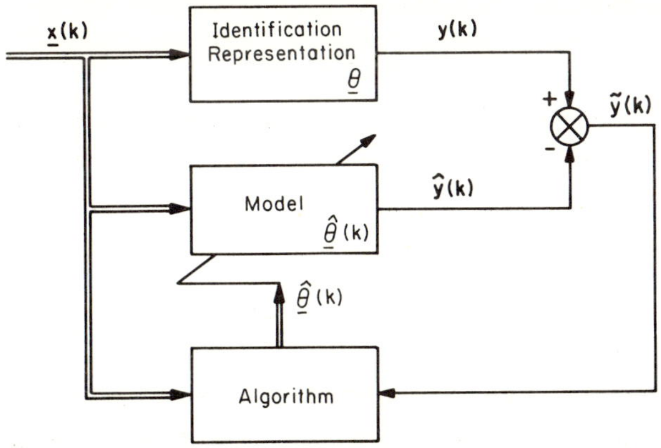

Fig. 4.2-1. Interpretation of scalar deterministic measurement equation and estimation equation as a scalar equation-error identification system.

Example 2. *Identification of Initial Condition Vector.* As in Example 2 of Section 2.2, we consider the problem of identifying the $n \times 1$ initial condition vector $\underline{x}(0)$ of the system

$$\underline{x}(k+1) = \Phi(k+1, k)\underline{x}(k) \quad . \tag{4.2-17}$$

Here, however, $\underline{x}(0)$ will be estimated from the perfect measurements

$$y(k) = \underline{h}'(k)\underline{x}(k) \quad , \tag{4.2-18}$$

$k \geq 0$. From Eqs. (4.2-17) and (1.3-58), it is clear, that

$$y(k) = \underline{h}'(k)\Phi(k, 0)\underline{x}(0) \quad . \tag{4.2-19}$$

Letting

$$\underline{\theta} = [x_1(0), x_2(0), \ldots, x_n(0)]' \tag{4.2-20}$$

and

$$\chi_1(k) = \sum_{i=1}^{n} h_i(k)\phi_{i1}(k) \tag{4.2-21}$$

$$\chi_2(k) = \sum_{i=1}^{n} h_i(k)\phi_{i2}(k) \tag{4.2-22}$$

$$\vdots$$

$$\chi_n(k) = \sum_{i=1}^{n} h_i(k)\phi_{in}(k) \quad , \tag{4.2-23}$$

it is clear that Eq. (4.2-19) can also be written as

$$y(k) = \underline{\chi}'(k)\underline{\theta} \tag{4.2-24}$$

where we have modified our usual notation for the input of the equation-error identification system from $\underline{x}(k)$ to $\underline{\chi}(k)$, in order to avoid possible confusion with $\underline{\chi}(k)$ and the components of $\underline{\theta}$. The signal $y(k)$ is modeled by $\hat{y}(k)$, where

$$\hat{y}(k) = \underline{\chi}'(k)\hat{\underline{\theta}}(k) \quad . \tag{4.2-25}$$

▲

Example 3. *Identification of Weights in a Superposition Summation.* As in Example 3 of Section 2.2, we wish to identify the weights f_1, f_2, ..., and f_N in the superposition summation

$$g(k) = \sum_{i=1}^{N} f_i w(k - i) \quad ; \tag{4.2-26}$$

however, here these weights are to be determined from perfect measurements of g(k), for $k \geq 1$ [the first measurable value of w(j) is assumed to be w(0)]. Letting

$$\underline{\theta} = (f_1, f_2, \ldots, f_N)' \quad , \tag{4.2-27}$$

$$\begin{aligned} x_1(k) &= w(k - 1) \\ x_2(k) &= w(k - 2) \\ &\vdots \\ x_N(k) &= w(k - N) \quad , \end{aligned} \tag{4.2-28}$$

and

$$y(k) = g(k) \quad , \tag{4.2-29}$$

it is clear that, once again, it has been possible to cast a problem into the scalar equation-error format. ▲

Example 4. *Identification of Coefficients in a Finite-Difference Equation.* As in Example 4 of Section 2.5, we shall be interested in identifying the coefficients a_1, a_2, ..., and a_n in the finite difference equation

$$y(k + n) + a_1 y(k + n - 1) + \cdots + a_n y(k) = \omega(k) \tag{4.2-30}$$

where it is assumed that y(k), y(k + 1), ..., and y(k + n) can be measured perfectly for all $k \geq 0$ [the first measurable value of y(j) is assumed to be y(0)].

If the driving function ω(k) is deterministic and perfectly measurable, we proceed to establish the scalar equation-error formulation by first rewriting Eq. (4.2-30) as

$$-y(k + n) + \omega(k) = a_1 y(k + n - 1) + \cdots + a_n y(k) \tag{4.2-31}$$

and then letting

$$\underline{\theta} = (a_1, a_2, \ldots, a_n)' \quad , \tag{4.2-32}$$

$$\begin{aligned} x_1(k) &= y(k + n - 1) \\ x_2(k) &= y(k + n - 2) \\ &\vdots \\ x_n(k) &= y(k) \quad , \end{aligned} \tag{4.2-33}$$

and

$$y^*(k) = -y(k + n) + \omega(k) \quad . \tag{4.2-34}$$

If, on the other hand, $\omega(k)$ is random, $y(k)$, $y(k + 1)$, ..., and $y(k + n)$ will also be random, since these quantities are obtained as solutions of the original finite-difference equation (cause and effect). The techniques of this chapter are inapplicable to this situation, since *all* measured signals are now random quantities. We will study this application in greater detail in Chapter 5, and will show that if the algorithm in Eq. (4.2-11) is applied to the identification of a_1, a_2, ..., and a_n, the estimates of these quantities will be biased. This bias is due solely to the random disturbance $\omega(k)$; however, fortunately for us, the bias can be removed. ▲

Example 5. *Identification of Aerodynamic Parameters.* As in Section 2.4.4, let us consider the identification of the two aerodynamic parameters, M_α and M_δ, that are associated with the high-performance aerospace control system described in Section 1.3.5. Assuming that $\ddot{\theta}(k)$, $\alpha(k)$, and $\delta(k)$ can be measured without errors, we shall estimate M_α and M_δ from the equation

$$\ddot{\theta}(k) = M_\alpha \alpha(k) + M_\delta \delta(k) \quad . \tag{4.2-35}$$

Clearly, by defining

$$\underline{\theta} = (M_\alpha, M_\delta)' \quad , \tag{4.2-36}$$

$$\begin{aligned} x_1(k) &= \alpha(k) \\ x_2(k) &= \delta(k) \quad , \end{aligned} \tag{4.2-37}$$

and

$$y(k) = \ddot{\theta}(k) \quad , \tag{4.2-38}$$

we have cast this problem into the scalar equation-error formulation. ▲

4.3 CONVERGENCE: CONTRACTION MAPPING APPROACH

4.3.1 Introduction

Now that we have decided to update the estimate of $\underline{\theta}$ according to Eq. (4.2-11), we must investigate whether or not this choice will accomplish the desired objective, minimization of $J[\hat{\underline{\theta}}(k)]$ with the minimum occurring for $\hat{\underline{\theta}}(k) \to \underline{\theta}$. The convergence of $\hat{\underline{\theta}}(k)$ to $\underline{\theta}$ when $\hat{\underline{\theta}}(k)$ is updated by means of Eq. (4.2-11) is the subject of this section.

It is well known (*4-10*, for example) that for a finite-dimensional Euclidean space

$$\lim_{k\to\infty} \hat{\underline{\theta}}(k) \to \underline{\theta} \qquad \text{if} \quad \lim_{k\to\infty} ||\tilde{\underline{\theta}}(k)||^2 \to 0 \quad . \tag{4.3-1}$$

Here, we shall obtain an expression for $||\tilde{\underline{\theta}}(k + 1)||^2$ and conditions under which $||\tilde{\underline{\theta}}(k + 1)||^2 \to 0$ as $k \to \infty$, when $\hat{\underline{\theta}}(k)$ is updated by means of the deterministic-gradient identification algorithm in Eq. (4.2-1).

4.3.2 Identification-Error System

To begin, let us show that the finite-difference equation for parameter identification error $\tilde{\underline{\theta}}(k)$ is

$$\tilde{\underline{\theta}}(k + 1) = [I - R(k)\underline{x}(k)\underline{x}'(k)]\tilde{\underline{\theta}}(k) \tag{4.3-2}$$

for $k \geq 0$. In the sequel, this equation is referred to as the *identification-error system*.† Equation (4.3-2) is obtained from Eqs. (4.2-11) and (4.2-4)-(4.2-6) as follows:

$$\begin{aligned}
\hat{\underline{\theta}}(k + 1) &= \hat{\underline{\theta}}(k) + R(k)\underline{x}(k)\tilde{y}(k) \\
&= \hat{\underline{\theta}}(k) + R(k)\underline{x}(k)[y(k) - \hat{y}(k)] \\
&= \hat{\underline{\theta}}(k) + R(k)\underline{x}(k)\underline{x}'(k)[\underline{\theta} - \hat{\underline{\theta}}(k)] \\
&= \hat{\underline{\theta}}(k) + R(k)\underline{x}(k)\underline{x}'(k)\tilde{\underline{\theta}}(k) \quad .
\end{aligned} \tag{4.3-3}$$

---

† We obtained a comparable system for a generalized least-squares estimation algorithm in Section 2.11.2.

Subtracting both sides of this equation from $\underline{\theta}$, we find

$$\underline{\theta} - \hat{\underline{\theta}}(k + 1) = \underline{\theta} - \hat{\underline{\theta}}(k) - R(k)\underline{x}(k)\underline{x}'(k)\tilde{\underline{\theta}}(k) \qquad (4.3\text{-}4)$$

or, finally,

$$\tilde{\underline{\theta}}(k + 1) = \tilde{\underline{\theta}}(k) - R(k)\underline{x}(k)\underline{x}'(k)\tilde{\underline{\theta}}(k) \qquad (4.3\text{-}5)$$

which is Eq. (4.3-2).

4.3.3 Identification-Error Norm-Squared System

Next, let us show that the square of the Euclidean norm of the identification error, $||\tilde{\underline{\theta}}(k + 1)||^2$, can be found recursively from the following scalar, first-order finite-difference equation:

$$||\tilde{\underline{\theta}}(k + 1)||^2 = [1 - \xi(k)]||\tilde{\underline{\theta}}(k)||^2 \qquad (4.3\text{-}6)$$

for $k \geq 0$, where

$$\xi(k) = \frac{2\tilde{y}(k)\tilde{\theta}'(k)R(k)\underline{x}(k) - \tilde{y}^2(k)||R(k)\underline{x}(k)||^2}{||\tilde{\underline{\theta}}(k)||^2} \quad . \qquad (4.3\text{-}7)$$

Recall that

$$||\tilde{\underline{\theta}}(k + 1)||^2 = \tilde{\underline{\theta}}'(k + 1)\tilde{\underline{\theta}}(k + 1) \quad ; \qquad (4.3\text{-}8)$$

hence, substituting Eq. (4.3-2) for $\tilde{\underline{\theta}}(k + 1)$ into Eq. (4.3-8), one finds that

$$\begin{aligned} ||\tilde{\underline{\theta}}(k + 1)||^2 &= \{\tilde{\underline{\theta}}'(k)[I - \underline{x}(k)\underline{x}'(k)R(k)]\} \\ &\quad \times \{[I - R(k)\underline{x}(k)\underline{x}'(k)]\tilde{\underline{\theta}}(k)\} \\ &= \tilde{\underline{\theta}}'(k)[I - R(k)\underline{x}(k)\underline{x}'(k) - \underline{x}(k)\underline{x}'(k)R(k) \\ &\quad + \underline{x}(k)\underline{x}'(k)R(k)R(k)\underline{x}(k)\underline{x}'(k)]\tilde{\underline{\theta}}(k) \\ &= ||\tilde{\underline{\theta}}||^2 - \tilde{\underline{\theta}}'R\underline{x}\underline{x}'\tilde{\underline{\theta}} - \tilde{\underline{\theta}}'\underline{x}\underline{x}'R\tilde{\underline{\theta}} \\ &\quad + \tilde{\underline{\theta}}'\underline{x}||R\underline{x}||^2\underline{x}'\tilde{\underline{\theta}} \end{aligned} \qquad (4.3\text{-}9)$$

where for convenience we have dropped the notational dependency of all terms on the right-hand side of this equation on k, and, where we have used the fact that R(k) is symmetric. It is clear from Eqs. (4.2-4)-(4.2-6) that

$$\tilde{y}(k) = \underline{x}'(k)\tilde{\underline{\theta}}(k) = \tilde{\underline{\theta}}'(k)\underline{x}(k) \quad ; \tag{4.3-10}$$

hence, Eq. (4.3-9) can also be written as

$$\begin{aligned} ||\tilde{\underline{\theta}}(k+1)||^2 &= ||\tilde{\underline{\theta}}(k)||^2 - \tilde{\underline{\theta}}'R\underline{x}\tilde{y}(k) - \tilde{y}(k)\underline{x}'R\tilde{\underline{\theta}} \\ &\quad + \tilde{y}^2(k)||R\underline{x}||^2 \\ &= ||\tilde{\underline{\theta}}(k)||^2 - 2\tilde{y}(k)\underline{x}'(k)R(k)\tilde{\underline{\theta}}(k) \\ &\quad + \tilde{y}^2(k)||R(k)\underline{x}(k)||^2 \end{aligned} \tag{4.3-11}$$

which is precisely Eq. (4.3-6) when $\xi(k)$ is defined as in Eq. (4.3-7).

4.3.4 Convergence

Theorem 4-1. *Given that* $\hat{\underline{\theta}}$ *is updated by means of the deterministic-gradient identification algorithm in Eq. (4.2-11), and* $\xi(k)$ *is defined in Eq. (4.3-7). If*

$$0 < \xi(k) < 2 \tag{4.3-12}$$

then

$$\lim_{k\to\infty} ||\tilde{\underline{\theta}}(k)||^2 \to 0 \quad . \tag{4.3-13}$$

Proof. Equation (4.3-6) is a first-order scalar finite-difference equation in $||\tilde{\underline{\theta}}(k)||^2$. We have already studied the stability of such an equation in Example 7 of Section 2.11. By analogy between Eqs. (4.3-6) and (2.11-1), we conclude that when $|1 - \xi(k)| < 1$, $\lim_{k\to\infty} ||\tilde{\underline{\theta}}(k)||^2 \to 0$; hence,

$$\lim_{k\to\infty} ||\tilde{\underline{\theta}}(k)||^2 \to 0 \quad \text{if} \quad 0 < \xi(k) < 2 \quad . \tag{4.3-14}$$

▲

If it is possible to choose $\xi(k)$ in this manner, $||\tilde{\underline{\theta}}(k)||^2$ will become continually smaller from iteration to iteration; that is to say, $||\tilde{\underline{\theta}}(k)||^2$ will *contract* from iteration to iteration, and, in this case, Eq. (4.3-6) is said to be a *contraction mapping*.

Theorem 4-2. *A* necessary condition *for* $0 < \xi(k) < 2$ *is*

$$\tilde{y}(k)\tilde{\underline{\theta}}'(k)R(k)\underline{x}(k) > 0 \tag{4.3-15}$$

for all k.

Proof. The second term on the right-hand side of Eq. (4.3-7) is always greater than or equal to zero; thus, unless the first term in that equation is positive, it is absolutely impossible for $\xi(k)$ to be positive. ▲

Example 6. In 1967, Nagumo and Noda (*4-6*) proposed the following choice for R(k):

$$R(k) = \alpha I/||\underline{x}(k)||^2 \quad . \tag{4.3-16}$$

Observe, first, that for this weighting matrix,

$$\tilde{y}(k)\underline{\tilde{\theta}}'(k)R(k)\underline{x}(k) = \frac{\alpha\tilde{y}(k)\underline{\tilde{\theta}}'(k)\underline{x}(k)}{||\underline{x}(k)||^2}$$

$$= \alpha\frac{\tilde{y}^2(k)}{||\underline{x}(k)||^2} \tag{4.3-17}$$

which is greater than zero, as long as

$$\alpha > 0 \quad . \tag{4.3-18}$$

Next, we shall show that

$$\alpha < 2 \quad . \tag{4.3-19}$$

This is obtained by substituting Eq. (4.3-16) into Eq. (4.3-7):

$$\xi(k) = \frac{1}{||\underline{\tilde{\theta}}(k)||^2}\left[\frac{2\tilde{y}(k)\underline{\tilde{\theta}}'(k)\underline{x}(k)\alpha}{||\underline{x}(k)||^2} - \frac{\tilde{y}^2(k)||\alpha\underline{x}(k)||^2}{||\underline{x}(k)||^4}\right]$$

$$= \frac{\tilde{y}^2(k)(2-\alpha)\alpha}{||\underline{\tilde{\theta}}(k)||^2\,||\underline{x}(k)||^2} \quad . \tag{4.3-20}$$

Obviously, for $\xi(k)$ to be positive, as it must be,

$$\alpha > 0 \quad \text{and} \quad (2-\alpha) > 0 \tag{4.3-21}$$

which means, of course, that

$$0 < \alpha < 2 \quad . \tag{4.3-22}$$

We leave, as an exercise to the reader (Problem 4-1), the proof of the result that $0 < \xi(k) < 2$ if $0 < \alpha < 2$. ▲

During an identification, $\tilde{\underline{\theta}}(k)$ is not physically available; hence, the condition in Eq. (4.3-15) generally is physically unrealizable. It is not known how to obtain general physically realizable conditions for $\xi(k)$ in Eq. (4.3-7) so that $0 < \xi(k) < 2$; hence, Theorem 4-1 is not of much practical value, since it cannot be used to indicate how R(k) should be chosen a priori so that convergence is assured. In the next section a different approach is taken to obtain physically realizable conditions on the elements of R(k), so that $\tilde{\underline{\theta}}(k)$ converges to zero.

4.4 CONVERGENCE: STABILITY THEORY APPROACH

4.4.1 Introduction

Instead of studying the convergence properties of the equation for $||\tilde{\underline{\theta}}(k + 1)||^2$, we shall turn directly to a study of the stability properties of the identification error system,

$$\tilde{\underline{\theta}}(k + 1) = [I - R(k)\underline{x}(k)\underline{x}'(k)]\tilde{\underline{\theta}}(k) \quad . \qquad (4.4\text{-}1)$$

This is a linear *time-varying* first-order vector difference equation. *It is the time-varying nature of this equation that makes a study of its stability properties challenging.* Since this is the first place where we have spoken about stability of time-varying systems, it is appropriate to digress briefly, to state some definitions of different types of stability.

4.4.2 Stability Definitions (*4-11* and *4-12*)

In this section, we direct our attention to the free (unforced) discrete-time dynamic system

$$\underline{x}(t_{k+1}) = \underline{f}[\underline{x}(t_k), t_k] \qquad (4.4\text{-}2)$$

where $\underline{x}$ is an $n \times 1$ state vector, t_k indicates discrete values of time (k = integer), $-\infty < \cdots < t_{k-1} < t_k < t_{k+1} < \cdots < +\infty$, and $t_k \to \infty$ when $k \to \infty$. Although our interest is mainly in linear discrete-time dynamic systems [Eq. (4.4-1)], we shall present definitions of different types of stability for the more general nonlinear discrete-time dynamic system in Eq. (4.4-2), since linear systems are a special case of nonlinear systems.

For any initial state $\underline{x}_0$, any initial time t_0, and any time t_k, we call $\underline{\phi}(t_k; \underline{x}_0, t_0)$ the *motion* of the system in Eq. (4.4-2) going through state $\underline{x}_0$ at t_0; that is to say, $\underline{\phi}(t_k; \underline{x}_0, t_0)$ denotes the solution (assumed to be unique) of Eq. (4.4-2) where $\underline{x}(t_0) = \underline{x}_0$ and t_k is observed time.

A state $\underline{x}_e$ of a free dynamic system is called an *equilibrium state* if

$$\underline{f}[\underline{x}_e, t_k] = \underline{0} \tag{4.4-3}$$

for all t_k. We shall assume that if there is more than one equilibrium state they are isolated from each other. Note that the determination of equilibrium states does not involve the solution of the system's difference equation, (4.4-2), but only the solution of the algebraic equation, (4.4-3).

Stability Definition 1: An equilibrium state $\underline{x}_e$ of the system in Eq. (4.4-2) is said to be *stable* if for any real number $\varepsilon > 0$ there corresponds a real number $\delta(\varepsilon, t_0) > 0$ such that if $||\underline{x}_0 - \underline{x}_e||$ $\leq \delta(\varepsilon, t_0)$ then $||\underline{\phi}(t_k; \underline{x}_0, t_0) - \underline{x}_e|| \leq \varepsilon$ for all $t \geq t_0$. ▲

Figure 4.4-1 illustrates the definition for a two-dimensional system. It is important to note that by this definition *stability is a local concept;* it refers to behavior near the equilibrium state since one does not know how small the δ in the definition may have to be chosen, ahead of time.

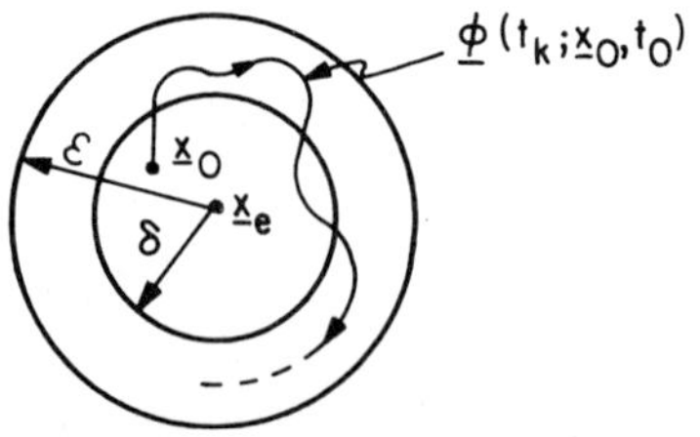

Fig. 4.4-1. Definition of stability (*4-11*).

Stability Definition 2: An equilibrium state $\underline{x}_e$ of the system in Eq. (4.4-2) is said to be *uniformly stable* if for any real number $\varepsilon > 0$ there corresponds a real number $\delta(\varepsilon)$ such that if $||\underline{x}_0 - \underline{x}_e|| \leq \delta(\varepsilon)$ then $||\underline{\phi}(t_k; \underline{x}_0, t_0) - \underline{x}_e|| \leq \varepsilon$ for all $t \geq t_0$. ▲

Obviously, the difference between uniform stability and stability is that δ is not dependent upon t_0 in the former, whereas it does depend upon t_0 in the latter. Uniform stability is an important concept when the system under investigation is time varying. Recall that for time-varying systems $\underline{\phi}$ is a function of t_0, whereas for time-invariant systems, $\underline{\phi}$ does not depend upon t_0 (*4-13* and *4-14*).

Stability Definition 3: An equilibrium state $\underline{x}_e$ of the system in Eq. (4.4-2) is said to be *asymptotically stable* if (1) it is stable, and (2) if every solution starting at $\underline{x}_0$ sufficiently near $\underline{x}_e$ converges to $\underline{x}_e$ as t_k increases indefinitely. A more precise statement of condition (2) is: given two real numbers $\delta > 0$ and $\mu > 0$, there are real numbers $\varepsilon > 0$ and $T(\mu, \delta, t_0)$ such that

$$||\underline{x}_0 - \underline{x}_e|| \leq \delta(\varepsilon, t_0)$$

implies

$$||\underline{\phi}(t_k; \underline{x}_0, t_0) - \underline{x}_e|| \leq \mu$$

for all

$$t_k \geq t_0 + T(\mu, \delta, t_0) \quad .$$ ▲

Figure 4.4-2 illustrates the definition for a second-order system. If the equilibrium state $\underline{x}_e$ is asymptotically stable, then every motion starting at a state $\underline{x}_0$ in the region $||\underline{x}_0 - \underline{x}_e|| \leq \delta(\varepsilon)$ converges, without leaving the region $||\underline{\phi}(t_k; \underline{x}_0, t_0) - \underline{x}_e|| \leq \mu$, to $\underline{x}_e$ as time increases indefinitely. Observe, also, that asymptotic stability is also a local concept; that is, one does not know ahead of time how small $\delta(\varepsilon)$ may have to be.

Stability Definition 4: An equilibrium state $\underline{x}_e$ of the system in Eq. (4.4-2) is said to be *uniformly asymptotically stable* (1) if it is uniformly stable, and (2) if $||\underline{x}_0 - \underline{x}_e|| \leq \delta(\varepsilon)$ implies $||\underline{\phi}(t_k; \underline{x}_0, t_0) - \underline{x}_e|| \leq \mu$ for all $t_k \geq t_0 + T(\mu, \delta)$. ▲

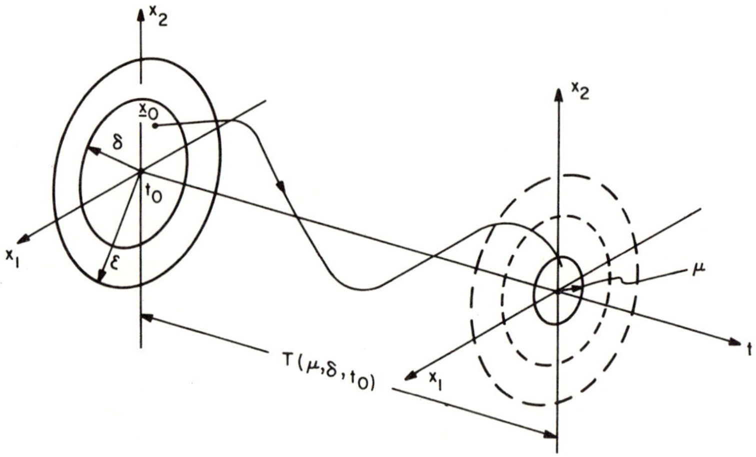

Fig. 4.4-2. Definition of asymptotic stability (*4-12*).

Observe that both δ and T do not depend upon the initial time t_0 in this definition.

Stability Definition 5: An equilibrium state $\underline{x}_e$ of the system in Eq. (4.4-2) is said to be *asymptotically stable in the large* if it is asymptotically stable for *all states* $\underline{x}_0$ from which motions may originate. ▲

The importance of this type of stability is that it is independent of the distance of the initial state from $\underline{x}_e$.

Stability Definition 6: An equilibrium state $\underline{x}_e$ of the system in Eq. (4.4-2) is said to be *uniformly asymptotically stable in the large* if it is uniformly asymptotically stable for all states $\underline{x}_0$ from which motions may originate. ▲

Let us relate these stability definitions to the identification-error system, Eq. (4.4-1). This system is (1) linear, (2) time varying, and (3) true for arbitrary initial errors $\underline{\tilde{\theta}}(0)$; hence, we shall direct our attention to determining conditions on the weighting matrix R(k) such that the equilibrium state, $\underline{\tilde{\theta}} = \underline{0}$, of the identification-error system is uniformly asymptotically stable in the large. Having done this, we will then feel quite confident about using the deterministic-gradient identification algorithm in Eq. (4.2-11) to estimate $\underline{\theta}$.

4.4.3 Second Method of Lyapunov (*4-11*, *4-12*, and *4-15*)

Because of the difficulty associated with obtaining a closed-form solution for the identification-error system in Eq. (4.4-1), it would be quite advantageous to determine its stability properties without having to obtain such a solution. One of the most important events in the theory of stability of dynamic systems was the publication in 1892 of the memoir of the Russian mathematician A. M. Lyapunov. The so-called "second method," which appeared in this memoir, gives information on the stability of equilibrium states of differential (difference) equations, utilizing the given form of the equations but without explicit knowledge of the solutions.

The principal idea of the second method is contained in the following reasoning (*4-11*): If the rate of change $dE(\underline{x})/dt$ of the energy $E(\underline{x})$ of an isolated physical system is negative for every possible state $\underline{x}$, except for a single equilibrium state $\underline{x}_e$, then the energy will continually decrease until it finally assumes its minimum value $E(\underline{x}_e)$. In other words, a dissipative system perturbed from its equilibrium state will always return to it.

The second method of Lyapunov consists of determining a fictitious energy function called a *Lyapunov function*. As a rule there is no natural way of defining energy when the equations of motion are given in a purely mathematical form; hence, the idea of the Lyapunov function is more general then that of energy and is more widely applicable. Lyapunov functions are functions of $\underline{x}(t_k)$ and t_k and are denoted below as $V(\underline{x}, t_k)$.

In the second method of Lyapunov the sign behaviors of $V(\underline{x}, t_k)$ and its rate of increase $\Delta V(\underline{x}, t_k)$ give information on uniform asymptotic stability in the large of the equilibrium state under consideration, without directly solving for the solution. We assume that the equilibrium state under consideration is at the origin of the state space. The essence of the second method of Lyapunov is contained in the following theorem.

Theorem 4-3 (Lyapunov's Main Stability Theorem). *Consider the discrete-time, free dynamic system*

$$\underline{x}(t_{k+1}) = \underline{f}[\underline{x}(t_k), t_k] \tag{4.4-4}$$

where

$$\underline{f}(\underline{0}, t_k) = \underline{0} \tag{4.4-5}$$

for all t_k. Suppose *there exists a scalar function* $V(\underline{x}, t_k)$ *such that*

$$V(\underline{0}, t_k) = 0 \tag{4.4-6}$$

for all t_k, *and*

(i) $V(\underline{x}, t_k)$ *is positive definite; i. e., there exists a continuous nondecreasing scalar function* α *such that* $\alpha(0) = 0$ *and, for all* t_k *and all* $\underline{x} \neq \underline{0}$,

$$V(\underline{x}, t_k) \geq \alpha(||\underline{x}||) > 0 \quad ; \tag{4.4-7}$$ †

(ii) there exists a continuous scalar function γ *such that* $\gamma(0) = 0$ *and, for all* t_k *and all* $\underline{x} \neq \underline{0}$,

$$\begin{aligned}\Delta V(\underline{x}, t_k) &= \text{rate of increase of } V \text{ along motion starting at } \underline{x}(t_k) \\ &= \{V[\underline{\phi}(t_{k+1}; \underline{x}, t_k), t_{k+1}] - V(\underline{x}, t_k)\} \\ &\quad \times (t_{k+1} - t_k)^{-1} \leq -\gamma(||\underline{x}||) < 0 \quad ;\end{aligned} \tag{4.4-8}$$ ‡

(iii) there exists a continuous, nondecreasing scalar function β *such that* $\beta(0) = 0$ *and, for all* t_k *and* $\underline{x} \neq \underline{0}$,

$$V(\underline{x}, t_k) \leq \beta(||\underline{x}||) \quad ; \text{ and} \tag{4.4-9}$$

---

† The notation $\alpha(||\underline{x}||)$ means α is a function of $||\underline{x}||$.

‡ When the discrete-time system is given abstractly (i. e., not by a discretization of a continuous-time system), there is no loss of generality in assuming $t_{k+1} - t_k = 1$ for all k.

(iv) $\alpha(\|\underline{x}\|) \to \infty$ *when* $\|\underline{x}\| \to \infty$.

Then *the equilibrium state* $\underline{x}_e = \underline{0}$ *is* uniformly asymptotically stable in the large *and* $V(\underline{x}, t_k)$ *is a Lyapunov function of the system in Eq. (4.4-4).* ▲

A proof of this important theorem is given in Appendix D. The reader unfamiliar with the second method of Lyapunov is strongly urged to study the proof, in order to obtain an understanding of the physical implications of the theorem's conditions.

4.4.4 A Choice for the Weighting Matrix R(k)

We shall assume that

$$R(k) = c(k)\,\mathrm{diag}[h_1(k), h_2(k), \ldots, h_n(k)] \qquad (4.4\text{-}10)$$

where

(i) $0 < h_L \leq h_i(k) \leq h_U < \infty$ for $i = 1, 2, \ldots, n$ and all $k \geq 0$;

(ii) there exists at least one of the n quantity h_i's, denoted $h_m(k)$, such that

$$\frac{h_m(k) - h_m(k+1)}{h_m(k)} \geq \frac{h_i(k) - h_i(k+1)}{h_i(k)} \qquad (4.4\text{-}11)$$

for all $i = 1, 2, \ldots, n$ and all $k \geq 0$; and

(iii) $c(k)$ is a scalar which remains to be specified.

For $R(k)$ in Eq. (4.4-10), the ith component of $\hat{\underline{\theta}}(k+1)$ in Eq. (4.2-11) becomes

$$\hat{\theta}_i(k+1) = \hat{\theta}_i(k) + c(k)h_i(k)x_i(k)\tilde{y}(k) \quad . \qquad (4.4\text{-}12)$$

In this way, only $x_i(k)$ directly affects $\hat{\theta}_i$ [suggested by the structure of the perfect measurement equation, Eq. (4.2-4)], and, by suitably choosing $h_i(k)$, it is possible to weight the effect of $x_i(k)$ as heavily as desired.

Most steepest-descent algorithms require $R(k) = c(k)I$, in which case $h_i(k) = 1$ for all i and k. These algorithms are often referred to as *first-order gradient algorithms* (*4-16*) (Problem 4-2). For weighting matrices of the more general form in Eq. (4.4-10), the algorithms are referred to as *second-order gradient algorithms* (see Problem 4-3). The rate of convergence is usually much faster for second-order gradient algorithms than for first-order gradient algorithms (Section 4.7).

Condition (i) of Section 4.4.4 merely means that the h_i's are bounded both from above and below. We do not need to know explicit values for h_L and h_U, only that they exist.

Condition (ii) of Section 4.4.4 is easily satisfied in practice. The reason for it will become clear in Eq. (4.4-28). Note that this condition can also be written as

$$\frac{h_m(k+1)}{h_m(k)} \le \frac{h_i(k+1)}{h_i(k)} \tag{4.4-13}$$

for $i = 1, 2, \ldots, n$ and all $k \ge 0$.

Example 7. Suppose that

$$h_i(k) = \mu^{ik} \tag{4.4-14}$$

where

$$0 < \mu < 1 \quad . \tag{4.4-15}$$

Then, from Eq. (4.4-13),

$$\mu^m \le \mu^i \qquad \text{for all } i = 1, 2, \ldots, n \tag{4.4-16}$$

and it is clear that

$$m = \max\{i\} = n \tag{4.4-17}$$

which means that

$$h_m(k) = h_n(k) = \mu^{nk} \quad . \tag{4.4-18}$$

The choice for the h_i's in Eq. (4.4-14) can be used to provide the identifier with a fading memory of the input data. Consider Example 3 in Section 4.2. The most recent input measurement, $w(k-1)$, would be weighted by μ^k, whereas the least recent measurement, $w(k-N)$, would be weighted by μ^{Nk}, and for $0 < \mu < 1$, $\mu^{Nk} \ll \mu^k$; thus, recent input measurements are weighted more heavily than past measurements. ▲

4.4.5 Convergence

Theorem 4-4. *If* $\hat{\underline{\theta}}(k)$ *is updated by means of the deterministic-gradient identification algorithm in Eq. (4.2-11), where the elements of* R(k) *in Eq. (4.4-10) are subject to the constraints in conditions (i) and (ii) of Section 4.4.4,* and

$$0 < c(k) < \frac{2}{\sum_{i=1}^{n} h_i(k)x_i^2(k)} \tag{4.4-19}$$

then almost always [unless $\tilde{\underline{\theta}}(k)$ *and* $\underline{x}(k)$ *are orthogonal*[†]*]* $\hat{\underline{\theta}}(k) \rightarrow \underline{\theta}$ *as* $k \rightarrow \infty$*, regardless of the initial estimate* $\hat{\underline{\theta}}(0)$.

Proof. A scalar function $V(\tilde{\underline{\theta}}, k)$ is defined as

$$V(\tilde{\underline{\theta}}, k) = h_m(k) \sum_{i=1}^{n} \frac{\left[\tilde{\theta}_i(k)\right]^2}{h_i(k)} \quad . \tag{4.4-20}$$

We shall demonstrate that, under the conditions of the Theorem, $V(\tilde{\underline{\theta}}, k)$ is a Lyapunov function for the identification-error system.

Comparing Eqs. (4.4-4) and (4.4-1), we associate $\tilde{\underline{\theta}}$ with $\underline{x}$, and $[I - R(k)\underline{x}(k)\underline{x}'(k)]\tilde{\underline{\theta}}(k)$ with $\underline{f}[\underline{x}(t_k), t_k]$. We leave it to the reader (Problem 4-5) to verify Eqs. (4.4-5), (4.4-6), and conditions (i), (iii), and (iv) in Theorem 4-3. We shall direct our attention to condition (ii) - the most difficult condition to verify.

Constraints on c(k) and the $h_i(k)$ $(i = 1, 2, \ldots, n)$ must be established so that

$$\Delta V(\tilde{\underline{\theta}}, k) = V(\tilde{\underline{\theta}}, k + 1) - V(\tilde{\underline{\theta}}, k) < 0 \tag{4.4-21}$$

for all $\tilde{\underline{\theta}} \neq \underline{0}$. Let

$$\Delta V_m(\tilde{\underline{\theta}}, k) \triangleq \frac{V(\tilde{\underline{\theta}}, k + 1)}{h_m(k + 1)} - \frac{V(\tilde{\underline{\theta}}, k)}{h_m(k)} \quad . \tag{4.4-22}$$

Our approach will be to obtain an explicit expression for $\Delta V_m(\tilde{\underline{\theta}}, k)$, bound it from above, and then relate things back to $\Delta V(\tilde{\underline{\theta}}, k)$.

[†] We shall discuss this special case separately in Section 4.5.

$$\Delta V_m(\underline{\tilde{\theta}}, k) = \sum_{i=1}^{n} \frac{[\tilde{\theta}_i(k+1)]^2}{h_i(k+1)} - \sum_{i=1}^{n} \frac{[\tilde{\theta}_i(k)]^2}{h_i(k)}$$

$$= \sum_{i=1}^{n} \left\{ \frac{[\tilde{\theta}_i(k+1)]^2}{h_i(k+1)} - \frac{[\tilde{\theta}_i(k)]^2}{h_i(k)} + \frac{[\tilde{\theta}_i(k+1)]^2}{h_i(k)} - \frac{[\tilde{\theta}_i(k+1)]^2}{h_i(k)} \right\}$$

$$= \sum_{i=1}^{n} \left\{ \frac{[\tilde{\theta}_i(k+1)]^2 - [\tilde{\theta}_i(k)]^2}{h_i(k)} \right\} + \sum_{i=1}^{n} [\tilde{\theta}_i(k+1)]^2 \left[\frac{1}{h_i(k+1)} - \frac{1}{h_i(k)} \right]$$

$$= \sum_{i=1}^{n} \frac{[-2\theta_i\hat{\theta}_i(k+1) + \hat{\theta}_i^2(k+1) + 2\theta_i\hat{\theta}_i(k) - \hat{\theta}_i^2(k)]}{h_i(k)} + \sum_{i=1}^{n} [\tilde{\theta}_i(k+1)]^2 \left[\frac{h_i(k) - h_i(k+1)}{h_i(k)h_i(k+1)} \right] . \qquad (4.4\text{-}23)$$

For the moment, let us direct our attention to the first term on the right-hand side of this expression. We denote it as T1, where

$$T1 = \sum_{i=1}^{n} \frac{1}{h_i(k)} \{2\theta_i[\hat{\theta}_i(k) - \hat{\theta}_i(k+1)] + [\hat{\theta}_i^2(k+1) - \hat{\theta}_i^2(k)]\}$$

$$= \sum_{i=1}^{n} \frac{1}{h_i(k)} [\hat{\theta}_i(k+1) - \hat{\theta}_i(k)][\hat{\theta}_i(k+1) + \hat{\theta}_i(k) - 2\theta_i] . \qquad (4.4\text{-}24)$$

At this point, we introduce the deterministic-gradient identification algorithm (which is the underlying basis for the identification-error system) into our analysis. Substituting Eq. (4.4-12) into Eq. (4.4-24), we find

$$T1 = \sum_{i=1}^{n} [c(k)x_i(k)\tilde{y}(k)][\hat{\theta}_i(k) + c(k)h_i(k)x_i(k)\tilde{y}(k) + \hat{\theta}_i(k) - 2\theta_i]$$

$$= c(k)\tilde{y}(k) \sum_{i=1}^{n} x_i(k)[c(k)h_i(k)x_i(k)\tilde{y}(k) - 2\tilde{\theta}_i(k)] .$$

$$T1 = c^2(k)\tilde{y}^2(k) \sum_{i=1}^{n} h_i(k)x_i^2(k) - 2c(k)\tilde{y}^2(k) \tag{4.4-25}$$

where we have made use of the fact, from Eq. (4.3-10), that

$$\tilde{y}(k) = \sum_{i=1}^{n} x_i(k)\tilde{\theta}_i(k) \quad . \tag{4.4-26}$$

Let us now return to an analysis of $\Delta V_m(\tilde{\underline{\theta}}, k)$, in Eq. (4.4-23), which we shall, for the time being, write as

$$\Delta V_m(\tilde{\underline{\theta}}, k) = T1 + \sum_{i=1}^{n} [\tilde{\theta}_i(k+1)]^2 \left[\frac{h_i(k) - h_i(k+1)}{h_i(k)h_i(k+1)}\right] \quad . \tag{4.4-27}$$

Using Eq. (4.4-11), we determine from Eq. (4.4-27) that

$$\Delta V_m(\tilde{\underline{\theta}}, k) \leq T1 + \left[\frac{h_m(k) - h_m(k+1)}{h_m(k)}\right] \sum_{i=1}^{n} \frac{[\tilde{\theta}_i(k+1)]^2}{h_i(k+1)}$$

$$\leq T1 + \left[\frac{h_m(k) - h_m(k+1)}{h_m(k)h_m(k+1)}\right] V(\tilde{\underline{\theta}}, k+1) \quad . \tag{4.4-28}$$

Now we shall rewrite this inequality using Eq. (4.4-22):

$$\frac{V(\tilde{\underline{\theta}}, k+1)}{h_m(k+1)} - \frac{V(\tilde{\underline{\theta}}, k)}{h_m(k)} + \left[\frac{h_m(k+1) - h_m(k)}{h_m(k)h_m(k+1)}\right] V(\tilde{\underline{\theta}}, k+1) \leq T1 \quad . \tag{4.4-29}$$

Combining the terms on the left-hand side of this equation, we find

$$\Delta V(\tilde{\underline{\theta}}, k) \leq h_m(k)T1 \quad ; \tag{4.4-30}$$

hence, $\Delta V(\tilde{\underline{\theta}}, k) < 0$ if $T1 < 0$, where

$$T1 = -c(k)\tilde{y}^2(k)[2 - c(k) \sum_{i=1}^{n} h_i(k)x_i^2(k)] \quad . \tag{4.4-31}$$

By means of a simple inequality analysis of T1 in this expression (Problem 4-6), it is easily shown that $T1 < 0$ if

$$0 < c(k) < \frac{2}{\sum_{i=1}^{n} h_i(k)x_i^2(k)} \tag{4.4-32}$$

which was to be proved.

Observe, also, that when $\tilde{\underline{\theta}}(k)$ and $\underline{x}(k)$ are orthogonal, the equation error $\tilde{y}(k)$ will be zero; and in this case, $\Delta V(\tilde{\underline{\theta}}, k) = 0$ for $\tilde{\underline{\theta}}(k) \neq \underline{0}$.

We see, therefore, that if $\tilde{\underline{\theta}}(k)$ and $\underline{x}(k)$ are never orthogonal, the identification error system is uniformly asymptotically stable in the large. If, on the other hand, $\tilde{\underline{\theta}}(k)$ and $\underline{x}(k)$ become orthogonal [for some $k > k_1$, say] then the identification error system is uniformly stable in the large. ▲

Example 8. If

$$c(k) \triangleq \frac{c}{\sum_{i=1}^{n} h_i(k)x_i^2(k)} \tag{4.4-33}$$

where

$$0 < c < 2 \quad , \tag{4.4-34}$$

then $\Delta V(\tilde{\underline{\theta}}, k)$ in Eq. (4.4-30) becomes

$$\Delta V(\tilde{\underline{\theta}}, k) \leq - \frac{c(c - 2)h_m(k)}{\sum_{i=1}^{n} h_i(k)x_i^2(k)} \tilde{y}^2(k) < 0 \tag{4.4-35}$$

for all $k \geq 0$. For this choice of $c(k)$,

$$R(k) = \frac{c}{\sum_{i=1}^{n} h_i(k)x_i^2(k)} \operatorname{diag}[h_1(k), \ldots, h_n(k)] \, , \tag{4.4-36}$$

which represents a generalization of the Nagumo and Noda weighting matrix in Eq. (4.3-16) (see Example 6 in Section 4.3.4).

4.5 ORTHOGONALITY OF PARAMETER ESTIMATION ERROR AND INPUT VECTORS

We have observed that when $\tilde{\underline{\theta}}(k)$ and $\underline{x}(k)$ are orthogonal, the identification-error system is uniformly stable in the large but is not uniformly, asymptotically stable in the large, meaning that $\hat{\underline{\theta}}(k)$ may not converge to $\underline{\theta}$ as $k \to \infty$. This represents a limitation of deterministic-gradient identification algorithms. In practice, however, convergence of $\hat{\underline{\theta}}(k)$ to $\underline{\theta}$ is almost always obtained. [This has been observed by the author and also by Shipley et al. (*4-17*) and Stefani (*4-18*).]

The condition $\tilde{y}(k) = \tilde{\underline{\theta}}'(k)\underline{x}(k) = 0$ when $\tilde{\underline{\theta}}(k) \neq \underline{0}$ is strongly dependent upon the input sequence $\underline{x}(k)$. Lion (*4-19*) relates the frequency content of $\underline{x}(k)$ to the nonorthogonality of $\underline{x}(k)$ and $\tilde{\underline{\theta}}(k)$. He concludes that *if* $\underline{x}(t)$ *contains at least* n/2 *independent frequencies when* n *is even, or* (n + 1)/2 *independent frequencies when* n *is odd, then* $\tilde{y}(k) = 0$ *only if* $\tilde{\underline{\theta}}(k) = \underline{0}$. Let us demonstrate the truth of this by means of a simple example.

Example 9. We shall examine the situation when $\underline{\theta} = (\theta_1, \theta_2, \theta_3)'$; hence, n = 3. To begin, let us assume that

$$x_1(k) = A_1 \sin(\omega k + \phi_1) \quad ,$$

$$x_2(k) = A_2 \sin(\omega k + \phi_2) \quad , \text{ and} \qquad (4.5\text{-}1)$$

$$x_3(k) = A_3 \sin(\omega k + \phi_3) \quad .$$

If $\tilde{y}(k) = 0$ for some $k > k_1$, then $\underline{x}'(k)\tilde{\underline{\theta}}(k) = 0$, which means

$$\tilde{\theta}_1(k)\, A_1 \sin(\omega k + \phi_1) + \tilde{\theta}_2\, A_2 \sin(\omega k + \phi_2) + \tilde{\theta}_3\, A_3 \sin(\omega k + \phi_3) = 0 \quad . \qquad (4.5\text{-}2)$$

This equation can also be written as

$$\sin \omega k[\tilde{\theta}_1(k)\, A_1 \cos \phi_1 + \tilde{\theta}_2(k)\, A_2 \cos \phi_2 + \tilde{\theta}_3(k)\, A_3 \cos \phi_3] + \cos \omega k[\tilde{\theta}_1(k)\, A_1 \sin \phi_1 + \tilde{\theta}_2(k)\, A_2 \sin \phi_2 + \tilde{\theta}_3(k)\, A_3 \sin \phi_3] = 0 \qquad (4.5\text{-}3)$$

for $k > k_1$; hence,

$$\tilde{\theta}_1(k)\, A_1 \cos \phi_1 + \tilde{\theta}_2(k)\, A_2 \cos \phi_2 + \tilde{\theta}_3(k)\, A_3 \cos \phi_3 = 0$$

and (4.5-4)

$$\tilde{\theta}_1(k)\, A_1 \sin \phi_1 + \tilde{\theta}_2(k)\, A_2 \sin \phi_2 + \tilde{\theta}_3(k)\, A_3 \sin \phi_3 = 0 \quad .$$

This underdetermined system of equations for $\tilde{\theta}_1(k)$, $\tilde{\theta}_2(k)$, and $\tilde{\theta}_3(k)$ does not possess unique solutions for these quantities; thus, although

$\tilde{\theta}_1(k) = \tilde{\theta}_2(k) = \tilde{\theta}_3(k) = 0$ is a solution that satisfies these equations, other solutions are possible. These other solutions represent solutions for which $\underline{\tilde{\theta}}(k)$ is orthogonal to $\underline{x}(k)$.

Next, let us assume that

$$x_1(k) = A_1 \sin(\omega_1 k + \phi_1) + B_1 \sin(\omega_2 k + \psi_1) \quad ,$$
$$x_2(k) = A_2 \sin(\omega_1 k + \phi_2) + B_2 \sin(\omega_2 k + \psi_2) \quad , \quad \text{and} \qquad (4.5\text{-}5)$$
$$x_3(k) = A_3 \sin(\omega_1 k + \phi_3) + B_3 \sin(\omega_2 k + \psi_3) \quad .$$

Repeating the proceding analysis, one easily determines that $\tilde{y}(k) = 0$, if

$$\tilde{\theta}_1(k)\, A_1 \cos\phi_1 + \tilde{\theta}_2(k)\, A_2 \cos\phi_2 + \tilde{\theta}_3(k)\, A_3 \cos\phi_3 = 0 \ ,$$
$$\tilde{\theta}_1(k)\, B_1 \cos\psi_1 + \tilde{\theta}_2(k)\, B_2 \cos\psi_2 + \tilde{\theta}_3(k)\, B_3 \cos\psi_3 = 0 \ ,$$
$$\tilde{\theta}_1(k)\, A_1 \sin\phi_1 + \tilde{\theta}_2(k)\, A_2 \sin\phi_2 + \tilde{\theta}_3(k)\, A_3 \sin\phi_3 = 0 \ , \qquad (4.5\text{-}6)$$

and

$$\tilde{\theta}_1(k)\, B_1 \sin\psi_1 + \tilde{\theta}_2(k)\, B_2 \cos\psi_2 + \tilde{\theta}_3(k)\, B_3 \cos\psi_3 = 0$$

which is an overdetermined system of equations for $\tilde{\theta}_1(k)$, $\tilde{\theta}_2(k)$, and $\tilde{\theta}_3(k)$. Clearly, the only way in which these four equations can be satisfied is if $\tilde{\theta}_1(k) = \tilde{\theta}_2(k) = \tilde{\theta}_3(k) = 0$.

Thus, when $\underline{x}(t)$ contains $(n + 1)/2 = 2$ independent frequencies, $\tilde{y}(k) = 0$ only if $\underline{\tilde{\theta}}(k) = \underline{0}$. ▲

In some applications, it is possible for signals to reach steady-state values. Suppose, for example, $\underline{x}(k)$, $y(k)$, and $\underline{\hat{\theta}}(k)$ have reached steady state; $\underline{x}^{ss}$, y^{ss}, and $\underline{\hat{\theta}}^{ss}$ denote these steady-state values.

Theorem 4-5. *(4-20)* If: *(1)* $\underline{\hat{\theta}}$ *achieves some final value* $\underline{\hat{\theta}}^{ss}$*;* *(2) the identifier is* performing well; *(3)* y *achieves a zero steady state (i. e.,* $y^{ss} = 0$*); and (4)* $\underline{x}$ *achieves a nonzero steady state (i. e.,* $\underline{x}^{ss} \neq \underline{0}$*),* then

$$\frac{\hat{\theta}_1^{ss}}{\theta_1} = \frac{\hat{\theta}_2^{ss}}{\theta_2} = \cdots = \frac{\hat{\theta}_n^{ss}}{\theta_n} \quad . \qquad (4.5\text{-}7)$$

Proof. If the identifier is *performing well*, then $\tilde{y}(k) \to 0$ for $k > k_1$; hence

$$y^{ss} - \hat{y}^{ss} = y^{ss} - (\underline{\hat{\theta}}^{ss})'\underline{x}^{ss} = 0 \quad . \tag{4.5-8}$$

From this equation, condition (3), and Eq. (4.2-4), it follows that

$$(\underline{\hat{\theta}}^{ss})'\underline{x}^{ss} = 0 \tag{4.5-9}$$

and

$$\underline{\theta}'\underline{x}^{ss} = 0 \quad . \tag{4.5-10}$$

Using Eq. (4.5-10) to obtain x_1^{ss}, we find

$$x_1^{ss} = -\frac{\theta_2}{\theta_1}x_2^{ss} - \cdots - \frac{\theta_n}{\theta_1}x_n^{ss} \quad . \tag{4.5-11}$$

Substituting this expression into Eq. (4.5-9), we determine that

$$x_2^{ss}\left(\hat{\theta}_2^{ss} - \frac{\theta_2}{\theta_1}\hat{\theta}_1^{ss}\right) + \cdots + x_n^{ss}\left(\hat{\theta}_n^{ss} - \frac{\theta_n}{\theta_1}\hat{\theta}_1^{ss}\right) = 0 \quad . \tag{4.5-12}$$

Because $\underline{x}^{ss} \neq \underline{0}$, it follows from this expression that

$$\frac{\hat{\theta}_i^{ss}}{\theta_i} = \frac{\hat{\theta}_1^{ss}}{\theta_1} \quad \text{for} \quad i = 2, 3, \ldots, n \quad . \tag{4.5-13}$$

Equation (4.5-7) is an immediate consequence of the n - 1 conditions in Eq. (4.5-13). ▲

If the four conditions of the preceding theorem are met, then the best that any identification algorithm can achieve is convergence of $\underline{\hat{\theta}}$ to $\underline{\theta}$ in the sense of Eq. (4.5-7). Naturally, this includes the desirable case when $\underline{\hat{\theta}} \to \underline{\theta}$ [for which Eq. (4.5-7) reduces to $1 = 1 = \cdots = 1$]; however, it also includes other cases for which $\underline{\hat{\theta}} \not\to \underline{\theta}$.

If $\underline{\hat{\theta}}^{ss} \neq \underline{\theta}$, then the result in Eq. (4.5-7) is equivalent to the statement: $\tilde{y}(k) \to 0$, because $\underline{x}^{ss}$ is orthogonal to the identification error $\underline{\tilde{\theta}}^{ss} = \underline{\theta} - \underline{\hat{\theta}}^{ss}$ [subtract Eq. (4.5-9) from Eq. (4.5-10)].

It is important to observe that none of the preceding results in this section are tied to a specific identification algorithm. *They represent general properties of the scalar equation-error approach to parameter identification.*

In some situations, the conditions in Eq. (4.5-7) represent an asset of the scalar equation-error approach, rather than a liability. This is demonstrated in Problem 4-9 for the high-performance aerospace control system that was described in Section 1.3.5.

4.6 AN OPTIMUM CHOICE FOR THE WEIGHTING MATRIX R(k)

In this section, a weighting matrix is derived from two completely different points of view: the Lyapunov stability theory and an error-correction property. Interestingly enough, the weighting matrices obtained from the two approaches can be made identical.

4.6.1 A Lyapunov-Optimum Weighting Matrix

The expression $\Delta V(\tilde{\underline{\theta}}, k)$ provides a measure of the rate at which $\hat{\underline{\theta}}(k + 1)$ converges to $\underline{\theta}$. Naturally, we desire rapid convergence. Setting the partial derivative of the right-hand side of Eq. (4.4-31), with respect to c(k), equal to zero, leads to the following optimum choice, $c^*(k)$, for c(k):

$$c^*(k) = \frac{1}{\sum_{i=1}^{n} h_i(k)x_i^2(k)} \quad . \tag{4.6-1}$$

Observe that $c^*(k)$ is of the form c(k) in Eq. (4.4-33), and that the condition in Eq. (4.4-34) is satisfied (in fact, $c^* = 1$) for $c^*(k)$; hence, convergence is assured for $c^*(k)$. $R^*(k)$, the *Lyapunov-optimum weighting matrix*, obtained by substituting Eq. (4.6-1) into Eq. (4.4-10), is

$$R^*(k) = \frac{\text{diag}[h_1(k), \ldots, h_n(k)]}{\sum_{i=1}^{n} h_i(k)x_i^2(k)} \quad . \tag{4.6-2}$$

4.6.2 An Error-Correction Weighting Matrix

The elements of R(k) shall be chosen to satisfy an error-correction property. In this property, which is based on the error-correction training procedure from pattern recognition,[†] *the output of the*

[†] The reader interested in pattern recognition and the error-correction training procedure can consult Refs. *4-21* and *4-22*.

adjusted model in Fig. 4.2-1 must equal the output of the representation block if the same input is applied at the next sampling instant. More precisely, we have the following error-correction property theorem:

Theorem 4-6. *If* $\hat{y}(k) \neq y(k)$ *and* $\underline{x}(k+1) = \underline{x}(k)$, *then* $\hat{y}(k+1) = y(k)$ *if and only if*

$$\underline{x}'(k)R(k)\underline{x}(k) = 1 \quad . \tag{4.6-3}$$

Proof. (1) Under the conditions of the theorem and from Eqs. (4.2-5), (4.2-6), and (4.2-11), it follows that

$$\begin{aligned}\hat{y}(k+1) &= \underline{\hat{\theta}}'(k+1)\underline{x}(k)\\ &= \underline{\hat{\theta}}'(k)\underline{x}(k) + \tilde{y}(k)\underline{x}'(k)R(k)\underline{x}(k)\\ &= \hat{y}(k) + \tilde{y}(k)\\ &= y(k)\end{aligned} \tag{4.6-4}$$

which proves the sufficiency of Eq. (4.6-3).

(2) Assume $\hat{y}(k) \neq y(k)$; then

$$\begin{aligned}\hat{y}(k+1) &= \hat{y}(k) + \tilde{y}(k)\underline{x}'(k)R(k)\underline{x}(k)\\ &= \hat{y}(k)[I - \underline{x}'(k)R(k)\underline{x}(k)]\\ &\quad + y(k)\underline{x}'(k)R(k)\underline{x}(k) \quad .\end{aligned} \tag{4.6-5}$$

Upon setting $\hat{y}(k+1) = y(k)$, one finds

$$[y(k) - \hat{y}(k)][1 - \underline{x}'(k)R(k)\underline{x}(k)] = 0 \tag{4.6-6}$$

which must hold for all values of k. Since $\hat{y}(k) \neq y(k)$, Eq. (4.6-6) can be satisfied for all k only if $\underline{x}'(k)R(k)\underline{x}(k) = 1$, which proves the necessity of Eq. (4.6-3). ▲

Many choices of weighting matrices that satisfy Eq. (4.6-3) are possible. In fact, $R^*(k)$ satisfies Eq. (4.6-3); hence, we have shown the validity of Theorem 4-7.

Theorem 4-7. *The Lyapunov-optimum weighting matrix for the deterministic-gradient identification algorithm in Eq. (4.2-11),*

$$R^*(k) = \frac{\text{diag}[h_1(k), h_2(k), \ldots, h_n(k)]}{\sum_{i=1}^{n} h_i(k)x_i^2(k)} \tag{4.6-7}$$

subject to conditions (i) and (ii) in Section 4.4.4, is also an error-correction weighting matrix for that algorithm. ▲

It is refreshing to observe that optimal step sizes which are derived from stability theory can also be derived from a physical principle, the error-correction property.† Such a property seems basic when identifying parameters from perfect measurements.

4.7 APPLICATIONS

The theoretical development of the preceding sections has advocated weighting each element of the input sequence $\underline{x}(k)$ by a different amount; thus, a matrix of weights R(k) appears in the identification algorithm rather than the more usual scalar gain. Why should one use a matrix of weights instead of a scalar weight? If it could be demonstrated that the rate of convergence or some other meaningful performance measure is improved when the matrix of weights is used, it would definitely be more advantageous to use it in an on-line identification. Analytical results for this problem have not been obtained (to the author's knowledge); however, in every example studied by the author, the performance of the identification algorithm with the matrix of weights was superior to the algorithm with the scalar weight. This is demonstrated in the examples below.

4.7.1 Impulse-Response Identification (*4-8*, *4-9*)

The sampled impulse response in Fig. 4.7-1 is to be identified (see Example 3 in Section 4.2). The input sequence g(k) is the realization of a random sequence of zeros and ones, both of which are assumed to be equally likely. Input and output measurements are

---

† This property was first pointed out by Nagumo and Noda (*4-6*) for the less general weighting matrix in Eq. (4.3-16).

summarized in Table 4-1. [The output sequence can be computed directly from Eq. (4.2-26) for $k \geq 4$.]

Estimates of f_1, f_2, and f_3 are to be obtained when $R(k) = I/||\underline{x}(k)||^2$ (scalar weight) and also when

$$R(k) = R^*(k) = \frac{\text{diag}[h_1(k),\ h_2(k),\ h_3(k)]}{\sum_{i=1}^{3} h_i(k)x_i^2(k)} \tag{4.7-1}$$

where

$$h_1(k) = 1 \ , \quad h_2(k) = \frac{1}{2} \ , \quad h_3(k) = \frac{1}{4} \quad . \tag{4.7-2}$$

The meaning of the weightings in Eq. (4.7-2) [see Example 7 in Section 4.4.4] is that past values of the input sequence are weighted less heavily than more recent values.

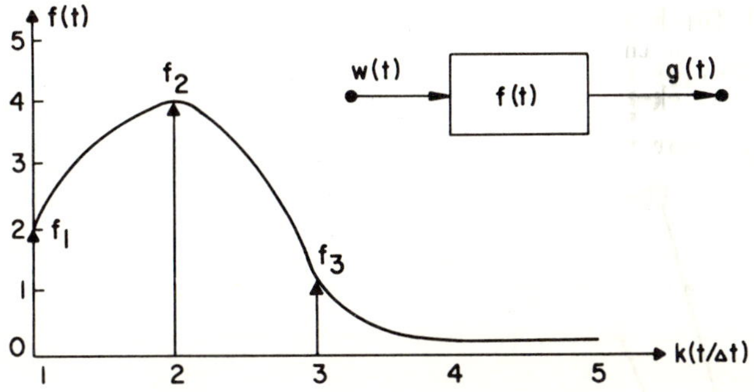

Fig. 4.7-1. Impulse response.

TABLE 4-1

Data for Impulse-Response Identification[a]

k:	1	2	3	4	5	6	7	8	9	10	11	12	13	14	15	16
w(k)	0	0	0	1	1	0	0	1	1	1	0	1	0	1	1	0
g(k)				2	6	5	1	2	6	7	5	3	4	3	6	5

[a] The relations between w(k - i) and x(i), and g(k) and y(k) are given in Example 3, Section 4.2.

Results for $R(k) = I/||\underline{x}(k)||^2$, when† $\hat{\underline{\theta}}(4) = \hat{\underline{f}}(4) = \underline{0}$, are depicted in Fig. 4.7-2. Convergence to the correct values of f_1, f_2, and f_3 occurs, and in relatively few iterations. Results for the matrix of weights in Eqs. (4.7-1) and (4.7-2) are compared with those for the scalar weight in Fig. 4.7-3; $||\tilde{\underline{\theta}}(k)||_M$ and $||\tilde{\underline{\theta}}(k)||_S$ are associated with the matrix and scalar weights, respectively. It is clear that for $k \geq 9$, the identification using the matrix of weights becomes superior to the results for the scalar weight.

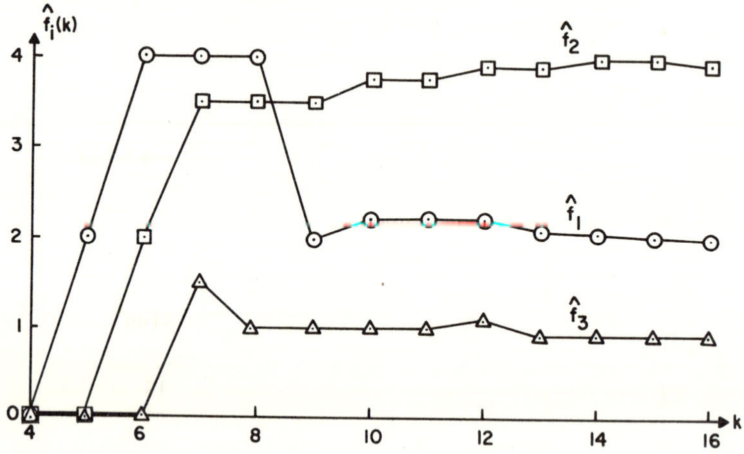

Fig. 4.7-2. Estimates of f_1, f_2, and f_3 for a scalar weight and for the input realization in Table 4-1.

4.7.2 Identification of Aerodynamic Parameters *(4-8, 4-9)*

Let us consider the identification of the two aerodynamic parameters M_α and M_δ that are associated with the high-performance, adaptive, aerospace control system that was described in Example 5, Section 4.2. Our approach will be to estimate M_α and M_δ from the equation

$$\ddot{\theta}(k) = M_\alpha \alpha(k) + M_\delta \delta(k) \qquad (4.7\text{-}3)$$

† Observe, in Table 4-1 that the first measured output occurs at $k = 4$ $(k \geq N + 1 = 4)$.

where, in the spirit of the present chapter, we are assuming that $\ddot{\theta}(k)$, $\alpha(k)$, and $\delta(k)$ can all be measured without measurement errors.

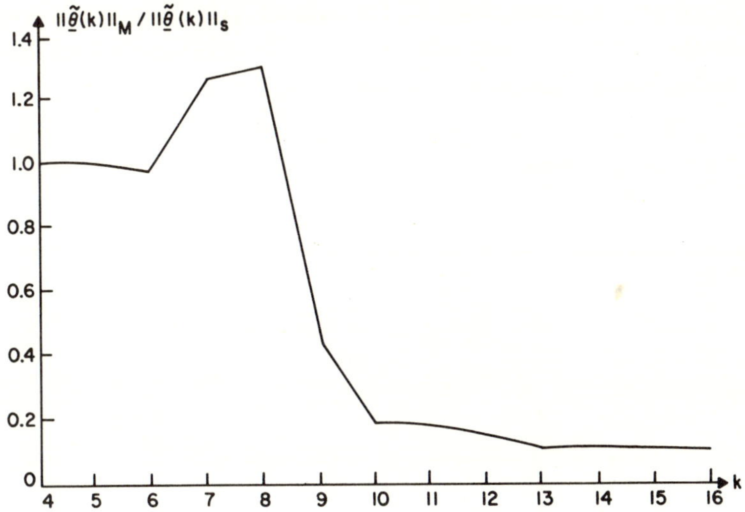

Fig. 4.7-3. Comparison of estimates for matrix of weights and scalar weight.

Estimates of M_α and M_δ are obtained from Eqs. (4.2-11) and (4.6-2). When these equations are combined, we obtain the following expressions for $\hat{M}_\alpha(k+1)$ and $\hat{M}_\delta(k+1)$:

$$\hat{M}_\alpha(k+1) = \hat{M}_\alpha(k) + \frac{h_1(k)\alpha(k)\tilde{y}(k)}{h_1(k)\alpha^2(k) + h_2(k)\delta^2(k)} \tag{4.7-4}$$

and

$$\hat{M}_\delta(k+1) = \hat{M}_\delta(k) + \frac{h_2(k)\delta(k)\tilde{y}(k)}{h_1(k)\alpha^2(k) + h_2(k)\delta^2(k)} \tag{4.7-5}$$

where

$$\tilde{y}(k) = \ddot{\theta}(k) - \hat{M}_\alpha(k)\alpha(k) - \hat{M}_\delta(k)\delta(k) \quad . \tag{4.7-6}$$

Observe that the estimates in Eqs. (4.7-4) and (4.7-5) depend upon only the ratio of $h_2(k)$ and $h_1(k)$ and not upon their absolute values.

The system in Fig. 1.3-2 was simulated along with Eqs. (4.7-4)-(4.7-6), and the identification of M_α and M_δ was accomplished for different settings of $h_2(k)/h_1(k)$. The objective of the study was to compare $\hat{M}_\alpha$ and $\hat{M}_\delta$ for different choices of $h_2(k)/h_1(k)$.

In order to remove some of the subjectivity from these comparisons, the following indexes of performance were used as the basis for rating the different identifiers:

$$(IP)_1 \triangleq J(N) \quad , \tag{4.7-7}$$

$$(IP)_2 \triangleq \sum_{k=1}^{N} J(k) \quad , \tag{4.7-8}$$

and

$$(IP)_3 \triangleq \sum_{k=1}^{N} W(k)J(k) \quad , \tag{4.7-9}$$

where

$$J(k) = \left[\left(\frac{\hat{M}_\alpha(k) - M_\alpha}{M_\alpha}\right)^2 + \left(\frac{\hat{M}_\delta(k) - M_\delta}{M_\delta}\right)^2\right]^{1/2} \tag{4.7-10}$$

and $W(k)$ is a weighting function given by the expression

$$W(k) = 1 + \left(\frac{N_a(k) - N_i}{N_i}\right)^2 \quad . \tag{4.7-11}$$

This weighting weights the performance of the identifier more heavily over the interval of time immediately preceding a command, N_i, and approaches unity as $N_a(k) \rightarrow N_i$. The index $(IP)_3$ provides a measure of performance for the identifier in operation in the control system, whereas $(IP)_1$ and $(IP)_2$ provide open-loop measures of performance for the identifier. A 0.5-sec flight was considered, with a command history as shown in Fig. 4.7-4. Measurements were made every 0.01 sec.

Some results of this study are tabulated in Table 4-2. In this table, the values of $(IP)_1$, $(IP)_2$, and $(IP)_3$ have each been normalized by the value of the respective index of performance for the unity $h_2(k)/h_1(k)$ case. Identifiers for which $\overline{IP} < 1$ are better than those for which $\overline{IP} \geq 1$. It is clear that weighting angle-of-attack (α) more heavily than control-surface deflection (δ) leads to better identifiers for this application.

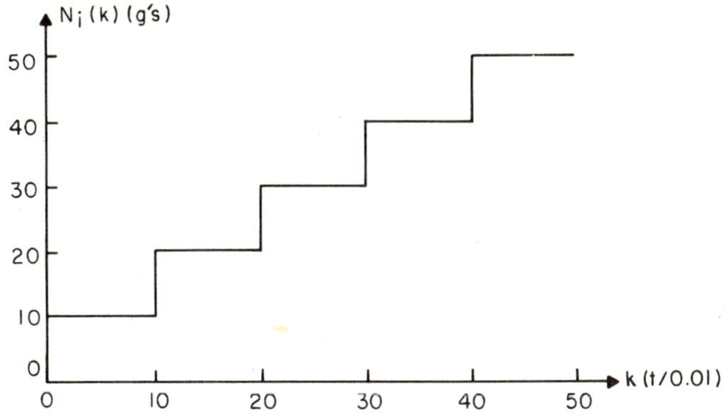

Fig. 4.7-4. Command history for input normal acceleration.

4.8 COMPUTATIONAL CONSIDERATIONS

One of the primary objectives of this chapter has been the development of an optimal algorithm for estimating $\underline{\theta}$ that has the structure New estimate = Old estimate + Linear transformation of equation error, and that has a transformation matrix that is easy to compute. The deterministic-gradient algorithm, with a Lyapunov-optimum weighting matrix, can be written as

$$\hat{\underline{\theta}}(k+1) = \hat{\underline{\theta}}(k) + K^{g}(k)\tilde{y}(k) \tag{4.8-1}$$

where

$$K^{g}(k) = \frac{\text{diag}[h_1(k), \ldots, h_n(k)]\underline{x}(k)}{\sum_{i=1}^{n} h_i(k)x_i^2(k)} \tag{4.8-2}$$

The matrix $K^g(k)$ is $n \times 1$, in which $h_1(k)$, ..., and $h_n(k)$ must be prespecified by the user, in much the same way that the weights $w(k)$ have to be prespecified by a user of the generalized least-squares estimation algorithm. Clearly, *the total computations to obtain* $K^g(k)$ *are very small*. In fact, $K^g(k)$ requires only 1 division, $(2n - 1)$ additions, and $2n$ multiplications, and (Problem 4-12) these are considerably fewer computations than are required to compute $K^*(k + 1)$, the transformation matrix for the sequential generalized least-squares estimation algorithm. Additional discussions on computational requirements for gradient identification algorithms are given by Mendel (*4-23*).

TABLE 4-2

Results for M_α - M_δ Identification

$\frac{h_2(k)}{h_1(k)}$	Normalized performance index $(\overline{IP})_i$			Ranking of identifier		
	$(\overline{IP})_1$	$(\overline{IP})_2$	$(\overline{IP})_3$	$(\overline{IP})_1$	$(\overline{IP})_2$	$(\overline{IP})_3$
$M_\alpha = 500$, $M_\delta = 1000$; $\hat{M}_\alpha(0) = 250$, $\hat{M}_\delta(0) = 500$						
0.01	0.236	0.367	0.443	2	2	2
0.10	0.200	0.219	0.232	1	1	1
1.00	1.000	1.000	1.000	5	5	5
10.00	1.340	1.425	1.440	6	6	6
100.00	0.526	0.840	0.956	3	3	4
$(0.9)^k$	0.928	0.926	0.930	4	4	3
$M_\alpha = -100$, $M_\delta = 500$; $\hat{M}_\alpha(0) = -250$, $\hat{M}_\delta(0) = 750$						
0.01	0.265	0.656	0.800	1	2	2
0.10	0.378	0.555	0.615	2	1	1
1.00	1.000	1.000	1.000	4	4	4
10.00	2.810	2.040	1.848	5	5	5
100.00	5.750	3.340	2.850	6	6	6
$(0.9)^k$	0.910	0.936	0.945	3	3	3

4.9 APPLICABILITY OF DETERMINISTIC-GRADIENT PARAMETER ESTIMATION ALGORITHMS

Deterministic-gradient parameter estimation algorithms are applicable whenever: (1) an application can be cast into a scalar equation-error formulation, and (2) all signals which must be used in the algorithms can be measured without measurement errors. In terms of the scalar equation-error identification system in Fig. 4.2-1, these signals are the input vector $\underline{x}(k)$ to the "identification representation" block, and the output $y(k)$ of the representation block.

The main disadvantage of deterministic-gradient parameter estimation algorithms is that even the slightest bit of measurement error will cause the estimate of $\underline{\theta}$ to be biased. We explore this and ways to remedy it in more detail in the next chapter.

PROBLEMS

4-1. In Example 6, Section 4.3, show that if $0 < \alpha < 2$ then $0 < \xi(k) < 2$. [*Hint:* Apply the Schwarz inequality - Eq. (G-2) - to $\tilde{y}^2(k)$.]

4-2. Demonstrate by graphical techniques that $\hat{\underline{\theta}}$ is adjusted in a direction equal to the negative gradient of $J[\tilde{\underline{\theta}}(k)] = \frac{1}{2}\tilde{y}^2(k)$ only if $R(k) = c(k)I$.

4-3. Let us consider a different approach for obtaining an optimum choice for $R(k)$. We shall view J as a function of $\hat{\underline{\theta}}(k)$ and begin by expanding $J[\hat{\underline{\theta}}(k)]$ about $\hat{\underline{\theta}}(k - 1)$ in the following Taylor's series:

$$J[\hat{\underline{\theta}}(k)] \cong J[\hat{\underline{\theta}}(k-1)] + [\operatorname{grad}_{\hat{\underline{\theta}}} J\big|_{\hat{\underline{\theta}}=\hat{\underline{\theta}}(k-1)}]'[\hat{\underline{\theta}}(k) - \hat{\underline{\theta}}(k-1)] + \frac{1}{2}[\hat{\underline{\theta}}(k) - \hat{\underline{\theta}}(k-1)]'G[\hat{\underline{\theta}}(k) - \hat{\underline{\theta}}(k-1)] \quad ,$$

where G is the Hessian matrix

$$G = \left.\begin{pmatrix} \dfrac{\partial^2 J}{\partial \hat{\theta}_1^2} & \cdots & \dfrac{\partial^2 J}{\partial \hat{\theta}_1 \partial \hat{\theta}_n} \\ \vdots & \ddots & \vdots \\ \dfrac{\partial^2 J}{\partial \hat{\theta}_n \partial \hat{\theta}_1} & \cdots & \dfrac{\partial^2 J}{\partial \hat{\theta}_n^2} \end{pmatrix}\right|_{\hat{\underline{\theta}}=\hat{\underline{\theta}}(k-1)} .$$

(a) Show that the value of $\hat{\underline{\theta}}(k)$ which minimizes $J[\hat{\underline{\theta}}(k)]$ is

$$\hat{\underline{\theta}}(k) = \hat{\underline{\theta}}(k-1) - G^{-1}\left[\operatorname{grad}_{\hat{\underline{\theta}}} J\Big|_{\hat{\underline{\theta}}=\hat{\underline{\theta}}(k-1)}\right] .$$

(b) Show that when $J[\hat{\underline{\theta}}(k)] = \frac{1}{2}\tilde{y}^2(k)$, where $y(k)$ is defined in Eq. (4.2-6),

$$G = -\underline{x}(k)\underline{x}'(k) .$$

(c) Is it possible to compute G^{-1} for G in (b)? Observe that the Lyapunov-optimum weighting matrix may be viewed as an approximation of G^{-1}. Explain.

4-4. Determine $h_m(k)$ when $h_i(k) = \mu^{ik}$ and $\mu > 1$, for $i = 1, 2, \ldots, n$, and all $k \geq 0$.

4-5. Demonstrate that for $V(\tilde{\underline{\theta}}, k)$ in Eq. (4.4-20) all conditions in Theorem 4.4, other than condition (ii) which was examined in the text, are satisfied when the elements of $R(k)$ in Eq. (4.4-10) are subject to the constraints in condition (i) in Section 4.4.4. Be sure to give explicit functions for $\alpha(||\tilde{\underline{\theta}}||)$ and $\beta(||\tilde{\underline{\theta}}||)$.

4-6. The T1, in Eq. (4.4-31) can be made negative in two distinctly different ways. Explain why only one of these ways is *admissible* (i. e., makes sense).

4-7. Prove the assertion, "If $\underline{x}(t)$ contains at least $n/2$ independent frequencies when n is even or $(n + 1)/2$ independent frequencies when n is odd, then $\tilde{y}(k) = 0$ only if $\tilde{\underline{\theta}}(k) = \underline{0}$."

4-8. What happens to the results in Theorem 4-1 when $\tilde{\underline{\theta}}(k)$ and $\underline{x}(k)$ are orthogonal?

4-9. Here we shall demonstrate that the orthogonality of $\tilde{y}(k)$ and $\underline{x}(k)$, which is usually a liability of the equation-error formulation, may be an asset in a specific application. We shall study the estimation of M_α and M_δ for the high-performance, aerodynamically controlled aerospace vehicle depicted in Fig. 1.3-2.

(a) Under the assumptions leading to Eq. (1.3-101), show that if $N_i(t)$ is a step function, then $\ddot{\theta}(t_k \to \infty) = 0$, $\alpha(t_k \to \infty) \neq 0$, and $\delta(t_k \to \infty) \neq 0$.

(b) Assuming that $\hat{M}_\alpha$ and $\hat{M}_\delta$ achieve steady-state values $\hat{M}_\alpha^{ss}$ and $\hat{M}_\delta^{ss}$, demonstrate that $\hat{M}_\alpha^{ss}/M_\alpha = \hat{M}_\delta^{ss}/M_\delta$.

(c) Because only estimates of M_α and M_δ are available for on-line computations of the control gains, the on-line gains will differ from their optimal counterparts in Eqs. (1.3-98) through (1.3-100). The on-line gains denoted $\hat{K}_{Ni}$, $\hat{K}_{\dot{\alpha}}$, and $\hat{K}_{Na}$ are obtained by setting $M_\alpha = \hat{M}_\alpha$ and $M_\delta = \hat{M}_\delta$ in Eqs. (1.3-98) through (1.3-100). Show that the on-line or *adaptive closed-loop transfer function* (i. e., the dynamic relationship between N_a and N_i in Fig. 1.3-2 when K_{Ni}, $K_{\dot{\alpha}}$, and K_{Na} are replaced by $\hat{K}_{Ni}$, $\hat{K}_{\dot{\alpha}}$, and $\hat{K}_{Na}$) is

$$\frac{N_a(s)}{N_i(s)} = \frac{C_2\, M_\delta/\hat{M}_\delta}{\{s^3 + 100s^2 + [C_1\, M_\delta/\hat{M}_\delta + 100Z_\alpha\, 1845/\mu\, (1 - M_\delta/\hat{M}_\delta) + \hat{M}_\alpha\, (M_\delta/\hat{M}_\delta - M_\alpha/\hat{M}_\alpha)]s + [C_2\, M_\delta/\hat{M}_\delta + \hat{M}_\alpha\, (M_\delta/\hat{M}_\delta - M_\alpha/\hat{M}_\alpha)]\}} \quad .$$

(d) Zero steady-state error occurs when the output acceleration N_a follows the input command N_i perfectly as $t \to \infty$. It is known that zero steady-state error occurs when $N_a(0)/N_i(0) = 1$ (J. G. Truxal, *Automatic Feedback Control System Synthesis*, McGraw-Hill, New York, 1955). Show that zero steady-state error occurs for the present system if $\hat{M}_\alpha^{ss}/M_\alpha = \hat{M}_\delta^{ss}/M_\delta$.

It is quite interesting to note that zero steady-state error can be achieved even if $\hat{M}_\alpha^{ss} \neq M_\alpha$ and $\hat{M}_\delta^{ss} \neq M_\delta$. Only the ratios $\hat{M}_\alpha^{ss}/M_\alpha$ and $\hat{M}_\delta^{ss}/M_\delta$ must be equal. Of course, identification errors will cause the adaptive system's bandwidth and damping ratios to deviate from their designed values, which may be undesirable.

4-10. Let x_1, x_2, ..., and x_n be independent random variables having the same symmetric distribution in time with respect to zero. Assume that $\underline{\theta}$ is estimated by Eq. (4.2-11) with $R(k) = I/||\underline{x}(k)||^2$.

(a) Show that

$$E\left\{\frac{x_i^2(k)}{||\underline{x}(k)||^2}\right\} = \frac{1}{n} \quad .$$

(*Hint:* $x_1^2/\sum x_i^2 + \cdots + x_n^2/\sum x_i^2 = 1$.)

(b) Show that

$$E\left\{\frac{x_i(k)x_j(k)}{||\underline{x}(k)||^2}\right\} = 0 \qquad \text{all} \quad i \neq j \quad .$$

(c) Show that

$$E\{||\tilde{\underline{\theta}}(k+1)||^2\} = (1 - 1/n)^k \, E\{||\tilde{\underline{\theta}}(0)||^2\} \quad .$$

(*Hint:* The following identity is useful, $E\{\tilde{\underline{\theta}}'\tilde{\underline{\theta}}\} = \mathrm{Tr}\, E\{\tilde{\underline{\theta}}\tilde{\underline{\theta}}'\}$.)

(d) Letting

$$\bar{e}(k) = \frac{\sqrt{E\{||\tilde{\underline{\theta}}(k)||^2\}}}{||\underline{\theta}||}$$

show that, if $\hat{\underline{\theta}}(0) \triangleq \underline{0}$, then

$$\bar{e}(k) = \varepsilon \qquad \text{if} \quad k^* = 1 + \frac{2 \cdot \ln \varepsilon}{\ln(1 - 1/n)} \quad .$$

This value of k is known as the *expected time for identification* (*4-6*).

4-11. Let us generalize the results of Problem 4-10 to the weighting matrix $R(k) = \alpha I/||\underline{x}(k)||^2$, where $0 < \alpha < 2$ (see Example 6 in Section 4.3).

(a) Show that

$$E\{||\tilde{\underline{\theta}}(k+1)||^2\} = [1 - \alpha(2-\alpha)n^{-1}]^k \, E\{||\tilde{\underline{\theta}}(0)||^2\} \quad .$$

(b) Defining $\bar{e}(k)$ as in Problem 4-10, show that if $\hat{\underline{\theta}}(0) = \underline{0}$, then $\bar{e}(k) = \varepsilon$, if

$$k^* = 1 + \frac{2 \cdot \ln \varepsilon}{\ln[1 - \alpha(2-\alpha)n^{-1}]} \quad .$$

4-12. Let us compare the computational requirements in the scalar measurement situation for deterministic-gradient and sequential generalized least-squares algorithms. We shall make this comparison on the basis of the total number of additions, subtractions, multiplications, and divisions required for the implementation of each algorithm (actually, a computer implementation of mathematical operations requires additional steps for properly controlling and sequencing the operations; for more details on the computation of the logic time associated with these additional steps, see *2-12*).

(a) Let A(n), M(n), and D(n) denote the total number of additions and subtractions, multiplications, and divisions, respectively. Compute A(n), M(n), and D(n) for the deterministic-gradient algorithm in Eqs. (4.2-11) and (4.6-2), and for the sequential generalized least-squares algorithm in Eqs. (2.7-46) through (2.7-48). These quantities will be explicit functions of the number of unknown parameters n.

(b) Assume that addition, multiplication, and division require 2, 6, and 12 execution times, respectively, and, that T_u is the basic execution time. Letting C(n) denote the computation time associated with A(n), M(n), and D(n), it is clear that

$$C(n) = [2A(n) + 6M(n) + 12D(n)]T_u \quad .$$

Plot $C(n)/T_u$ for both estimation algorithms for n = 1, 2, ..., 20.

(c) Let $C_g(n)$ and $C_L(n)$ denote C(n) for the gradient and sequential generalized least-squares algorithms. Plot $C_L(n)/C_g(n)$ for n = 1, 2, ..., 20.

4-13. Refer to Problem 2-3. Find the equation of a straight line fitting the data in Table P2-3 using a deterministic-gradient algorithm.

4-14. Refer to Problem 2-8. Find the equation of a parabola fitting the data in Table P2-8 using a deterministic-gradient algorithm.

REFERENCES

4-1. H. Lass, *Vector and Tensor Analysis*, McGraw-Hill, New York, 1950.

4-2. C. C. Blaydon, *Recursive Algorithms for Pattern Classification*, Rept. No. 520, Div. Eng. Appl. Phys., Harvard Univ., Cambridge, Mass., March 1967.

4-3. C. B. Tompkins, Methods of steep descent, in *Modern Mathematics for the Engineer* (E. F. Beckenbach, ed.), Chapter 18, pp. 448-479, McGraw-Hill, New York, 1956.

4-4. M. Aoki, *Introduction to Optimization Techniques*, MacMillan, New York, 1971.

4-5. E. M. Braverman, Determination of a plant's differential equation during its normal operation, *Automation and Remote Control*, 27, 425-431 (1966).

4-6. J. Nagumo and A. Noda, A learning method for system identification, *IEEE Trans. Auto. Control*, 12, 282-287 (1967).

4-7. R. Roy and J. Sherman, System identification and pattern recognition, International Federation on Automatic Control Symposium on Identification in Automatic Control Systems, Prague, Czechoslovakia, 1967, Paper 1.6.

4-8. J. M. Mendel, Gradient, error-correction identification algorighms, *Inform. Sci.*, 1, 23-42 (1968).

4-9. J. M. Mendel, Gradient identification for linear systems, in *Adaptive, Learning, and Pattern Recognition Systems: Theory and Applications* (J. M. Mendel and K. S. Fu, eds.), Chapter 6, pp. 209-242, Academic Press, New York, 1970.

4-10. B. Z. Vulikh, *Introduction to Functional Analysis for Scientists and Technologists*, Pergamon Press, New York, 1963.

4-11. R. E. Kalman and J. E. Bertram, Control system analysis and design via the second method of Liapunov: I. Continuous-time systems, *ASME J. Basic Eng.*, Series D, 82, 371-393 (1960).

4-12. K. Ogata, *State Space Analysis of Control Systems*, Prentice-Hall, Englewood Cliffs, New Jersey, 1967.

4-13. L. A. Zadeh and C. A. Desoer, *Linear System Theory, The State Space Approach*, McGraw-Hill, New York, 1963.

4-14. R. J. Schwarz and B. Friedland, *Linear Systems*, McGraw-Hill, New York, 1965.

4-15. R. E. Kalman and J. E. Bertram, Control system analysis and design via the second method of Liapunov: II. Discrete-time systems, *ASME J. Basic Eng.*, Series D, 82, 394-400 (1960).

4-16. G. A. Bekey, System identification - an introduction and a survey, *Simulation,* pp. 151-166, October (1970).

4-17. R. P. Shipley, A. G. Engel, and J. W. Hung, *Self-Adaptive Flight Control by Multivariable Parameter Identification,* Rept. No. AFFDL-TR-65-90, Wright-Patterson Air Force Base, Dayton, Ohio, May 1965.

4-18. R. T. Stefani, *Design and Simulation of a High Performance Digital, Adaptive, Normal Acceleration Control System Using Modern Parameter Estimation Techniques,* Rept. No. DAC-60637, Douglas Aircraft Co., Santa Monica, Cal., May 1967.

4-19. P. N. Lion, Rapid identification of linear and nonlinear systems, *AIAA J.,* 5, 1835-1842 (1967).

4-20. J. M. Mendel, Property of the equation-error approach to parameter identification, *IEEE Trans. Auto. Control,* 15, 676-678 (1970).

4-21. N. J. Nilsson, *Learning Machines: Foundations of Trainable Pattern-Classifying Systems,* McGraw-Hill, New York, 1965.

4-22. J. M. Mendel and K. S. Fu (eds.), *Adaptive, Learning, and Pattern Recognition Systems: Theory and Applications,* Academic Press, New York, 1970, Chapters 1-5.

4-23. J. M. Mendel, Sequential identification by means of gradient learning algorithms, in *Pattern Recognition and Machine Learning* (K. S. Fu, ed.), pp. 70-78, Plenum Press, New York, 1971.

5

Stochastic–Gradient Parameter Estimation

5.1 FORMULATION AND STATEMENT OF THE ESTIMATION PROBLEM

Suppose that a measurement of a scalar signal y(t) is made at time t_k where $k \geq 0$. The times t_k, t_{k+1}, ..., etc., do not have to be uniformly spaced. *A basic assumption in this chapter is that the measurement of* y(t), z(t), *is made with measurement error;* that is to say,

$$z(k) = y(k) + v(k) \qquad (5.1\text{-}1)$$

for $k \geq 0$, where v(k) is a zero-mean discrete random sequence.

The signal y(t) is assumed to be a linear combination of n parameters, θ_1, θ_2, ..., θ_n; that is to say,

$$y(t) = x_1(t)\theta_1 + x_2(t)\theta_2 + \cdots + x_n(t)\theta_n \qquad (5.1\text{-}2)$$

where it is also assumed that $x_1(t)$, $x_2(t)$, ..., $x_n(t)$ *cannot be measured perfectly*. Letting $r_i(t)$ denote the measured value of $x_i(t)$, this means that only

$$r_i(k) = x_i(k) + n_i(k) \qquad i = 1, 2, \ldots, n \qquad (5.1\text{-}3)$$

for $k \geq 0$ are available to the analyst for purposes of identification of $\theta_1, \theta_2, \ldots, \theta_n$. The noise $n_i(k)$ is assumed to be a zero-mean, independent discrete random sequence of known variance σ_i^2.

Letting

$$\underline{x}(t) = [x_1(t), x_2(t), \ldots, x_n(t)]' \quad , \tag{5.1-4}$$

$$\underline{\theta} = (\theta_1, \theta_2, \ldots, \theta_n)' \quad , \tag{5.1-5}$$

$$\underline{r}(t) = [r_1(t), r_2(t), \ldots, r_n(t)]' \quad , \tag{5.1-6}$$

and

$$\underline{n}(t) = [n_1(t), n_2(t), \ldots, n_n(t)]' \quad , \tag{5.1-7}$$

the actual signal y(t) and the measured x_i's can be written more compactly as

$$y(t) = \underline{x}'(t)\underline{\theta} \tag{5.1-8}$$

and

$$\underline{r}(k) = \underline{x}(k) + \underline{n}(k) \quad , \tag{5.1-9}$$

where

$$E\{\underline{n}(k)\} = \underline{0} \tag{5.1-10}$$

and

$$E\{\underline{n}(k)\underline{n}'(k)\} = \Sigma_n \quad . \tag{5.1-11}$$

The covariance matrix Σ_n is assumed known a priori, and is diagonal.

Substituting Eq. (5.1-8) into Eq. (5.1-1), we obtain

$$z(k) = \underline{x}'(k)\underline{\theta} + v(k) \tag{5.1-12}$$

for $k \geq 0$. This equation is the starting point for all of the analyses in this chapter. It was also the starting point for much of the analyses in Chapters 2 and 3 (scalar measurement situation).

The following approximation (model) to z(k) is assumed:

$$\hat{z}(k) = \underline{r}'(k)\hat{\underline{\theta}}(k) \quad , \tag{5.1-13}$$

where

$$\hat{\underline{\theta}}(k) = [\hat{\theta}_1(k), \hat{\theta}_2(k), \ldots, \hat{\theta}_n(k)]' \tag{5.1-14}$$

is the kth approximation of $\underline{\theta}$. If $\underline{x}(k)$ could be measured without measurement errors, the model to z(k) would be $\hat{z}(k) = \underline{x}'(k)\hat{\underline{\theta}}(k)$ -

as was the situation in Chapter 4. We have, however, assumed that this is not possible; thus, under the given assumptions, the best we can do is to assume $\hat{z}(k) = \underline{r}'(k)\hat{\underline{\theta}}(k)$.

The error between z(k) and $\hat{z}(k)$, $\tilde{z}(k)$, is

$$\tilde{z}(k) = z(k) - \hat{z}(k) \quad . \tag{5.1-15}$$

Our approach for obtaining $\hat{\underline{\theta}}(k)$ is to choose it in a manner that minimizes some measure of the error $\tilde{z}(k)$, and to then demonstrate that, for that choice, $\hat{\underline{\theta}}(k)$ converges to $\underline{\theta}$. Here we define an instantaneous quadratic measure of the error $\tilde{z}(k)$ as

$$J[\hat{\underline{\theta}}(k)] = \frac{1}{2}\tilde{z}^2(k) \tag{5.1-16}$$

for $k \geq 0$, and obtain $\hat{\underline{\theta}}(k)$ by minimizing $J[\hat{\underline{\theta}}(k)]$ with respect to each of the n components of $\hat{\underline{\theta}}(k)$. Observe that $J[\hat{\underline{\theta}}(k)]$ is a *stochastic criterion function.*

As in Chapter 3, where we also dealt directly with noisy measurements and their known statistics, we shall require the estimate of $\underline{\theta}$ to be *unbiased.* Our approach in this chapter will be first to synthesize an algorithm that is guaranteed to give unbiased estimates of $\underline{\theta}$, and then to study the convergence properties of the identification error system. The study of convergence is complicated here by the stochastic nature of signals which appear in the algorithm.

Important observation. The signal $\tilde{z}(k)$ is a scalar stochastic equation error of exactly the type that was discussed in Chapter 1, Section 1.2.2. The z(k) in Eq. (5.1-12) represents the measured output of the identification system in Fig. 1.2-2c, and $\hat{z}(k)$ represents the output of the model of the identification system. We depict these relationships in Fig. 5.1-1.

We conclude this section with two examples that illustrate how some identification problems can be cast into the formulation just described.

Example 1. *Identification of Aerodynamic Parameters.* We shall once again consider the identification of the two aerodynamic parameters

M_α and M_δ that are associated with the high-performance aerospace control system described in Section 1.3.5. Here, however, $\ddot{\theta}(k)$, $\alpha(k)$, and $\delta(k)$ are not available directly; only measured values of these quantities, $\ddot{\theta}_m(k)$, $\alpha_m(k)$, and $\delta_m(k)$, are available, where

$$\ddot{\theta}_m(k) = \ddot{\theta}(k) + v_{\ddot{\theta}}(k) \quad , \tag{5.1-17}$$

$$\alpha_m(k) = \alpha(k) + n_\alpha(k) \quad , \tag{5.1-18}$$

and

$$\delta_m(k) = \delta(k) + n_\delta(k) \quad . \tag{5.1-19}$$

The measurement equation from which M_α and M_δ will be estimated is [Eqs. (5.1-17) and (4.2-35)]

$$\ddot{\theta}_m(k) = M_\alpha \alpha(k) + M_\delta \delta(k) + v_{\ddot{\theta}}(k) \quad . \tag{5.1-20}$$

The assumed model for $\ddot{\theta}_m(k)$ is

$$\hat{\ddot{\theta}}_m(k) = \hat{M}_\alpha(k)\alpha_m(k) + \hat{M}_\delta(k)\delta_m(k) \quad . \tag{5.1-21}$$

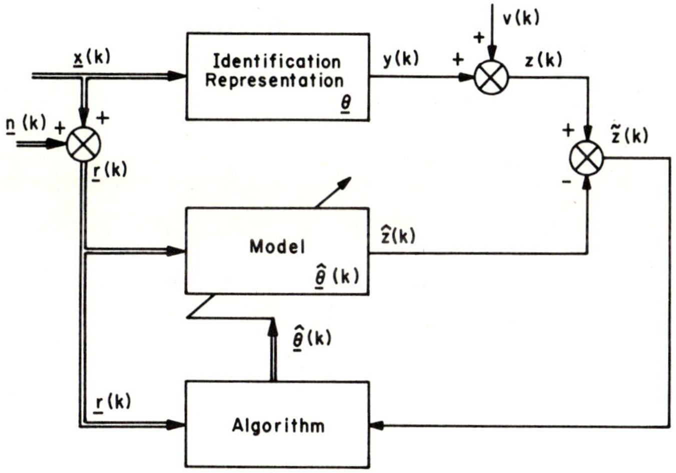

Fig. 5.1-1. Interpretation of scalar measurement equation and estimation equation as a scalar equation-error identification system.

By defining

$$\underline{\theta} = (M_\alpha, M_\delta)' \quad , \tag{5.1-22}$$

$$\underline{x}(k) = [\alpha(k), \delta(k)]' \quad , \tag{5.1-23}$$

$$\underline{r}(k) = [\alpha_m(k), \delta_m(k)]' \quad , \tag{5.1-24}$$

$$\underline{n}(k) = [n_\alpha(k), n_\delta(k)]' \quad , \tag{5.1-25}$$

$$v(k) = v_{\ddot{\theta}}(k) \quad , \tag{5.1-26}$$

and

$$z(k) = \ddot{\theta}_m(k) \quad , \tag{5.1-27}$$

we have cast this problem into the scalar equation-error formulation depicted in Fig. 5.1-1. ▲

Example 2. *Identification of Coefficients in a Finite-Difference Equation.* As in Example 4 of Section 4.2, we shall be interested in identifying the coefficients $a_1, a_2, \ldots, a_n$ in the finite-difference equation

$$y(k + n) + a_1y(k + n - 1) + \cdots + a_ny(k) = \omega(k) \quad . \tag{5.1-28}$$

Here, however, $y(k + n)$ and $y(k + n - j)$ $(j = 1, 2, \ldots, n)$ are not available directly; only measured values $y_m(k + n)$ and $y_m(k + n - j)$ $(j = 1, 2, \ldots, n)$ are available, where

$$y_m(k + n) = y(k + n) + n_y(k + n) \tag{5.1-29}$$

and

$$y_m(k + n - j) = y(k + n - j) + n_y(k + n - j) \tag{5.1-30}$$

for $j = 1, 2, \ldots, n$, and all $k \geq 0$. The measurement equation from which $a_1, a_2, \ldots, a_n$ will be estimated [Eqs. (5.1-29) and (5.1-28)] is

$$y_m(k + n) = -a_1y(k + n - 1) - a_2y(k + n - 2) - \cdots - a_ny(k) + \omega(k) + n_y(k + n) \quad . \tag{5.1-31}$$

The model of $y_m(k + n)$ is[†]

[†] The argument of $\hat{a}_i$, $k + n$, is chosen so that it corresponds to the argument of $\hat{y}(k + n)$ [see discussion following Eq. (1.3-32)].

$$\hat{y}_m(k + n) = -\hat{a}_1(k + n)y_m(k + n - 1) - \hat{a}_2(k + n)y_m(k + n - 2) - \cdots$$

$$- \hat{a}_n(k + n)y_m(k) \quad . \qquad (5.1\text{-}32)$$

By defining

$$\underline{\theta} = (-a_1, -a_2, \ldots, -a_n)' \quad , \qquad (5.1\text{-}33)$$

$$\underline{x}(k) = [y(k + n - 1), y(k + n - 2), \ldots, y(k)]' \quad , \qquad (5.1\text{-}34)$$

$$\underline{r}(k) = [y_m(k + n - 1), y_m(k + n - 2), \ldots, y_m(k)]' \quad , \qquad (5.1\text{-}35)$$

$$\underline{n}(k) = [n_y(k + n - 1), n_y(k + n - 2), \ldots, n_y(k)]' \quad , \qquad (5.1\text{-}36)$$

$$v(k) = \omega(k) + n_y(k + n) \quad , \qquad (5.1\text{-}37)$$

and

$$z(k+n) = y_m(k+n) \quad , \qquad (5.1\text{-}38)$$

we have again cast this problem into the desired formulation.

We shall examine this example, as well as generalizations of it to more complicated forcing functions, in great detail in this chapter, for none of our previously studied estimation techniques can be used to provide unbiased estimates of $a_1, a_2, \ldots, a_n$. ▲

5.2 THREE TYPES OF STOCHASTIC IDENTIFICATION PROBLEMS

By making different assumptions about the statistics of the input vector $\underline{x}(k)$ (Fig. 5.1-1) and the (measurement) error process $v(k)$, we can obtain three quite different and important identification problems. We present these problems next, in order of increasing difficulty (*5-1*).

5.2.1 Class-1 Identification Problem

In this first class of identification problems, the identification system is as depicted in Fig. 5.1-1. The following assumptions are made concerning the statistics of $\underline{x}(k)$ and $v(k)$:

(i) Terms $\underline{x}(k)$ and $v(k)$ are uncorrelated, so that

$$E\{\underline{x}(k)v(k)\} = E\{\underline{x}(k)\}E\{v(k)\} \quad ; \tag{5.2-1}$$

(ii) The conditional covariance of $\underline{x}(k)$ is constant and independent of $\hat{\underline{\theta}}(k)$; that is to say,†

$$E\{\underline{x}(k)\underline{x}'(k)|\hat{\underline{\theta}}(k)\} = \Omega \tag{5.2-2}$$

where Ω *is positive definite, but is not known explicitly.*‡

Assumptions (i) and (ii) mean that there is no coupling between $\underline{x}(k)$ and $v(k)$, and $\underline{x}(k)$ and $\hat{\underline{\theta}}(k)$.

The following applications fall within the framework of the Class-1 identification problem: (1) identification of coefficients in algebraic equations, when some or all of the signals can only be measured with measurement error - as in Example 1 in Section 5.1; and (2) identification of coefficients of finite-difference equations, when random forcing functions are not present, and when measurements are corrupted by measurement noise (Problem 5-1).

5.2.2 Class-2 Identification Problem

For Class-2 identification problems, the scalar equation-error identification system in Fig. 5.1-1 is modified to the structure depicted in Fig. 5.2-1. Now

$$v(k) = v_m(k) + v_d(k) \tag{5.2-3}$$

where $v_m(k)$ and $v_d(k)$ denote measurement and disturbance sequences. Observe that $v_d(k)$ excites a dynamic system, the outputs of which comprise the input vector $\underline{x}(k)$ to the "Identification representation" block.

The following assumptions are made concerning the statistics of $\underline{x}(k)$ and $v(k)$:

(i) The terms $\underline{x}(k)$ and $v(k)$ are correlated.

---

† The signal $\underline{x}(k)$ is also assumed to be zero mean.

‡ Explicit knowledge of Ω is not needed to obtain an asymptotically unbiased estimate of $\underline{\theta}$. This is demonstrated in Section 5.4.

(ii) The conditional covariance of $\underline{x}(k)$ is constant and independent of $\hat{\underline{\theta}}(k)$; that is to say,

$$E\{\underline{x}(k)\underline{x}'(k)|\hat{\underline{\theta}}(k)\} = \Omega \tag{5.2-4}$$

where Ω *is positive definite, but is not known explicitly.*

Now there is coupling between $\underline{x}(k)$ and $v(k)$.

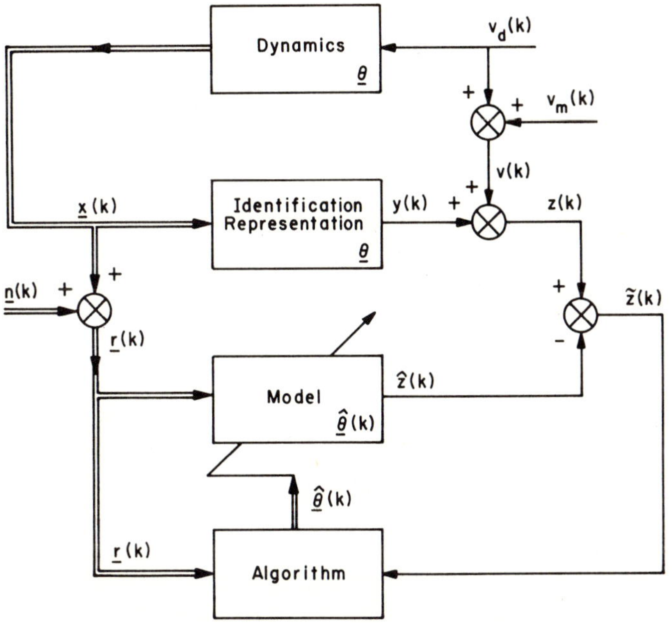

Fig. 5.2-1. Scalar stochastic equation-error identification system for Class-2 identification problems.

An application that falls within the scope of the Class-2 identification problem is identification of coefficients of finite-difference equations when random forcing functions are present, and when measurements are corrupted by measurement noise - as in Example 2 in Section 5.1 [refer specifically to Eq. (5.1-37), to see that $v(k)$ is of the structure in Eq. (5.2-3)].

5.2.3 Class-3 Identification Problem

In this, the most difficult of the three classes of identification problems, the scalar equation-error identification system in

Fig. 5.2-1 is modified further to the structure depicted in Fig. 5.2-2. Now, not only is there coupling between $\underline{x}(k)$ and $v(k)$, but there is coupling between $\underline{x}(k)$ and $\hat{\underline{\theta}}(k)$; that is to say,

(i) $\underline{x}(k)$ and $v(k)$ are correlated;

(ii) $E\{\underline{x}(k)\underline{x}'(k)|\hat{\underline{\theta}}(k)\} = \Omega[\hat{\underline{\theta}}(k)]$.

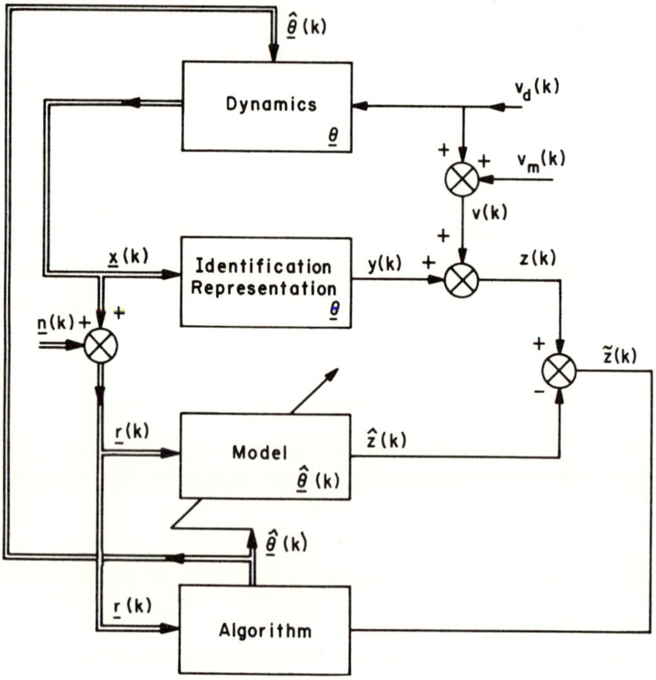

Fig. 5.2-2. Scalar stochastic equation-error identification system for Class-3 identification problems.

An application that falls within the scope of the Class-3 identification problem is, the identification of coefficients of a finite-difference equation when random forcing terms are present, when measurements are corrupted by measurement noise, *and* when feedback control gains, which are embedded in the unknown coefficients, are adjusted on-line by expressions which are explicit functions of $\hat{\underline{\theta}}(k)$ - as in *adaptive control* (see, for example, Section 1.3.5 in Chapter 1).

The difficulty in analyzing the Class-3 identification problem is that the identification-error system for it is nonlinear. We shall present algorithms and supporting analyses only for Class-1 and Class-2 identification problems.

5.3 A STOCHASTIC-GRADIENT ALGORITHM: FIRST ATTEMPT

5.3.1 Development of Algorithm

By analogy with our development for the deterministic situation in Section 4.2, we shall update the kth estimate of $\underline{\theta}$ by means of the following gradient descent algorithm:

$$\hat{\underline{\theta}}(k + \ell) = \hat{\underline{\theta}}(k) - R(k)\ \text{grad}_{\hat{\underline{\theta}}(k)}\ J[\hat{\underline{\theta}}(k)] \quad . \tag{5.3-1}$$

Paralleling the gradient computation for the deterministic situation, it is easily shown that

$$\text{grad}_{\hat{\underline{\theta}}(k)}\ J[\hat{\underline{\theta}}(k)] = -\underline{r}(k)\tilde{z}(k) \quad ; \tag{5.3-2}$$

hence, the *stochastic-gradient parameter identification algorithm* in Eq. (5.3-1) becomes

$$\hat{\underline{\theta}}(k + \ell) = \hat{\underline{\theta}}(k) + R(k)\underline{r}(k)\tilde{z}(k) \tag{5.3-3}$$

for $k \geq 0$.

The estimate $\hat{\underline{\theta}}(k + \ell)$ is spaced ℓ time units from $\hat{\underline{\theta}}(k)$, for reasons which will become clear in Section 5.4.3.

5.3.2 Asymptotic Bias in Estimate

As stated in Section 5.1, we desire the estimate of $\underline{\theta}$ to be unbiased. Because of the sequential nature of our estimation algorithm, only asymptotic unbiasedness is possible (see Example 9 in Section 2.11.1, and, Section 2.11.4); thus, we shall investigate whether or not

$$\lim_{k\to\infty} E\{\hat{\underline{\theta}}(k)\} = \underline{\theta} \quad , \tag{5.3-4}$$

when $\hat{\underline{\theta}}(k)$ is updated by means of Eq. (5.3-3).

To begin, let us demonstrate that Eq. (5.3-3) can also be written as

$$\hat{\underline{\theta}}(k + \ell) = \hat{\underline{\theta}}(k) + R(k)[-\underline{r}(k)\underline{r}'(k)\hat{\underline{\theta}}(k) + \underline{r}(k)v(k) + \underline{r}(k)\underline{x}'(k)\underline{\theta}] \quad . \tag{5.3-5}$$

This is obtained by substituting Eqs. (5.1-15), (5.1-13), and (5.1-12) into Eq. (5.3-3) as follows:

$$\begin{aligned}\hat{\underline{\theta}}(k + \ell) &= \hat{\underline{\theta}}(k) + R(k)\underline{r}(k)[z(k) - \hat{z}(k)] \\ &= \hat{\underline{\theta}}(k) + R(k)\underline{r}(k)[\underline{x}'(k)\underline{\theta} + v(k) - \underline{r}'(k)\hat{\underline{\theta}}(k)] \quad .\end{aligned} \tag{5.3-6}$$

▲

Our approach will be to take the expected value of the entire Eq. (5.3-5), assume a steady state (final value) exists for $E\{\hat{\underline{\theta}}(k)\}$, and solve for its value. The question of existence of the steady-state value will be treated separately in the section entitled "Convergence" (Section 5.5). Before proceeding with this analysis, let us list the major assumptions that are required to make the analysis meaningful and tractable.

Major assumptions

(i) The terms $\underline{x}(k)$ and $\hat{\underline{\theta}}(k)$ are independent, and

$$E\{\underline{x}(k)\underline{x}'(k)|\hat{\underline{\theta}}(k)\} = \Omega \tag{5.3-7}$$

where Ω is constant and positive definite; its value does not have to be known.

(ii) The terms $\underline{x}(k)$ and $\underline{n}(k)$ are statistically independent.

(iii) The terms $\underline{n}(k)$ and $v(k)$ are statistically independent.†

(iv) The following statistics of $\underline{n}(k)$ and $v(k)$ are known a priori:

$$E\{\underline{n}(k)\} = \underline{0} \quad , \tag{5.3-8}$$

† In some applications, $\underline{n}(k)$ and $v(k)$ may be dependent. It is a simple matter to modify the results of these assumptions for this situation.

$$E\{\underline{n}(k)\underline{n}'(k)\} = \Sigma_n \quad , \tag{5.3-9}$$

and

$$E\{v(k)\} = 0 \quad . \tag{5.3-10}$$

Assumption (i) means that we are limiting our attention to Class-1 and Class-2 identification problems. Assumptions (ii) and (iv), and (iii) and (iv) mean that

$$E\{\underline{x}(k)\underline{n}'(k)\} = E\{\underline{x}(k)\}E\{\underline{n}'(k)\} = \underline{0} \tag{5.3-11}$$

and

$$E\{v(k)\underline{n}(k)\} = E\{v(k)\}E\{\underline{n}(k)\} = \underline{0} \quad . \tag{5.3-12}$$

We shall rely quite heavily on conditional expectations; for, clearly, $\hat{\underline{\theta}}(k + \ell)$ *is conditioned on the preceding estimate* $\hat{\underline{\theta}}(k)$. Some facts about conditional expectations are summarized in Appendix E. The following applications of Eqs. (E-13) and (E-14) will play prominent parts in our analyses:

$$E\{\hat{\underline{\theta}}(k + \ell)\} = E\{E\{\hat{\underline{\theta}}(k + \ell)|\hat{\underline{\theta}}(k)\}\} \tag{5.3-13}$$

where the outer expectation is with respect to $\hat{\underline{\theta}}(k)$; and

$$E\{A(k)\hat{\underline{\theta}}(k)\} = E\{E\{A(k)|\hat{\underline{\theta}}(k)\}\hat{\underline{\theta}}(k)\} \tag{5.3-14}$$

where the outer expectation is also with respect to $\hat{\underline{\theta}}(k)$. ▲

Taking the expected value on both sides of Eq. (5.3-5), and using Eqs. (5.3-13) and (5.3-14), we find

$$\begin{aligned}E\{\hat{\underline{\theta}}(k + \ell)\} &= E\{\hat{\underline{\theta}}(k)\} - R(k)[E\{\underline{r}(k)\underline{r}'(k)\hat{\underline{\theta}}(k)\} \\ &\qquad - E\{\underline{r}(k)v(k)\} - E\{\underline{r}(k)\underline{x}'(k)\underline{\theta}\}] \\ &= E\{\hat{\underline{\theta}}(k)\} - R(k)[E\{E\{\underline{r}(k)\underline{r}'(k)|\hat{\underline{\theta}}(k)\}\hat{\underline{\theta}}(k)\} \\ &\qquad - E\{E\{\underline{r}(k)v(k)|\hat{\underline{\theta}}(k)\}\} \\ &\qquad - E\{E\{\underline{r}(k)\underline{x}'(k)\underline{\theta}|\hat{\underline{\theta}}(k)\}\}]\end{aligned} \tag{5.3-15}$$

or

$$E\{\hat{\underline{\theta}}(k + \ell)\} = E\{\hat{\underline{\theta}}(k)\} - R(k)[\underline{TA} - \underline{TB} - \underline{TC}] \quad , \tag{5.3-16}$$

where

$$\underline{TA} = E\{E\{\underline{r}(k)\underline{r}'(k)|\underline{\hat{\theta}}(k)\}\underline{\hat{\theta}}(k)\} \quad , \tag{5.3-17}$$

$$\underline{TB} = E\{E\{\underline{r}(k)v(k)|\underline{\hat{\theta}}(k)\}\} \quad , \tag{5.3-18}$$

and

$$\underline{TC} = E\{E\{\underline{r}(k)\underline{x}'(k)\underline{\theta}|\underline{\hat{\theta}}(k)\}\} \quad . \tag{5.3-19}$$

We shall consider $\underline{TA}$, $\underline{TB}$, and $\underline{TC}$ separately.

Calculation of $\underline{TA}$

Let us compute the inner expectation first, using Eqs. (5.1-9), (5.3-7), (5.3-9), and (5.3-11):

$$\begin{aligned} E\{\underline{r}(k)\underline{r}'(k)|\underline{\hat{\theta}}(k)\} &= E\{[\underline{x}(k)\underline{x}'(k) + \underline{x}(k)\underline{n}'(k) \\ &\quad + \underline{n}(k)\underline{x}'(k) + \underline{n}(k)\underline{n}'(k)]|\underline{\hat{\theta}}(k)\} \\ &= \Omega + \Sigma_n \quad . \end{aligned} \tag{5.3-20}$$

Hence

$$\underline{TA} = E\{(\Omega + \Sigma_n)\underline{\hat{\theta}}(k)\} = (\Omega + \Sigma_n)E\{\underline{\hat{\theta}}(k)\} \quad . \tag{5.3-21}$$

Calculation of $\underline{TB}$

Since $\underline{x}(k)$ is not a function of $\underline{\hat{\theta}}(k)$,

$$\begin{aligned} \underline{TB} &= E\{E\{\underline{r}(k)v(k)|\underline{\hat{\theta}}(k)\}\} \\ &= E\{\underline{r}(k)v(k)\} \\ &= E\{\underline{x}(k)v(k)\} + E\{\underline{n}(k)v(k)\} \\ &= E\{\underline{x}(k)v(k)\} \end{aligned} \tag{5.3-22}$$

where we have made use of Eqs. (5.1-9) and (5.3-12). This is as far as we can proceed in the general development ot $\underline{TB}$ without making additional assumptions about the statistics between $\underline{x}(k)$ and $v(k)$. For example, *in Class-1 identification problems* $\underline{TB} = \underline{0}$, *whereas in Class-2 identification problems* $\underline{TB} \neq \underline{0}$ *and must be computed by making use of system dynamics* (see Fig. 5.2-1). We shall examine the computation of $\underline{TB}$ for certain applications in Section 5.6. Here, however, we prefer not to restrict our attention to either Class-1 or Class-2 identification problems; thus, we leave $\underline{TB}$ as in Eq. (5.3-22).

Calculation of $\underline{\underline{TC}}$

Again, since $\underline{x}(k)$ is not a function of $\hat{\underline{\theta}}(k)$,

$$\underline{\underline{TC}} = E\left\{E\{\underline{r}(k)\underline{x}'(k)\underline{\theta}|\hat{\underline{\theta}}(k)\}\right\}$$

$$= E\{\underline{r}(k)\underline{x}'(k)\underline{\theta}\}$$

$$= E\{\underline{x}(k)\underline{x}'(k)\}\underline{\theta} + E\{\underline{n}(k)\underline{x}'(k)\}\underline{\theta}$$

$$= \Omega\underline{\theta} \tag{5.3-23}$$

where we have made use of Eqs. (5.1-9) and (5.3-11). ▲

Returning to $E\{\hat{\underline{\theta}}(k + \ell)\}$ in Eq. (5.3-16), we have

$$E\{\hat{\underline{\theta}}(k + \ell)\} = E\{\hat{\underline{\theta}}(k)\} - R(k)(\Omega + \Sigma_n)E\{\hat{\underline{\theta}}(k)\}$$

$$+ R(k)\underline{\underline{TB}} + R(k)\Omega\underline{\theta} \tag{5.3-24}$$

for $k \geq 0$. This expression is a vector first-order, forced, time-varying finite-difference equation from which we can solve for $E\{\hat{\underline{\theta}}(k + j)\}$. *We are interested only in the steady-state solution;* that is to say, we are interested only in the situation when

$$\lim_{k\to\infty} E\{\hat{\underline{\theta}}(k + \ell)\} = \lim_{k\to\infty} E\{\hat{\underline{\theta}}(k)\} \quad . \tag{5.3-25}$$

Taking the limit as $k \to \infty$ on both sides of Eq. (5.3-24), and using the condition for a steady state in Eq. (5.3-25), we determine that

$$R(k)(\Omega + \Sigma_n) \lim_{k\to\infty} E\{\hat{\underline{\theta}}(k)\} = R(k)[\underline{\underline{TB}} + \Omega\underline{\theta}] \quad ; \tag{5.3-26}$$

hence

$$\lim_{k\to\infty} E\{\hat{\underline{\theta}}(k)\} = (\Omega + \Sigma_n)^{-1}(\underline{\underline{TB}} + \Omega\underline{\theta}) \quad . \tag{5.3-27}$$

We have proven Theorem 5-1.

Theorem 5-1. *Estimates of* $\underline{\theta}$ *obtained from the stochastic-gradient identification algorithm*

$$\hat{\underline{\theta}}(k + \ell) = \hat{\underline{\theta}}(k) + R(k)\underline{r}(k)\tilde{z}(k) \tag{5.3-28}$$

are asymptotically biased, and

$$\lim_{k\to\infty} E\{\hat{\underline{\theta}}(k)\} = (\Omega + \Sigma_n)^{-1}(\underline{TB} + \Omega\underline{\theta}) \quad , \tag{5.3-29}$$

where $\underline{TB} = E\{\underline{x}(k)v(k)\}$. ▲

Observe that if $\Sigma_n = 0$ *and* $\underline{TB} = \underline{0}$ then

$$\lim_{k\to\infty} E\{\hat{\underline{\theta}}(k)\} = \Omega^{-1}\Omega\underline{\theta} = \underline{\theta} \tag{5.3-30}$$

and the estimate is asymptotically unbiased.

Corollary 5-1. *For Class-1 identification problems in which* $\underline{x}(k)$ *can be measured perfectly, estimates of* $\underline{\theta}$ *obtained from the stochastic-gradient identification algorithm in Eq. (5.3-28) are asymptotically unbiased.* ▲

This is an interesting result; for it demonstrates that under certain conditions unbiased estimates of $\underline{\theta}$ are possible regardless of how much measurement noise may be present on the output y(k) of the "Identification representation" block in the scalar stochastic equation-error identification system.

5.4 A STOCHASTIC-GRADIENT ALGORITHM: SECOND ATTEMPT

The troublesome terms in Eq. (5.3-24) are $-R(k)\Sigma_n E\{\hat{\underline{\theta}}(k)\}$ and $R(k)\underline{TB}$. If, for example, these terms were not present, then Eq. (5.3-24) would be

$$E\{\hat{\underline{\theta}}(k + \ell)\} = E\{\hat{\underline{\theta}}(k)\} - R(k)\Omega E\{\hat{\underline{\theta}}(k)\} + R(k)\Omega\underline{\theta} \tag{5.4-1}$$

and, now when we take the limit as $k \to \infty$ on both sides of this equation, and assume the existence of a steady state, we find that

$$R(k)\Omega \lim_{k\to\infty} E\{\hat{\underline{\theta}}(k)\} = R(k)\Omega\underline{\theta} \tag{5.4-2}$$

or

$$\lim_{k\to\infty} E\{\hat{\underline{\theta}}(k)\} = \underline{\theta} \tag{5.4-3}$$

as desired.

5.4.1 Unbiased Algorithm for Class-1 Identification Problem

Recall that, for Class-1 identification problems (see Section 5.2.1), $\underline{x}(k)$ and v(k) are uncorrelated; hence,

$$\underline{TB} = E\{\underline{x}(k)v(k)\} = E\{\underline{x}(k)\}E\{v(k)\} = \underline{0} \quad , \qquad (5.4\text{-}4)$$

since v(k) is zero mean.

In this case there is only one troublesome term in Eq. (5.3-24), and it is $-R(k)\Sigma_n E\{\hat{\underline{\theta}}(k)\}$. Clearly, in order to remove it we must modify our original stochastic-gradient identification algorithm by adding a term to it whose expected value exactly cancels $-R(k)\Sigma_n E\{\hat{\underline{\theta}}(k)\}$. We add $R(k)\Sigma_n \hat{\underline{\theta}}(k)$, since

$$E\{R(k)\Sigma_n \hat{\underline{\theta}}(k)\} = R(k)\Sigma_n E\{\hat{\underline{\theta}}(k)\} \quad . \qquad (5.4\text{-}5)$$

In this manner, we have arrived at the following (asymptotically) *unbiased stochastic-gradient identification algorithm for Class-1 identification problems:*

$$\hat{\underline{\theta}}(k + \ell) = [I + R(k)\Sigma_n]\hat{\underline{\theta}}(k) + R(k)\underline{r}(k)\tilde{z}(k) \qquad (5.4\text{-}6)$$

for $k \geq 0$.

5.4.2 Unbiased Algorithm for Class-2 Identification Problems

For Class-2 identification problems, $E\{\underline{x}(k)v(k)\} \neq \underline{0}$, and we must remove $-R(k)\Sigma_n E\{\hat{\underline{\theta}}(k)\}$ and $R(k)\underline{TB}$ from Eq. (5.3-24) in order to obtain the desired Eq. (5.4-1). The structure of these terms suggests the addition of

$$R(k)\Sigma_n \hat{\underline{\theta}}(k) - R(k)\underline{TB}$$

to the algorithm in Eq. (5.3-3), since

$$E\{R(k)\Sigma_n \hat{\underline{\theta}}(k) - R(k)\underline{TB}\} = R(k)\Sigma_n E\{\hat{\underline{\theta}}(k)\} - R(k)\underline{TB} \quad . \qquad (5.4\text{-}7)$$

Unfortunately, $\underline{TB}$ *is not physically realizable.* Examination of Fig. 5.2-1 reveals the reason for this dilemma. It is because $\underline{x}(k)$, the output from the "Dynamics" box, will always be a function of the unknown parameters $\underline{\theta}$. Thus, $\underline{TB}$ *must also be an explicit function of* $\underline{\theta}$; and since $\underline{\theta}$ is not known to us a priori, the best we can do is to replace the troublesome term $R(k)\underline{TB}$ by $R(k)\hat{\underline{TB}}(k)$, where $\hat{\underline{TB}}(k)$ denotes the kth estimate of $\underline{TB}$.

Assumption.† It is assumed that $\underline{TB}$ is a linear function of $\underline{\theta}$; i. e.,

$$\underline{TB} = E\{\underline{x}(k)v(k)\} = \underline{\lambda} + ME\{\underline{\theta}\} = \underline{\lambda} + M\underline{\theta} \quad , \qquad (5.4\text{-}8)$$

where $\underline{\lambda}$ and M are application dependent.

We shall compute $\underline{\lambda}$ and M for the problem of identifying the coefficients of a finite-difference equation in Section 5.6. ▲

The term $\underline{TB}$ is approximated by $\underline{\hat{TB}}(k)$, where

$$\underline{\hat{TB}}(k) = \underline{\lambda} + ME\{\underline{\hat{\theta}}(k)\} \quad . \qquad (5.4\text{-}9)$$

Let us now assume that the stochastic-gradient algorithm can be modified so that Eq. (5.3-24) contains two additional terms whose expected values are $-R(k)\underline{\hat{TB}}(k)$ and $R(k)\Sigma_n E\{\underline{\hat{\theta}}(k)\}$, in which case

$$E\{\underline{\hat{\theta}}(k + \ell)\} = E\{\underline{\hat{\theta}}(k)\} - R(k)\Omega E\{\underline{\hat{\theta}}(k)\}$$

$$+ R(k)\Omega\underline{\theta} + R(k)[\underline{TB} - \underline{\hat{TB}}(k)] \quad . \qquad (5.4\text{-}10)$$

Under the preceding assumption on $\underline{TB}$, and by Eq. (5.4-9), Eq. (5.4-10) becomes

$$E\{\underline{\hat{\theta}}(k + \ell)\} = E\{\underline{\hat{\theta}}(k)\} - R(k)\Omega E\{\underline{\hat{\theta}}(k)\} + R(k)\Omega\underline{\theta}$$

$$+ R(k)M\underline{\theta} - R(k)ME\{\underline{\hat{\theta}}(k)\}$$

$$= E\{\underline{\hat{\theta}}(k)\} - R(k)(\Omega + M)E\{\underline{\hat{\theta}}(k)\}$$

$$+ R(k)(\Omega + M)\underline{\theta} \quad . \qquad (5.4\text{-}11)$$

Taking the limit as $k \to \infty$ (assuming the existence of a steady state), we find that

$$R(k)(\Omega + M) \lim_{k\to\infty} E\{\underline{\hat{\theta}}(k)\} = R(k)(\Omega + M)\underline{\theta} \qquad (5.4\text{-}12)$$

or

$$\lim_{k\to\infty} E\{\underline{\hat{\theta}}(k)\} = \underline{\theta} \qquad (5.4\text{-}13)$$

as desired. This is a rather remarkable result, considering that $\underline{TB}$ has been approximated by $\underline{\hat{TB}}(k)$ given in Eq. (5.4-9).

---

† In practice, this assumption must be verified.

The structure of the terms $R(k)\Sigma_n E\{\hat{\underline{\theta}}(k)\}$ and $-R(k)\widehat{\underline{TB}}(k)$, which have been added to Eq. (5.3-24), suggests we modify the stochastic-gradient identification algorithm in Eq. (5.3-3) by adding the terms $R(k)\Sigma_{n}\hat{\underline{\theta}}(k)$ and $-R(k)[\underline{\lambda} + M\hat{\underline{\theta}}(k)]$ for

$$E\{R(k)\Sigma_{n}\hat{\underline{\theta}}(k) - R(k)[\underline{\lambda} + M\hat{\underline{\theta}}(k)]\} = R(k)\Sigma_n E\{\hat{\underline{\theta}}(k)\} - R(k)\widehat{\underline{TB}}(k) \tag{5.4-14}$$

as required.

In this manner, we have arrived at the following (asymptotically) *unbiased stochastic-gradient identification algorithm for Class-2 identification problems:*

$$\hat{\underline{\theta}}(k + \ell) = [I + R(k)\Sigma_n - R(k)M]\hat{\underline{\theta}}(k) + R(k)\underline{r}(k)\tilde{z}(k) - R(k)\underline{\lambda} \tag{5.4-15}$$

for $k \geq 0$, in which $\underline{\lambda}$ and M must be computed ahead of time, and are application dependent.

5.4.3 Choice of the Spacing Parameter, ℓ

The value of ℓ is chosen so that major assumption (i) is satisfied [see Eq. (5.3-7)]. From Eq. (5.4-15), observe that[†]

$$\begin{aligned}\hat{\underline{\theta}}(k) &= [I + R(k - \ell)(\Sigma_n - M)]\hat{\underline{\theta}}(k - \ell) \\ &\quad + R(k - \ell)\underline{r}(k - \ell)\tilde{z}(k - \ell) - R(k - \ell)\underline{\lambda} \\ &= g[\underline{x}(k - \ell)] \quad ;\end{aligned} \tag{5.4-16}$$

hence,

$$E\{\underline{x}(k)\underline{x}'(k)\,|\,\hat{\underline{\theta}}(k)\} = E\{\underline{x}(k)\underline{x}'(k)\,|\,\underline{x}(k - \ell)\} \quad . \tag{5.4-17}$$

We choose ℓ so that

$$\underline{x}(k) \text{ and } \underline{x}(k - \ell) \text{ are statistically independent} \tag{5.4-18}$$

which means, of course, that

$$E\{\underline{x}(k)\underline{x}'(k)\,|\,\underline{x}(k - \ell)\} = E\{\underline{x}(k)\underline{x}'(k)\} \triangleq \Omega \quad . \tag{5.4-19}$$

[†] Both $\underline{r}(k - \ell)$ and $\tilde{z}(k - \ell)$ are viewed here as functions of $\underline{x}(k - \ell)$.

Example 2 (Continued). We determine ℓ for the finite-difference, equation-coefficient, identification problem that was described in Section 5.1. Using Eq. (5.1-34), we see that ℓ must be chosen so that

$$\underline{x}(k) = [y(k + n - 1), y(k + n - 2), \ldots, y(k)]' \quad (5.4\text{-}20)$$

and

$$\underline{x}(k - \ell) = [y(k + n - 1 - \ell), y(k + n - 2 - \ell), \ldots, y(k - \ell)]' \quad (5.4\text{-}21)$$

are statistically independent. Clearly, $\underline{x}(k)$ and $\underline{x}(k - \ell)$ are independent if they do not contain any common elements; thus, the first element in $\underline{x}(k - \ell)$ must occur *at least* one time unit earlier than the last element in $\underline{x}(k)$, which means that

$$k + n - 1 - \ell \leq k - 1 \quad . \quad (5.4\text{-}22)$$

Hence

$$\ell \geq n \quad . \quad (5.4\text{-}23)$$

It is customary to choose ℓ at its smallest admissible value, so that the identification can be completed in the shortest possible time; thus, we would normally choose

$$\ell = n \quad . \quad (5.4\text{-}24)$$

Since ℓ is a spacing parameter, the algorithm for estimating $\underline{\theta}$ must be updated for k = 0, n, 2n, ... ▲

Example 1 (Continued). The reader should show that $\ell \geq 1$ for the aerodynamic-parameter identification problem described in Section 5.1 (Problem 5-4).

5.5 CONVERGENCE

5.5.1 Introduction

Recall that in our study of bias (Section 5.3.2), we assumed the existence of a steady-state value of $\hat{\underline{\theta}}(k)$. This assumption is examined here by studying conditions under which convergence of $\hat{\underline{\theta}}(k)$ to $\underline{\theta}$ occurs. For a stochastic system (we show below that the identification-error system is stochastic), distinctly different types of convergence are possible (see *5-2* for a discussion and comparison of different types of stochastic convergence). Our attention will be directed for the most part to mean-square convergence of $\underline{\theta}(k)$ to $\underline{\theta}$.

The estimate $\hat{\underline{\theta}}(k)$ *converges in mean square* to $\underline{\theta}$, if

$$\lim_{k\to\infty} E\{||\hat{\underline{\theta}}(k) - \underline{\theta}||^2\} \to 0 \quad . \tag{5.5-1}$$

It is well known that if $\hat{\underline{\theta}}(k)$ converges in mean square to $\underline{\theta}$, then it also *converges in probability* to $\underline{\theta}$; that is to say,

$$\lim_{k\to\infty} [\text{Prob}|\hat{\theta}_i(k) - \theta_i| \geq \varepsilon] \to 0 \quad . \tag{5.5-2}$$

Additionally, if $\hat{\underline{\theta}}(k)$ converges in probability to $\underline{\theta}$, as $k \to \infty$, $\hat{\underline{\theta}}(k)$ is called a *consistent estimate* of $\underline{\theta}$. A consistent estimate is always asymptotically unbiased; however, an unbiased estimate need not be consistent (*5-3* and *5-4*, and discussions at end of Sections 2.5.2 and 2.11.4) .

We have purposely designed our identification algorithm so that estimates of $\underline{\theta}$ are asymptotically unbiased. Obviously, from the line of reasoning just presented, this in no way means that $\hat{\underline{\theta}}(k)$ converges in mean square to $\underline{\theta}$. We shall determine conditions on R(k) whereby convergence occurs. Having removed the biases ahead of time, we can be certain that

$$\lim_{k\to\infty} E\{||\hat{\underline{\theta}}(k) - \underline{\theta}||^2\} \to C \neq 0$$

will not occur.

Our approach will be, (1) to obtain the stochastic identification-error system for $\tilde{\underline{\theta}}(k)$; (2) to obtain the stochastic identification-error norm-squared system for $||\tilde{\underline{\theta}}(k)||^2$; (3) to obtain $E\{||\tilde{\underline{\theta}}(k)||^2\}$ and bound it from above; and (4) to establish conditions under which mean-square convergence occurs [after Holmes (*5-5*)].

Motivation for bounding $E\{||\tilde{\underline{\theta}}(k)||^2\}$ from above derives from the following very interesting theorem, first proved by Venter (*5-6*).

Theorem 5-2 (Venter's Theorem). *If*

$$\beta(k + 1) \leq [1 - \alpha(k)]\beta(k) + w(k) \tag{5.5-3}$$

and

(i) $\sum_{k=0}^{\infty} \alpha(k) \to \infty$,

(ii) $\alpha(k) \to 0$ *as* $k \to \infty$,

(iii) $\alpha(k)$, $w(k)$, *and* $\beta(k) \geq 0$, *and*

(iv) $\sum_{k=0}^{\infty} w(k) < \infty$,

then

$$\beta(k) \to 0 \quad \text{as } k \to \infty \quad . \tag{5.5-4}$$

A proof of this important theorem is given in Appendix F. ▲

In our analysis $E\{||\tilde{\underline{\theta}}(k)||^2\}$ will play the role of $\beta(k)$; $\alpha(k)$ and $w(k)$ will be related to $R(k)$.

5.5.2 Stochastic Identification-Error System

We shall demonstrate that the identification-error system for $\tilde{\underline{\theta}}(k)$ is given by the expression†

$$\begin{aligned}\tilde{\underline{\theta}}(k + \ell) = {} & [I - R(k)\underline{r}(k)\underline{r}'(k) + R(k)\Sigma_n - R(k)M]\tilde{\underline{\theta}}(k) \\ & + R(k)[-\underline{r}(k)v(k) - \underline{r}(k)\underline{x}'(k)\underline{\theta} - \Sigma_n\underline{\theta} \\ & + \underline{r}(k)\underline{r}'(k)\underline{\theta}] + R(k)E\{\underline{x}(k)v(k)\}\end{aligned} \tag{5.5-5}$$

for $k \geq 0$. Observe that this is a vector, forced, linear, time-varying finite-difference equation, and that the forcing function is stochastic; hence, Eq. (5.5-5) is referred to herein as the *stochastic identification-error system*. It should be compared with the (deterministic) identification-error system in Eq. (4.3-2). The system in

---

† Equation (5.5-5) includes both Class-1 and Class-2 identification problems.

Eq. (5.5-5) is vastly more complicated than its deterministic counterpart; however, we shall proceed undaunted by complexity. Remember, the stochastic identification problem we have formulated in this Chapter is *a fortiori* more difficult than the deterministic identification problem that was formulated in Chapter 4.

Equation (5.5-5) is obtained from Eqs. (5.4-15), (5.1-15), (5.1-13), and (5.1-12) in the following manner:

$$\begin{aligned}
\hat{\underline{\theta}}(k + \ell) &= [I + R(k)\Sigma_n - R(k)M]\hat{\underline{\theta}}(k) \\
&\quad + R(k)\underline{r}(k)[z(k) - \hat{z}(k)] - R(k)\underline{\lambda} \\
&= [I + R(k)\Sigma_n - R(k)M]\hat{\underline{\theta}}(k) \\
&\quad + R(k)\underline{r}(k)[\underline{x}'(k)\underline{\theta} + v(k) - \underline{r}'(k)\hat{\underline{\theta}}(k)] - R(k)\underline{\lambda} \\
&= [I - R(k)\underline{r}(k)\underline{r}'(k) + R(k)\Sigma_n - R(k)M]\hat{\underline{\theta}}(k) \\
&\quad + R(k)\underline{r}(k)[\underline{x}'(k)\underline{\theta} + v(k)] - R(k)\underline{\lambda} \quad . \qquad (5.5\text{-}6)
\end{aligned}$$

Adding and subtracting $[I - R(k)\underline{r}(k)\underline{r}'(k) + R(k)\Sigma_n - R(k)M]\underline{\theta}$ to the right-hand side, and $\underline{\theta}$ to the left-hand side of this equation, gives

$$\begin{aligned}
-\tilde{\underline{\theta}}(k + \ell) + \underline{\theta} &= -[I - R(k)\underline{r}(k)\underline{r}'(k) + R(k)\Sigma_n - R(k)M]\tilde{\underline{\theta}}(k) \\
&\quad + \underline{\theta} + R(k)[-\underline{r}(k)\underline{r}'(k) + \Sigma_n - M]\underline{\theta} \\
&\quad + R(k)\underline{r}(k)[\underline{x}'(k)\underline{\theta} + v(k)] - R(k)\underline{\lambda}
\end{aligned}$$

$$\begin{aligned}
-\tilde{\underline{\theta}}(k + \ell) &= -[I - R(k)\underline{r}(k)\underline{r}'(k) + R(k)\Sigma_n - R(k)M]\tilde{\underline{\theta}}(k) \\
&\quad + R(k)[-\underline{r}(k)\underline{r}'(k) + \Sigma_n]\underline{\theta} \\
&\quad + R(k)\underline{r}(k)[\underline{x}'(k)\underline{\theta} + v(k)] - R(k)E\{\underline{x}(k)v(k)\} \quad . \\
&\qquad (5.5\text{-}7)
\end{aligned}$$

Hence

$$\underline{\tilde{\theta}}(k + \ell) = [I - R(k)\underline{r}(k)\underline{r}'(k) + R(k)\Sigma_n - R(k)M]\underline{\tilde{\theta}}(k)$$

$$+ R(k)[\underline{r}(k)\underline{r}'(k)\underline{\theta} - \Sigma_n\underline{\theta} - \underline{r}(k)\underline{x}'(k)\underline{\theta}$$

$$- \underline{r}(k)v(k)] + R(k)E\{\underline{x}(k)v(k)\} \tag{5.5-8}$$

which is precisely Eq. (5.5-5).

5.5.3 Stochastic Identification-Error Norm Squared System

For the rest of Section 5.5, our developments shall be for Class-1 identification problems, in which case we set M and $E\{\underline{x}(k)v(k)\}$ equal to zero in Eq. (5.5-8). This will simplify our analyses somewhat, since there will not be as many terms in $||\underline{\tilde{\theta}}(k + \ell)||^2$ as there would be in the general case when M and $E\{\underline{x}(k)v(k)\}$ are not equal to zero. Extensions of all results obtained below to the more general situation are straightforward, and are left to the reader (Problem 5-6).

The term $||\underline{\tilde{\theta}}(k + \ell)||^2$, found directly from Eq. (5.5-8) by forming $\underline{\tilde{\theta}}'(k + \ell)\underline{\tilde{\theta}}(k + \ell)$, is

$$||\underline{\tilde{\theta}}(k + \ell)||^2 = ||\underline{\tilde{\theta}}(k)||^2 + T2 + T3 + T4 + 2(T5 - T6)$$

$$+ 2T7 - 2T8 + 2T9 - 2T10 \quad , \tag{5.5-9}$$

where

$$T2 = ||R(k)\Sigma_n\underline{\tilde{\theta}}(k)||^2 \quad , \tag{5.5-10}$$

$$T3 = ||R(k)\underline{r}(k)\underline{r}'(k)\underline{\tilde{\theta}}(k)||^2 \quad , \tag{5.5-11}$$

$$T4 = ||R(k)\underline{f}(k)||^2 \quad , \tag{5.5-12}$$

$$\underline{f}(k) = [\underline{r}(k)\underline{r}'(k) - \Sigma_n - \underline{r}(k)\underline{x}'(k)]\underline{\theta} - \underline{r}(k)v(k) \quad , \tag{5.5-13}$$

$$T5 = \underline{\tilde{\theta}}'(k)R(k)\Sigma_n\underline{\tilde{\theta}}(k) \quad , \tag{5.5-14}$$

$$T6 = \underline{\tilde{\theta}}'(k)R(k)\underline{r}(k)\underline{r}'(k)\underline{\tilde{\theta}}(k) \quad , \tag{5.5-15}$$

$$T7 = \underline{\tilde{\theta}}'(k)R(k)\underline{f}(k) \quad , \tag{5.5-16}$$

$$T8 = \underline{\tilde{\theta}}'(k)\Sigma_n R^2(k)\underline{r}(k)\underline{r}'(k)\underline{\tilde{\theta}}(k) \quad , \tag{5.5-17}$$

$$T9 = \underline{\tilde{\theta}}'(k)\Sigma_n R^2(k)\underline{f}(k) \quad , \tag{5.5-18}$$

and

$$T10 = \underline{\tilde{\theta}}'(k)\underline{r}(k)\underline{r}'(k)R^2(k)\underline{f}(k) \quad . \tag{5.5-19}$$

In the derivation of these results, we have assumed that R(k) *is a symmetric matrix* (as was assumed in Chapter 4).

5.5.4 Upper Bound on $E\{||\underline{\tilde{\theta}}(k + \ell)||^2\}$

For notational convenience, we write $\overline{(\cdot)}$ to mean $E\{(\cdot)\}$, U.B.(·) to mean the upper bound of (·), and L.B.(·) to mean the lower bound of (·).

Numerous upper bounds can be found for $\overline{||\underline{\tilde{\theta}}(k + \ell)||^2}$; however, the following bound appears to be most useful:

$$\begin{aligned}\overline{||\underline{\tilde{\theta}}(k + \ell)||^2} \le\ & \overline{||\underline{\tilde{\theta}}(k)||^2} + \text{U.B.}(\overline{T2}) + \text{U.B.}(\overline{T3}) \\ & + \text{U.B.}(\overline{T4}) - 2\ \text{L.B.}(|\overline{T5} - \overline{T6}|) \\ & + 2\ \text{U.B.}(\overline{T7}) + 2\ \text{U.B.}(\overline{T8}) \\ & + 2\ \text{U.B.}(\overline{T9}) + 2\ \text{U.B.}(\overline{T10}) \quad .\end{aligned} \tag{5.5-20}$$ †

The details associated with determining all of these bounds are given in Appendix G. It is shown there that [see Eq. (G-49)]

$$\begin{aligned}\overline{||\underline{\tilde{\theta}}(k + \ell)||^2} \le\ & [1 + c_3||R(k)||^2 - 2\lambda_2]\overline{||\underline{\tilde{\theta}}(k)||^2} \\ & + c_1^*||R(k)||^2\end{aligned} \tag{5.5-21}$$

where c_3, c_1^*, and λ_2 are defined in Eqs. (G-51), (G-50), and (G-34). In order to obtain L.B.($|\overline{T5} - \overline{T6}|$), the *additional assumption*, that

$$R(k)\Omega \text{ is positive definite} \tag{5.5-22}$$

was made.

---

† In Appendix G, it is shown that a minus sign is factorable from $\overline{T5} - \overline{T6}$.

To proceed further, we must *assume a specific structure for* R(k). Motivated by the choice made in Section 4.4.4, we shall choose

$$R(k) = \rho(k)\ \mathrm{diag}[h_1(k), h_2(k), \ldots, h_n(k)] \tag{5.5-23}$$

where

$$0 < h_L \leq h_i(k) \leq h_U < \infty \tag{5.5-24}$$

for i = 1, 2, ..., n, and all $k \geq 0$, and

$$\rho(k) > 0 \quad . \tag{5.5-25}$$

We leave it as an exercise for the reader to demonstrate that for R(k) chosen in this manner $R(k)\Omega$ is positive definite (Problem 5-8).

From matrix theory, it is known that one norm of a matrix is its largest eigenvalue (*5-7 - 5-9*); hence,

$$\begin{aligned} ||R(k)|| &= \text{max eigenvalue } \{\rho(k)\ \mathrm{diag}[h_1(k), \ldots, h_n(k)]\} \\ &= c_4\rho(k) \end{aligned} \tag{5.5-26}$$

where

$$c_4 = \max_{i,k} \{h_i(k)\} \leq h_U \quad . \tag{5.5-27}$$

Now λ_2 has been defined in Eq. (G-34) to be the smallest eigenvalue of $R(k)\Omega$; thus,

$$\lambda_2 = \rho(k)\lambda_2^* \tag{5.5-28}$$

where†

$$\lambda_2^* = \text{smallest eigenvalue } \{\mathrm{diag}[h_1(k), \ldots, h_n(k)]\Omega\} \quad . \tag{5.5-29}$$

Substituting Eqs. (5.5-26) and (5.5-28) into Eq. (5.5-21), we find

$$\begin{aligned} \overline{||\underline{\tilde{\theta}}(k + \ell)||^2} \leq{}& [1 + c_3c_4^2\rho^2(k) - 2\lambda_2^*\rho(k)]\overline{||\underline{\tilde{\theta}}(k)||^2} \\ &+ c_1^*c_4^2\rho^2(k) \end{aligned} \tag{5.5-30}$$

or, finally, that

---

† The quantity $\lambda_2^* > 0$, since the eigenvalues of a positive definite matrix are positive.

$$\overline{||\tilde{\underline{\theta}}(k + \ell)||^2} \leq \{1 - 2\rho(k)[\lambda_2^* - \frac{1}{2} c_5\rho(k)]\}\overline{||\tilde{\underline{\theta}}(k)||^2} + c_6\rho^2(k) \quad , \tag{5.5-31}$$

where

$$c_5 = c_3c_4^2 \tag{5.5-32}$$

and

$$c_6 = c_1^*c_4^2 \quad . \tag{5.5-33}$$

5.5.5 Conditions for Mean-Square Convergence

We shall show that

$$\lim_{k\to\infty} \overline{||\tilde{\underline{\theta}}(k)||^2} \to 0 \tag{5.5-34}$$

by applying Venter's theorem to Eq. (5.5-31). We begin by establishing the following associations between the various quantities in Eqs. (5.5-31) and (5.5-3):

$$w(k) = c_6\rho^2(k) \tag{5.5-35}$$

and

$$\alpha(k) = c_7\rho(k) + O[\rho^2(k)] \tag{5.5-36}$$

where

$$c_7 = 2\lambda_2^* \tag{5.5-37}$$

and $O[\rho^2(k)]$ is used to mean that there is an extra term in $\alpha(k)$ which, as we shall demonstrate below, goes to zero much faster than $\rho(k)$ does; hence, since we are interested in the limiting behavior of $\overline{||\tilde{\underline{\theta}}(k)||^2}$, the $O[\rho^2(k)]$ term can be neglected.

From Eqs. (5.5-35) and (5.5-36) and conditions (i) through (iv) in the statement of Venter's theorem, we see that the sequence $\rho(k)$ $(k \geq 0)$ must be chosen such that

$$\text{(i)} \quad \sum_{k=0}^{\infty} \rho(k) \to \infty \tag{5.5-38a}$$

$$\text{(ii)} \quad \rho(k) \to 0 \quad \text{as} \quad k \to \infty \tag{5.5-38b}$$

(iii)† $\rho(k) > 0$ (5.5-38c)

(iv) $\sum_{k=0}^{\infty} \rho^2(k) < \infty$. (5.5-38d)

Recall, from the theory of infinite series (*5-24* and *5-25*), that the series

$$\frac{1}{1^p} + \frac{1}{2^p} + \cdots + \frac{1}{k^p} + \cdots$$

is convergent if $p > 1$ and divergent if $p \leq 1$. When $p = 1$ we obtain the well-known *harmonic series*

$$1 + \frac{1}{2} + \cdots + \frac{1}{k} + \cdots .$$

Recall, also, that the series

$$\frac{1}{1^{2p}} + \frac{1}{2^{2p}} + \cdots + \frac{1}{k^{2p}} + \cdots$$

is convergent if $p > 1/2$. Putting these facts together, we conclude that one choice for $\rho(k)$, is

$$\rho(k) = \frac{1}{k^p} \quad \text{where} \quad \frac{1}{2} < p \leq 1 \quad . \tag{5.5-39}$$

For this choice $k > 0$. Other choices are examined in Problem 5-9.

For $\rho(k)$ in Eq. (5.5-39),

$$O[\rho^2(k)] = \frac{1}{k^{2p}} \tag{5.5-40}$$

and, as we stated above in our discussion of Eq. (5.5-36), $1/k^{2p}$ does go to zero much faster than $1/k^p$; thus, in the limit as $k \to \infty$

$$\alpha(k) \to c_7\rho(k) \tag{5.5-41}$$

as we assumed.

Before summarizing our results, let us provide some physical meaning to the conditions on $\rho(k)$ in Eq. (5.5-38) (*5-10*). The $\rho(k)$

---

† We exclude the possibility of zero $\rho(k)$, since $\rho(k) = 0$ would make it difficult to satisfy condition (i).

can be interpreted as the correction effort at t_k. Condition (i) in Eq. (5.5-38) tells us that an unlimited amount of correction effort is available, if necessary. If $\rho(k)$ were a convergent series then a limited amount of correction effort would be available, and we would run the risk of stopping short of the goal - driving $||\tilde{\underline{\theta}}(k)||^2$ to zero. Randomness introduces the possibility of moving in the wrong direction; hence, to compensate for such a possibility, we must have an unlimited correction effort.

5.5.6 Summary

We summarize all of the preceding results in the following theorems.

Theorem 5-3 (Class-1 Algorithm). *The sequence* $\hat{\underline{\theta}}(k + \ell)$, *defined by*

$$\hat{\underline{\theta}}(k + \ell) = [I + R(k)\Sigma_n]\hat{\underline{\theta}}(k) + R(k)\underline{r}(k)\tilde{z}(k) \qquad (5.5\text{-}42)$$

for $k \geq 0$, *converges in mean square (and, subsequently, in probability) to* $\underline{\theta}$, *if the following occur*

(i) $R(k) = \rho(k)\,\mathrm{diag}[h_1(k), h_2(k), \ldots, h_n(k)]$, *where:* $0 < h_L \leq h_i(k) \leq h_U < \infty$, *for* $i = 1, 2, \ldots, n$ *and all* $k \geq 0$; $\rho(k) > 0$; $\sum_{k=0}^{\infty} \rho(k) \to \infty$; *and* $\sum_{k=0}^{\infty} \rho^2(k) < \infty$.

(ii) $E\{\underline{x}(k)\underline{x}'(k)|\hat{\underline{\theta}}(k)\} = \Omega$, *where* Ω *is constant and positive definite, but does not have to be known a priori.*

(iii) $\underline{x}(k)$ *and* $\underline{n}(k)$, *and* $\underline{n}(k)$ *and* $v(k)$ *are statistically independent.*

(iv) $E\{\underline{n}(k)\} = \underline{0}$, $E\{\underline{n}(k)\underline{n}'(k)\} = \Sigma_n$ *where* Σ_n *is known a priori, and* $E\{v(k)\} = 0$.

(v)† $E\{v^2(k)\} < \infty$, $E\{||\underline{r}(k)||^2\underline{r}(k)\underline{r}'(k)|\hat{\underline{\theta}}(k)\} < \infty$, *and* $E\{||\underline{r}(k)||^2 v^2(k)|\hat{\underline{\theta}}(k)\} < \infty$.

(vi) $E\{\underline{x}(k)v(k)\} = \underline{0}$. ▲

---

† The bounds in this condition are required to bound $||\underline{f}(k)||^2$ in Appendix G, Eq. (G-23).

Obviously, because of condition (vi), this theorem is limited to Class-1 identification problems. We leave it to the reader to show that no additional conditions are required for mean-square convergence of $\hat{\underline{\theta}}(k)$ to $\underline{\theta}$ when $\hat{\underline{\theta}}(k)$ is obtained by means of the algorithm in Eq. (5.4-15) and $E\{\underline{x}(k)v(k)\} \neq \underline{0}$ (Problem 5-6).

Theorem 5-4 (Class-2 Algorithm). *The sequence* $\hat{\underline{\theta}}(k + \ell)$, *defined by*

$$\hat{\underline{\theta}}(k + \ell) = [I + R(k)\Sigma_n - R(k)M]\hat{\underline{\theta}}(k) + R(k)\underline{r}(k)\tilde{z}(k) - R(k)\underline{\lambda} \tag{5.5-43}$$

for $k \geq 0$, *converges in mean square (and, subsequently, in probability) to* $\underline{\theta}$, *if conditions (i)-(v) of Theorem 5-3 are satisfied.* ▲

Obviously, Theorem 5-3 is a special case of Theorem 5-4; however, because we are leaving the few additional steps required to prove Theorem 5-4 to the reader, we feel justified in distinguishing between Class-1 and Class-2 identification problems and their associated estimation algorithms in two distinct theorems.

5.5.7 Convergence with Probability One

A sequence of random vectors $\tilde{\underline{\theta}}(0)$, $\tilde{\underline{\theta}}(1)$, ..., $\tilde{\underline{\theta}}(k)$, ... is said to converge to $\underline{0}$ *with probability one* if

$$\text{Prob}[\lim_{k\to\infty} \tilde{\underline{\theta}}(k) = \underline{0}] \to 1 \quad . \tag{5.5-44}$$

Convergence with probability one neither implies nor is implied by convergence in mean square (*5-2* and *5-11*); however, *if* $\tilde{\underline{\theta}}(k)$ *converges in mean square to* $\underline{0}$ *in such a way that*

$$\sum_{k=0}^{\infty} E\{||\tilde{\underline{\theta}}(k)||^2\} < \infty \tag{5.5-45}$$

then it follows that $\tilde{\underline{\theta}}(k)$ *converges to* $\underline{0}$ *with probability one* [*5-11, Theorem 1A, page 416*].

Theorem 5-5. *The sequence* $\hat{\underline{\theta}}(k + \ell)$ *defined by Eq. (5.5-43) for* $k \geq 0$ *converges to* $\underline{\theta}$ *with probability one.*

The proof of this theorem is left as an exercise (Problem 5-7). ▲

The unbiased stochastic-gradient identification algorithms we have obtained in this chapter fall within the broad framework of stochastic approximation.

Stochastic approximation is concerned with schemes (algorithms) converging to some sought value when, owing to the stochastic nature of the problem, the observations involve errors. The interesting schemes are those which are self correcting, that is, in which a mistake always tends to be wiped out in the limit, and in which convergence to the desired value is of some specified nature, such as mean-square or with probability one convergence.

Stochastic approximation is not limited to linear models. It was invented by Robbins and Monro (*5-12*) in connection with the problem of finding the unique root of a noisy function in one variable; applied by Kiefer and Wolfowitz (*5-13*) to the problem of finding the unique extremum of a noisy function in one variable; extended by Blum (*5-14*) to multivariate situations; and generalized by Dvoretzky (*5-15*) to nonlinear algorithms that have the structure

New estimate = Nonlinear transformation on some or all preceding estimates + Noise

It has been applied to many diverse applications (*5-16* - *5-18*), including the parameter identification problem (*5-19* - *5-22*); however, its generality is not really needed for the linear identification problem, and in this book we are, for the most part, limiting our attention to linear identification problems. The results we have obtained in Theorems 5-3, 5-4, and 5-5 can also be obtained from an application of Dvoretzky's generalized stochastic approximation theory for multivariate functions. This approach is taken by Saridis and Stein, for example (*5-19*).

5.6 IDENTIFICATION OF COEFFICIENTS IN FINITE-DIFFERENCE EQUATIONS

We show, by means of examples, how the results of this chapter can be applied to identify the coefficients in a finite-difference equation. Our interest is directed to only those finite-difference equation applications that are classified as Class-2 identification problems. We shall show that for these specific applications,

$$E\{\underline{x}(k)v(k)\} = \underline{\lambda} + M\underline{\theta} \quad ; \tag{5.6-1}$$

and we shall obtain explicit expressions for $\underline{\lambda}$ and M.

Example 3. *Abstract Digital System.* Consider the following third-order digital system:

$$\begin{pmatrix} x_1(k+1) \\ x_2(k+1) \\ x_3(k+1) \end{pmatrix} = \begin{pmatrix} 0 & 1 & 0 \\ 0 & 0 & 1 \\ -a_3 & -a_2 & -a_1 \end{pmatrix} \begin{pmatrix} x_1(k) \\ x_2(k) \\ x_3(k) \end{pmatrix} + \begin{pmatrix} b_1 \\ b_2 \\ b_3 \end{pmatrix} m(k) + \begin{pmatrix} d_1 \\ d_2 \\ d_3 \end{pmatrix} \omega(k) \tag{5.6-2}$$

where

$$y(k) = x_1(k) \quad , \tag{5.6-3}$$

$$y_m(k) = y(k) + n_y(k) \quad , \tag{5.6-4}$$

$$m_m(k) = m(k) + n_m(k) \quad , \tag{5.6-5}$$

and $\omega(k)$, $n_y(k)$, and $n_m(k)$ are zero-mean independent discrete white processes for which

$$E\{\omega^2(k)\} = q \quad , \tag{5.6-6}$$

$$E\{n_y^2(k)\} = \sigma_y^2 \quad , \tag{5.6-7}$$

and

$$E\{n_m^2(k)\} = \sigma_m^2 \quad . \tag{5.6-8}$$

The reader is referred to Appendix A, where it is proven that the equivalent finite-difference equation representation of Eq. (5.6-2) is

$$y(k + 3) + a_1 y(k + 2) + a_2 y(k + 1) + a_3 y(k) = b_3^o m(k + 2) + b_2^o m(k + 1) + b_1^o m(k) + d_3^o \omega(k + 2) + d_2^o \omega(k + 1) + d_1^o \omega(k) \tag{5.6-9}$$

where

$$\underline{b}^o = T\underline{b} \quad , \tag{5.6-10}$$

$$\underline{d}^o = T\underline{d} \quad , \tag{5.6-11}$$

and

$$T = \begin{pmatrix} 1 & 0 & 0 \\ a_1 & 1 & 0 \\ a_2 & a_1 & 1 \end{pmatrix} \quad . \tag{5.6-12}$$

The vector $\underline{d} = (d_1, d_2, d_3)'$ is assumed known. The coefficients a_1, a_2, a_3, b_1, b_2, and b_3 are identified by: (1) identifying a_1, a_2, a_3, b_1^o, b_2^o, and b_3^o, using the finite-difference equation representation; and (2) obtaining $\hat{b}_1$, $\hat{b}_2$, and $\hat{b}_3$ from Eq. (5.6-10),

$$\hat{\underline{b}} = (\hat{T})^{-1}\hat{\underline{b}}^o \tag{5.6-13}$$

where

$$\hat{T} = \begin{pmatrix} 1 & 0 & 0 \\ \hat{a}_1 & 1 & 0 \\ \hat{a}_2 & \hat{a}_1 & 1 \end{pmatrix} \quad . \tag{5.6-14}$$

To begin, we formulate the scalar stochastic measurement equation from which the a and b^o parameters will be estimated. It is obtained from Eqs. (5.6-4) and (5.6-9), as

$$y_m(k + 3) = -a_1 y(k + 2) - a_2 y(k + 1) - a_3 y(k) + b_3^o m(k + 2) + b_2^o m(k + 1) + b_1^o m(k) + d_3^o \omega(k + 2) + d_2^o \omega(k + 1) + d_1^o \omega(k) + n_y(k + 3) \quad . \tag{5.6-15}$$

The model of $y_m(k + 3)$ is[†]

$$\hat{y}_m(k + 3) = -\hat{a}_1(k + 3)y_m(k + 2) - \hat{a}_2(k + 3)y_m(k + 1) - \hat{a}_3(k + 3)y_m(k) + \hat{b}_3^o(k + 3)m_m(k + 2) + \hat{b}_2^o(k + 3)m_m(k + 1) + \hat{b}_1^o(k + 3)m_m(k) \quad . \quad (5.6\text{-}16)$$

By defining

$$\underline{\theta} = (-a_1, -a_2, -a_3, b_3^o, b_2^o, b_1^o)' \quad , \quad (5.6\text{-}17)$$

$$\underline{x}(k) = [y(k+2), y(k+1), y(k), m(k+2), m(k+1), m(k)]' \quad , \quad (5.6\text{-}18)$$

$$\underline{r}(k) = [y_m(k+2), y_m(k+1), y_m(k), m_m(k+2), m_m(k+1), m_m(k)]' \quad , \quad (5.6\text{-}19)$$

$$\underline{n}(k) = [n_y(k+2), n_y(k+1), n_y(k), n_m(k+2), n_m(k+1), n_m(k)]' \quad , \quad (5.6\text{-}20)$$

$$v(k) = d_3^o\omega(k+2) + d_2^o\omega(k+1) + d_1^o\omega(k) + n_y(k+3) \quad , \quad (5.6\text{-}21)$$

and

$$z(k) = y_m(k+3) \quad , \quad (5.6\text{-}22)$$

this problem is cast into the stochastic scalar equation-error form.[‡]

Along the way, we have clearly defined $\underline{x}(k)$ and $v(k)$, so that we may now proceed to the computation of $E\{\underline{x}(k)v(k)\}$.

Clearly, $m(k + 2)$, $m(k + 1)$, and $m(k)$ are independent of $v(k)$; thus,

$$E\{m(k + 2)v(k)\} = E\{m(k + 1)v(k)\} = E\{m(k)v(k)\} = 0 \quad . \quad (5.6\text{-}23)$$

[†] See discussion following Eq. (1.3-32) for reason why arguments of $\hat{a}_i$ and $\hat{b}_i^o$ are k + 3 rather than k.

[‡] It is important to remember that the symbols on the left-hand sides of Eqs. (5.6-18)-(5.6-22) are generic. The discrepancies between the time arguments on the two sides of these expressions have been explained in Section 1.3.3.

Vector $\underline{x}(k)$, the solution to the state equation in Eq. (5.6-2), was derived in Section 1.3.3. From Eqs. (5.6-3), (5.6-2), and (1.3-58), we have that

$$y(k) = (1, 0, 0)\underline{x}(k) \tag{5.6-24}$$

and

$$\underline{x}(k) = \Phi^k \underline{x}(0) + \sum_{i=1}^{k} \Phi^{k-i}[\underline{b}m(i - 1) + \underline{d}\omega(i - 1)] , \tag{5.6-25}$$

where

$$\Phi = \begin{pmatrix} 0 & 1 & 0 \\ 0 & 0 & 1 \\ -a_3 & -a_2 & -a_1 \end{pmatrix} . \tag{5.6-26}$$

We proceed first with the computation of $E\{y(k)v(k)\}$.

$$E\{y(k)v(k)\} = (1, 0, 0)E\{\Phi^k \underline{x}(0)v(k)\} + (1, 0, 0) E\{ \sum_{i=1}^{k} \Phi^{k-i}[\underline{b}m(i - 1) + \underline{d}\omega(i - 1)]v(k)\} . \tag{5.6-27}$$

Because $\underline{x}(0)$ and all $m(j)$ are independent of $v(k)$, and $E\{v(k)\} = 0$, this expression reduces to

$$E\{y(k)v(k)\} = (1, 0, 0)E\{ \sum_{i=1}^{k} \Phi^{k-i}\underline{d}\omega(i - 1)[d_3^o\omega(k + 2) + d_2^o\omega(k + 1) + d_1^o\omega(k) + n_y(k + 3)]\} . \tag{5.6-28}$$

The summation contains terms in $\omega(0)$, $\omega(1)$, ..., and $\omega(k - 1)$, none of which appear in $v(k)$; hence,

$$E\{y(k)v(k)\} = 0 . \tag{5.6-29}$$

Next, let us compute $E\{y(k + 1)v(k)\}$. Arguing as in the preceding, we need not bother to include the $\underline{x}(0)$ and $m(j)$ terms in our analysis; thus,

$$E\{y(k + 1)v(k)\} = (1, 0, 0)E\{ \sum_{i=1}^{k+1} \Phi^{k+1-i}\underline{d}\omega(i - 1)[\underline{d}_3^o\omega(k + 2) + d_2^o\omega(k + 1) + d_1^o\omega(k) + n_y(k + 3)]\} . \tag{5.6-30}$$

This summation contains terms in $\omega(0), \omega(1), \ldots, \omega(k)$; hence,

$$E\{y(k + 1)v(k)\} = (1, 0, 0)E\{\Phi^0 \underline{d}\omega(k)d_1^o\omega(k)\}$$

$$= d_1 d_1^o q \quad . \tag{5.6-31}$$

Proceeding similarly for $E\{y(k + 2)v(k)\}$, it is easily shown that

$$E\{y(k + 2)v(k)\} = (d_2 d_1^o + d_1 d_2^o)q \quad . \tag{5.6-32}$$

Before collecting all these results together, we must express d_1^o and d_2^o in terms of the known values of d_1 and d_2. From Eqs. (5.6-11) and (5.6-12), we find

$$d_1^o = d_1 \quad \text{and}$$
$$d_2^o = a_1 d_1 + d_2 \quad ; \tag{5.6-33}$$

hence,

$$E\{y(k + 1)v(k)\} = d_1^2 q \quad \text{and} \tag{5.6-34}$$

$$E\{y(k + 2)v(k)\} = (a_1 d_1^2 + 2d_1 d_2)q \quad . \tag{5.6-35}$$

Collecting our results, we conclude that

$$E\{\underline{x}(k)v(k)\} = \left(\begin{array}{ccc|c} -d_1^2 & 0 & 0 & \\ 0 & 0 & 0 & 0_3 \\ 0 & 0 & 0 & \\ \hline & 0_3 & & 0_3 \end{array}\right)\left(\begin{array}{c} -a_1 \\ -a_2 \\ -a_3 \\ \hline \underline{b}^o \end{array}\right)q + \left(\begin{array}{c} 2d_1 d_2 \\ d_1^2 \\ 0 \\ \hline \underline{0}_3 \end{array}\right)q \tag{5.6-36}$$

where 0_3 and $\underline{0}_3$, are the 3 × 3 null matrix and 3 × 1 zero vector, respectively.

Equation (5.6-36) verifies our assumption that $E\{\underline{x}(k)v(k)\} = \underline{\lambda} + M\underline{\theta}$. Here,

$$\underline{\lambda} = (2d_1 d_2, d_1^2, 0 \mid \underline{0}_3')'q \tag{5.6-37}$$

and

$$M = \left(\begin{array}{ccc|c} -d_1^2 & 0 & 0 & \\ 0 & 0 & 0 & 0_3 \\ 0 & 0 & 0 & \\ \hline & 0_3 & & 0_3 \end{array}\right)q \quad . \tag{5.6-38}$$

The algorithm for estimating $\underline{\theta}$ is

$$\hat{\underline{\theta}}(k + \ell) = [I + R(k)\Sigma_n - R(k)M]\hat{\underline{\theta}}(k)$$

$$+ R(k)\underline{r}(k)\tilde{z}(k) - R(k)\underline{\lambda} \tag{5.6-39}$$

where M and $\underline{\lambda}$ are given above,

$$\Sigma_n = \left(\begin{array}{c|c} \sigma_y^2 I_3 & 0_3 \\ \hline 0_3 & \sigma_m^2 I_3 \end{array}\right) \quad , \tag{5.6-40}$$

$\ell \geq 3$, and $k = 0, \ell, 2\ell, \ldots$. Observe that for $k = 0$, $\hat{\underline{\theta}}(\ell)$ requires $\hat{\underline{\theta}}(0)$, which must be specified a priori, and $\underline{r}(0)$, which includes the first measured values of $y(k)$ and $m(k)$, $y_m(0)$ and $m_m(0)$, respectively.

The results in this example can be generalized to systems of arbitrary order (Problems 5-10 and 5-11). Saridis and Stein have done this, but for a different state space/finite-difference equation representation (*4-19*).

Example 2 (Continued). *Identification of Coefficients in a Finite-Difference Equation.* Throughout this book we have examined the problem of identifying the coefficients $a_1, a_2, \ldots, a_n$ in the finite-difference equation

$$y(k + n) + a_1 y(k + n - 1) + \cdots + a_n y(k) = \omega(k) \quad . \tag{5.6-41}$$

In Chapters 2 and 3 we showed that it is not possible to obtain unbiased estimates of these coefficients from either generalized least-squares or unbiased minimum-variance estimation algorithms. In the present chapter we showed that this identification problem is Class-2; and we obtained an unbiased stochastic-gradient identification algorithm for estimating parameters for Class-2 problems. The algorithm is given in Eq. (5.4-15).

The reader should refer back to Example 2 in Section 5.1, where we showed how to formulate this identification application into scalar equation-error form. All that remains to be done before estimating $\underline{\theta} = (-a_1, -a_2, \ldots, -a_n)'$ by means of Eq. (5.4-15) is to compute $\underline{\lambda}$ and M.

We shall compute $\underline{\lambda}$ and M for the case when n = 3, and leave generalizations to the reader (Problems 5-15 and 5-16). Our solution is obtained by comparing Eqs. (5.6-41) and (5.6-9), and using $\underline{\lambda}$ and M in Eqs. (5.6-37) and (5.6-38). Comparing the two equations, we see that

$$b_3^o = b_2^o = b_1^o = 0 \quad , \tag{5.6-42}$$

$$d_3^o = d_2^o = 0 \quad , \quad \text{and} \tag{5.6-43}$$

$$d_1^o = 1 \quad . \tag{5.6-44}$$

In order to compute $\underline{\lambda}$ and M, we must first determine d_1 and d_2. From Eqs. (5.6-33), (5.6-43), and (5.6-44), we find

$$d_1 = 1 \quad \text{and} \tag{5.6-45}$$

$$d_2 = -a_1 \quad ; \tag{5.6-46}$$

thus,

$$\underline{\lambda} = (-2a_1, 1, 0, 0, 0, 0)'q \tag{5.6-47}$$

and

$$M = q \left(\begin{array}{ccc|c} -1 & 0 & 0 & \\ 0 & 0 & 0 & 0_3 \\ 0 & 0 & 0 & \\ \hline & 0_3 & & 0_3 \end{array}\right) \quad . \tag{5.6-48}$$

A somewhat surprising event has occurred: $\underline{\lambda}$ is a function of $\underline{\theta}$ since, from Eq. (5.6-47),

$$\underline{\lambda} = q \left(\begin{array}{ccc|c} 2 & 0 & 0 & \\ 0 & 0 & 0 & 0_3 \\ 0 & 0 & 0 & \\ \hline & 0_3 & & 0_3 \end{array}\right) \underline{\theta} + \left(\begin{array}{c} 0 \\ 1 \\ 0 \\ \hline \underline{0}_3 \end{array}\right) q \quad . \tag{5.6-49}$$

This occurrence does not invalidate our assumption in Eq. (5.4-8). Clearly, if

$$\underline{\lambda} = M_1 \underline{\theta} + \underline{\lambda}^* \tag{5.6-50}$$

then

$$\underline{\underline{TB}} = \underline{\lambda}^* + (M + M_1)\underline{\theta} \tag{5.6-51}$$

or

$$\underline{\underline{TB}} = \underline{\lambda}^* + M^*\underline{\theta} \quad , \tag{5.6-52}$$

and the right-hand side of this expression has exactly the same structure as the right-hand side of Eq. (5.4-8). Now, however,

$$\underline{\underline{\hat{TB}}}(k) = \underline{\lambda}^* + M^* E\{\underline{\hat{\theta}}(k)\} \tag{5.6-53}$$

and we must replace M and $\underline{\lambda}$ in Eq. (5.4-15) by M* and $\underline{\lambda}^*$.

We see from Eq. (5.6-49) that

$$\underline{\lambda}^* = (0 \quad 1 \quad 0 \mid \underline{0}_3')' q \tag{5.6-54}$$

and

$$M_1 = q \left(\begin{array}{ccc|c} 2 & 0 & 0 & \\ 0 & 0 & 0 & 0_3 \\ 0 & 0 & 0 & \\ \hline & 0_3 & & 0_3 \end{array}\right) \quad ; \tag{5.6-55}$$

therefore,

$$M^* = M + M_1 = q \left(\begin{array}{ccc|c} 1 & 0 & 0 & \\ 0 & 0 & 0 & 0_3 \\ 0 & 0 & 0 & \\ \hline & 0_3 & & 0_3 \end{array}\right) \tag{5.6-56}$$

and the algorithm for estimating a_1, a_2, and a_3 is

$$\underline{\hat{\theta}}(k + \ell) = [I + R(k)\Sigma_n - R(k)M^*]\underline{\hat{\theta}}(k) + R(k)\underline{r}(k)\tilde{z}(k) - R(k)\underline{\lambda}^* \qquad (5.6\text{-}57)$$

for k = 0, 3, 6, ... Since no test signal forcing functions appear in Eq. (5.6-41), we set $\sigma_m^2 = 0$ in Eq. (5.6-40), so that

$$\Sigma_n = \left(\begin{array}{c|c} \sigma_y^2 I_3 & 0_3 \\ \hline 0_3 & 0_3 \end{array}\right) \quad . \qquad (5.6\text{-}58)$$

It is clear from this example, that unbiased estimates of a_1, a_2, and a_3 are obtained by introducing the statistics of the driving noise $\omega(k)$ into the estimation algorithm in the proper manner.

Example 4. *Finite-Difference Equation Representation of a Continuous-Time System.* We are given the following third-order differential equation:

$$\dddot{y}(t) + a_1\ddot{y}(t) + a_2\dot{y}(t) + a_3 y(t) = b_1\ddot{m}(t) + b_2\dot{m}(t) + b_3 m(t) + d_1\ddot{\omega}(t) + d_2\dot{\omega}(t) + d_3\omega(t) \qquad (5.6\text{-}59)$$

where ω is a zero-mean random process having the additional property that

$$E\{\omega(t)\omega(\tau)\} = q\delta(t - \tau) \quad ; \qquad (5.6\text{-}60)$$

Coefficients d_1, d_2, and d_3 are known; and a_1, a_2, a_3, b_1, b_2, and b_3 are unknown and are to be identified from measurements of y(t) made at $t = t_j$.

From Theorem B-5 in Appendix B, we find the equivalent finite-difference equation of Eq. (5.6-59) to be

$$y(k + 3) + \alpha_1 y(k + 2) + \alpha_2 y(k + 1) + \alpha_3 y(k) = \beta_1 m(k + 2) + \beta_2 m(k + 1) + \beta_3 m(k) + \gamma_1\omega(k + 2) + \gamma_2\omega(k + 1) + \gamma_3\omega(k) \qquad (5.6\text{-}61)$$

where α_i, β_i, and γ_i (i = 1, 2, 3) are given in Eqs. (B.3-7), (B.3-8), and (B.3-24), respectively. Observe from these equations that: the γ_i are completely determined, since d_1, d_2, and d_3 are known; a_1, a_2, and a_3 can be computed once α_1, α_2, and α_3 are known; and b_1, b_2, and b_3 can be computed once β_1, β_2, and β_3 are known. Our approach, therefore, is to estimate the α and β parameters and then to solve for the a and b parameters by means of Eqs. (B.3-7) and (B.3-8).

As always, we begin by formulating the measurement equation from which the α and β parameters will be estimated. It is obtained from Eq. (5.6-61), and the fact that

$$y_m(k+3) = y(k+3) + n_y(k+3) \tag{5.6-62}$$

where $n_y(j)$ is zero mean, and

$$E\{n_y^2(j)\} = \sigma_y^2 \tag{5.6-63}$$

for all j. The scalar stochastic measurement equation is

$$\begin{aligned} y_m(k+3) = &-\alpha_1 y(k+2) - \alpha_2 y(k+1) - \alpha_3 y(k) + \beta_1 m(k+2) \\ &+ \beta_2 m(k+1) + \beta_3 m(k) + \gamma_1 \omega(k+2) \\ &+ \gamma_2 \omega(k+1) + \gamma_3 \omega(k) + n_y(k+3) \quad . \end{aligned} \tag{5.6-64}$$

The model of $y_m(k+3)$ is

$$\begin{aligned} \hat{y}_m(k+3) = &-\hat{\alpha}_1(k+3)y_m(k+2) - \hat{\alpha}_2(k+3)y_m(k+1) \\ &- \hat{\alpha}_3(k+3)y_m(k) + \hat{\beta}_1(k+3)m_m(k+2) \\ &+ \hat{\beta}_2(k+3)m_m(k+1) + \hat{\beta}_3(k+3)m_m(k) \end{aligned} \tag{5.6-65}$$

where

$$m_m(j) = m(j) + n_m(j) \tag{5.6-66}$$

and $n_m(j)$ is a zero-mean noise process, with

$$E\{n_m^2(j)\} = \sigma_m^2 \quad . \tag{5.6-67}$$

In order to put this problem into the scalar equation-error formulation, we define

$$\underline{\theta} = (-\alpha_1, -\alpha_2, -\alpha_3, \beta_1, \beta_2, \beta_3)' \quad , \tag{5.6-68}$$

and $\underline{x}(k)$, $\underline{r}(k)$, and $\underline{n}(k)$ exactly as in Eqs. (5.6-18), (5.6-19), and (5.6-20), respectively,

$$v(k) = \gamma_1\omega(k+2) + \gamma_2\omega(k+1) + \gamma_3\omega(k) + n_y(k+3) \quad , \tag{5.6-69}$$

and

$$z(k) = y_m(k+3) \quad . \tag{5.6-70}$$

As in Example 3, m(k + 2), m(k + 1), and m(k) are independent of v(k); thus,

$$E\{m(k+2)v(k)\} = E\{m(k+1)v(k)\} = E\{m(k)v(k)\} = 0 \quad . \tag{5.6-71}$$

In order to compute the remaining components in $E\{\underline{x}(k)v(k)\}$, we will need to determine y(k), y(k + 1), and y(k + 2). This is readily accomplished by means of the state-space representation of our finite-difference equation given in Theorem B-5. Applying Eqs. (B.3-16)-(B.3-22) to our third-order difference equation, we find that

$$y(k) = \underline{b}'\underline{x}_1(k) + \underline{d}'\underline{x}_2(k) \tag{5.6-72}$$

where

$$\underline{b} = (b_3 \;\; b_2 \;\; b_1)' \quad , \tag{5.6-73}$$

$$\underline{d} = (d_3 \;\; d_2 \;\; d_1)' \quad , \tag{5.6-74}$$

and

$$\begin{pmatrix} \underline{x}_1(k+1) \\ \hline \underline{x}_2(k+1) \end{pmatrix} = \left(\begin{array}{c|c} \Phi & 0_3 \\ \hline 0_3 & \Phi \end{array}\right) \begin{pmatrix} \underline{x}_1(k) \\ \hline \underline{x}_2(k) \end{pmatrix} + \begin{pmatrix} \underline{e}_3 \\ \hline \underline{0}_3 \end{pmatrix} m(k) + \begin{pmatrix} \underline{0}_3 \\ \hline \underline{e}_3 \end{pmatrix} \omega(k) \tag{5.6-75}$$

in which

$$\Phi = \begin{pmatrix} 1 & T & 0 \\ 0 & 1 & T \\ -a_3T & -a_2T & 1 - a_1T \end{pmatrix} \tag{5.6-76}$$

and

$$\underline{e}_3 = (0, 0, 1)' \quad . \tag{5.6-77}$$

For notational convenience we write Eqs. (5.6-75) and (5.6-72) as

$$\underline{x}(k + 1) = \Phi_1\underline{x}(k) + \underline{e}_m m(k) + \underline{e}_\omega \omega(k) \tag{5.6-78}$$

and

$$y(k) = (\underline{b}' \mid \underline{d}')\underline{x}(k) \tag{5.6-79}$$

where

$$\Phi_1 = \begin{pmatrix} \Phi & \vdots & 0_3 \\ \cdots & & \cdots \\ 0_3 & \vdots & \Phi \end{pmatrix} , \tag{5.6-80}$$

$$\underline{e}_m = \begin{pmatrix} \underline{e}_3 \\ -- \\ \underline{0}_3 \end{pmatrix} , \tag{5.6-81}$$

and

$$\underline{e}_\omega = \begin{pmatrix} \underline{0}_3 \\ -- \\ \underline{e}_3 \end{pmatrix} . \tag{5.6-82}$$

The solution to Eq. (5.6-78) is

$$\underline{x}(k) = \Phi_1^k\underline{x}(0) + \sum_{i=1}^{k} \Phi_1^{k-i}[\underline{e}_m m(i - 1) + \underline{e}_\omega \omega(i - 1)] . \tag{5.6-83}$$

We are now ready to compute $E\{y(k)v(k)\}$.

$$E\{y(k)v(k)\} = (\underline{b}' \mid \underline{d}')E\{\Phi_1^k\underline{x}(0)v(0) + \sum_{i=1}^{k} \Phi_1^{k-i}[\underline{e}_m m(i - 1) + \underline{e}_\omega \omega(i - 1)]v(k)\} . \tag{5.6-84}$$

As in Example 3, this expression reduces quite a bit since $\underline{x}(0)$ and all $m(j)$ are independent of $v(k)$; thus,

$$E\{y(k)v(k)\} = (\underline{b}' \mid \underline{d}')E\left(\sum_{i=1}^{k} \Phi_1^{k-i}\underline{e}_\omega \omega(i - 1)v(k)\right) . \tag{5.6-85}$$

The summation does not contain any term that is common to $v(k)$, which means that [since $E\{v(k)\} = 0$]

$$E\{y(k)v(k)\} = 0 . \tag{5.6-86}$$

Proceeding to the calculation of $E\{y(k+1)v(k)\}$, we find

$$E\{y(k+1)v(k)\} = (\underline{b}' \,\vdots\, \underline{d}')E\left(\sum_{i=1}^{k+1} \Phi_1^{k+1-i}\underline{e}_\omega\,\omega(i-1)v(k)\right)$$

$$= (\underline{b}' \,\vdots\, \underline{d}')\underline{e}_\omega \gamma_3 E\{\omega^2(k)\}$$

$$= d_1\gamma_3 q \frac{1}{T} \tag{5.6-87}$$

where we have used Eq. (B.2-22) to determine $E\{\omega^2(k)\}$ from q in Eq. (5.6-60). In a similar manner, we compute $E\{y(k+2)v(k)\}$:

$$E\{y(k+2)v(k)\} = (\underline{b}' \,\vdots\, \underline{d}')E\left(\sum_{i=1}^{k+2} \Phi_1^{k+2-i}\underline{e}_\omega\,\omega(i-1)v(k)\right)$$

$$= (\underline{b}' \,\vdots\, \underline{d}')\underline{e}_\omega \gamma_2 E\{\omega^2(k+1)\}$$

$$+ (\underline{b}' \,\vdots\, \underline{d}')\Phi_1\underline{e}_\omega \gamma_3 E\{\omega^2(k)\}$$

$$= d_1\gamma_2 q \frac{1}{T} + [d_2T + d_1(1 - a_1T)]\gamma_3 q \frac{1}{T} \quad . \tag{5.6-88}$$

Before bringing all of these results together, we must express a_1 in Eq. (5.6-88) in terms of α_1, since it is α_1 and not a_1 that is going to be estimated [see Eq. (5.6-68)]. The coefficient α_1 is determined from Eq. (B.3-7), by setting i = n - 1. For n = 3, we find

$$\alpha_1 = a_1T - 3 \tag{5.6-89}$$

whereupon

$$a_1T = \alpha_1 + 3 \quad . \tag{5.6-90}$$

Substituting this expression into Eq. (5.6-88), we find

$$E\{y(k+2)v(k)\} = \frac{q}{T}[d_1(\gamma_2 - 2\gamma_3) + d_2T\gamma_3 - d_1\gamma_3\alpha_1] \quad . \tag{5.6-91}$$

Bringing all these results together, we find

$$E\{\underline{x}(k)v(k)\} = \frac{q}{T}\left(\begin{array}{ccc|c} d_1\gamma_3 & 0 & 0 & \\ 0 & 0 & 0 & 0_3 \\ 0 & 0 & 0 & \\ \hline & 0_3 & & 0_3 \end{array}\right)\begin{pmatrix} -\alpha_1 \\ -\alpha_2 \\ -\alpha_3 \\ --- \\ \beta_1 \\ \beta_2 \\ \beta_3 \end{pmatrix} + \frac{q}{T}\begin{pmatrix} d_1(\gamma_2 - 2\gamma_3) + d_2T\gamma_3 \\ d_1\gamma_3 \\ 0 \\ ------ \\ 0 \\ 0 \\ 0 \end{pmatrix} . \tag{5.6-92}$$

Once again, we have demonstrated that $E\{\underline{x}(k)v(k)\} = \underline{\lambda} + M\underline{\theta}$, where in this case

$$\underline{\lambda} = \frac{q}{T}\,[d_1(\gamma_2 - 2\gamma_3) + d_2T\gamma_3,\ d_1\gamma_3,\ 0 \mid \underline{0}_3']' \qquad (5.6\text{-}93)$$

and

$$M = \frac{q}{T}\begin{pmatrix} d_1\gamma_3 & 0 & 0 & \vdots & \\ 0 & 0 & 0 & \vdots & 0_3 \\ 0 & 0 & 0 & \vdots & \\ \hdashline & 0_3 & & \vdots & 0_3 \end{pmatrix} . \qquad (5.6\text{-}94)$$

As in Example 3, Σ_n is given by Eq. (5.6-40).

The algorithm for estimating $\underline{\theta}$ is

$$\hat{\underline{\theta}}(k + \ell) = [I + R(k)\Sigma_n - R(k)M]\hat{\underline{\theta}}(k)$$

$$+ R(k)\underline{r}(k)\tilde{z}(k) - R(k)\underline{\lambda} \qquad (5.6\text{-}95)$$

where $\underline{\lambda}$ and M are given above, $k = 0, \ell, 2\ell, \ldots, \ell \geq 3$.

The estimates $\hat{a}_1$, $\hat{a}_2$, $\hat{a}_3$, $\hat{b}_1$, $\hat{b}_2$, and $\hat{b}_3$ are found from the equations:

$$\begin{pmatrix} \hat{a}_1T \\ \hat{a}_2T^2 \\ \hat{a}_3T^3 \end{pmatrix} = \begin{pmatrix} 1 & 0 & 0 \\ 2 & 1 & 0 \\ 1 & 1 & 1 \end{pmatrix}\begin{pmatrix} \hat{\alpha}_1 \\ \hat{\alpha}_2 \\ \hat{\alpha}_3 \end{pmatrix} + \begin{pmatrix} 3 \\ 3 \\ 1 \end{pmatrix} \qquad (5.6\text{-}96)$$

and

$$\begin{pmatrix} \hat{b}_1T \\ \hat{b}_2T^2 \\ \hat{b}_3T^3 \end{pmatrix} = \begin{pmatrix} 1 & 0 & 0 \\ 2 & 1 & 0 \\ 1 & 1 & 1 \end{pmatrix}\begin{pmatrix} \hat{\beta}_1 \\ \hat{\beta}_2 \\ \hat{\beta}_3 \end{pmatrix} . \qquad (5.6\text{-}97)$$

▲

5.7 RATE OF CONVERGENCE AND CHOICE OF R(k)

From Eqs. (5.5-23) and (5.5-39), we see that one admissible choice (i. e., a choice for which mean-square convergence occurs) for R(k) is

$$R(k) = \frac{1}{k^p} \operatorname{diag}[h_1(k), h_2(k), \ldots, h_n(k)] \tag{5.7-1}$$

where $\frac{1}{2} < p \leq 1$.

Looking at the right-hand side of Eq. (5.5-43) we see, loosely speaking, that

$$\hat{\underline{\theta}}(k + \ell) = O\left(\frac{1}{k^p}\right) \quad . \tag{5.7-2}$$

In addition, we see from Eq. (5.5-31) that

$$\overline{||\tilde{\underline{\theta}}(k + \ell)||^2} \leq O\left(\frac{1}{k^p}\right) \quad . \tag{5.7-3}$$

This last result is disconcerting, for it means that *convergence is very slow* (the larger k becomes, the smaller $1/k^p$ becomes).

The sequence $1/k^p$ ($\frac{1}{2} < p \leq 1$) provides too much attenuation for the identification algorithm. It is important to remember, however, that *such attenuation is only needed in the limit as* $k \to \infty$.

Because of excellent results achieved in practical applications using the Lyapunov-optimum weighting matrix, $R^*(k)$ in Eq. (4.6-2), we suggest the following choice for R(k) in the stochastic situation:

For $k \leq k'$, use

$$R(k) = R^s(k) = \frac{\operatorname{diag}[h_1(k), \ldots, h_n(k)]}{\sum_{i=1}^n h_i(k) r_i^2(k)} \tag{5.7-4}$$

and, for $k > k'$, use

$$R(k) = \frac{1}{k^p} \operatorname{diag}[h_1(k), \ldots, h_n(k)] \quad . \tag{5.7-5}$$

The matrix $R^s(k)$ is the *stochastic analog* of $R^*(k)$. Often k' can be chosen rather arbitrarily.

Example 1 (Continued).

The aerodynamic parameters M_α and M_δ, associated with the high-performance aerospace control system that was described in Section 5.1, have been estimated by Mendel (*5-23*) using the stochastic analogs to Eqs. (4.7-4) and (4.7-5), obtained from Eqs. (5.4-6), (5.1-20) through (5.1-27), (5.7-4), and (5.7-5). When these equations are combined, the following expressions are obtained for $\hat{M}_\alpha(k + 1)$ and $\hat{M}_\delta(k + 1)$ (we showed that $\ell = 1$ in Example 1 of Section 5.4.3):

$$\hat{M}_\alpha(k + 1) = \left[1 + \frac{h_1\sigma_\alpha^2}{d(k)}\right]\hat{M}_\alpha(k) + \frac{h_1\alpha_m(k)\tilde{z}(k)}{d(k)} \tag{5.7-6}$$

and

$$\hat{M}_\delta(k + 1) = \left[1 + \frac{h_2\sigma_\delta^2}{d(k)}\right]\hat{M}_\delta(k) + \frac{h_2\delta_m(k)\tilde{z}(k)}{d(k)} \tag{5.7-7}$$

for k = 0, 1, 2, ..., where

$$d(k) = \begin{cases} h_1\alpha_m^2(k) + h_2\delta_m^2(k) , & k \leq k' \\ k , \quad k > k' \end{cases} \tag{5.7-8}$$

and

$$\tilde{z}(k) = \ddot{\theta}_m(k) - \hat{\ddot{\theta}}_m(k) \quad . \tag{5.7-9}$$

Estimates of $M_\alpha = 500$ and $M_\delta = 1000$ were obtained using these algorithms for the input normal acceleration command history depicted in Fig. 4.7-4, and for $h_2/h_1 = 10$ (see Section 4.7.2), $\sigma_\alpha^2 = 0.061$, $\sigma_\delta^2 = 0.016$, and $\sigma_{\ddot{\theta}}^2 = 3$. Results for M (k) are depicted in Fig. 5.7-1. Clearly, bias is indeed removed when the measurement noise statistics are included in the algorithm [the biased algorithm for $\hat{M}_\alpha(k + 1)$ is obtained from Eq. (5.7-6) by eliminating the term $[h_1\sigma_\alpha^2/d(k)]\hat{M}_\alpha(k)]$. It is also clear that for $k > k' = 52$, ripples in $\hat{M}_\alpha(k)$ decay to zero because of the attenuating factor k which appears in the denominator of two terms in Eq. (5.7-6).

5.8 COMPUTATIONAL CONSIDERATIONS

Unbiased stochastic-gradient identification algorithms are more complicated than their deterministic counterparts that were studied in Chapter 4; but as we pointed out in Section 5.5.2, the stochastic identification problem we have formulated in this chapter is *a fortiori* more difficult than the deterministic identification problem that was formulated in Chapter 4.

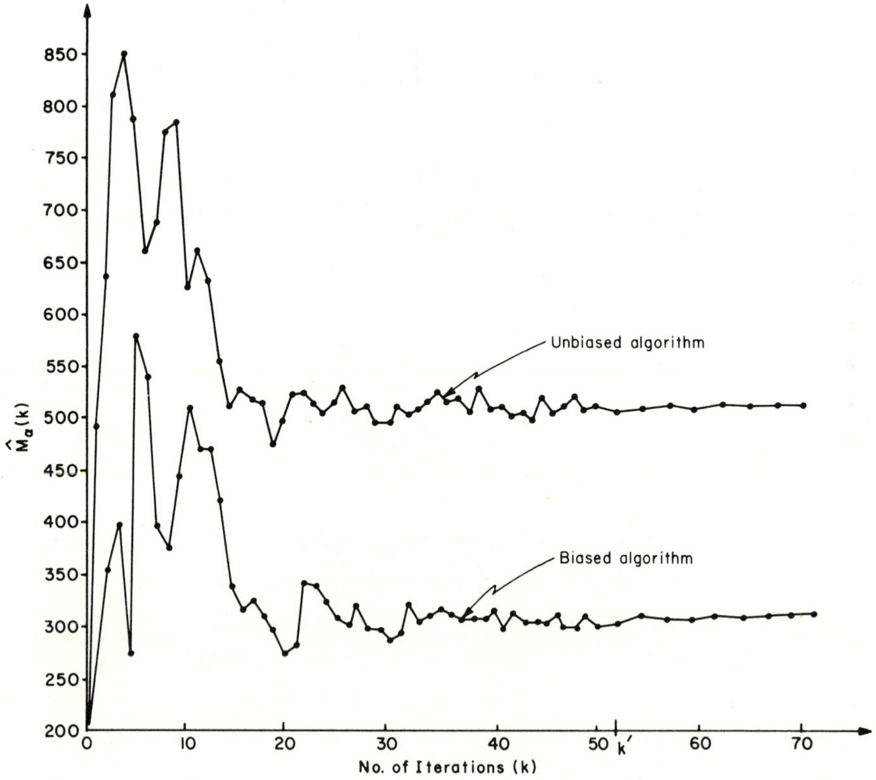

Fig. 5.7-1. Realizations of estimates of $M_\alpha = 500$.

We do not propose to compare the unbiased stochastic-gradient identification algorithms with the sequential unbiased minimum-variance identification algorithm derived in Chapter 3, since these algorithms would not be used for the same application - they are noncompetitive (see Section 5.9 for additional discussions).

Determination of total computational times for the algorithms in Eqs. (5.4-6) and (5.4-15) is left as an exercise for the reader (Problem 5-21). He will find that they are rather small.

In order to implement both Class-1 and Class-2 algorithms, Σ_n must be known a priori. Just as we studied the problem associated with mismatches in the measurement-noise covariance matrix in Section 3.10, it would be useful to obtain comparable results for the algorithms in this chapter. To date, no analytical results have been obtained. One reason for this is that we do not have an expression for the parameter estimation-error covariance matrix. In Chapter 3, on the other hand, this covariance matrix was computed as part of the estimation algorithm. We propose the analytical study of the sensitivity problem for unbiased stochastic-gradient identification algorithms as a research project, in Problem 5-27. The author has observed, by means of numerous sensitivity simulations, that estimates of the aerodynamic parameters M_α and M_δ (see Example 1 in Section 5.7) are quite insensitive to mismatches between actual values of σ_α^2 and σ_δ^2, and the values used in Eqs. (5.7-6) and (5.7-8) (*5-23*); however, these results are too specific to be generalized.

5.9 APPLICABILITY OF UNBIASED STOCHASTIC-GRADIENT PARAMETER ESTIMATION ALGORITHMS

Unbiased stochastic-gradient parameter estimation algorithms are applicable, whenever (1) an application can be cast into a scalar equation-error formulation; (2) some or all of the signals which must be used in the algorithms cannot be measured without errors, or if some of the signals used for identification are correlated with *output noise* (which may be a combination of output measurement noise and random forcing function, or it may be just a random forcing function - see Fig. 5.2-1); and (3) first- and second-order statistics of random measurement errors and/or forcing functions are known ahead of time.

We have distinguished three stochastic identification problems, and have directed our attention to the first two, since their underlying identification-error systems are linear and are, therefore, amenable to analysis. Any application that can be formulated as either a Class-1 or Class-2 identification problem, in the spirit of Figs. 5.1-1 and 5.2-1, respectively, is a suitable candidate for the algorithms obtained in this chapter. In closing, the reader should compare these figures with Fig. 2.2-1, which depicts our interpretation of the problem formulation associated with both generalized least-squares and unbiased minimum-variance estimation. Although Fig. 2.2-1 is structurally similar to Fig. 5.1-1, there is one very important difference between them, and that is that there is no measurement noise on the input signals to model and algorithm in the Fig. 2.2-1 identification system. The elements of $H(k)$ must be known without errors; hence, we can at best associate the Fig. 2.2-1 identification system with *very limited Class-2 problems*. An unbiased stochastic-gradient estimation algorithm for an analogous problem - one in which $\underline{x}(k)$ can be measured perfectly - is presented in Corollary 5.1 (Section 5.3.2). It is

$$\underline{\hat{\theta}}(k + \ell) = \underline{\hat{\theta}}(k) + R(k)\underline{x}(k)\tilde{z}(k) \tag{5.9-1}$$

for $k \geq 0$. This algorithm can be used wherever the sequential unbiased minimum-variance algorithm [Eq. (3.7-5), for example] can be, as long as the measurement $z(k)$ is a scalar. We have not treated the vector measurement case in this chapter. If one has a limited Class-1 identification problem and a vector of measurements, he should use the Chapter 3 algorithms.

It is important to remember that the algorithms of this chapter require a priori knowledge of first- and second-order statistics [e. g., Σ_n in Eq. (5.4-15)]. If this information is not available, it may have to be estimated. The maximum-likelihood technique (Section 3.9) can be applied to this situation. Generally, the solutions for the unknown parameters and statistics, obtained from the maximum-likelihood principle, must be found by means of nonlinear programming. These solutions are not within the scope of this book.

PROBLEMS

5-1. In Example 2 of Section 5.1, suppose that $\omega(k)$ is a *measurable test signal*. How must the equation-error formulation be modified to accommodate the measurability of $\omega(k)$? Show that this is a Class-1 identification problem.

5-2. Demonstrate that the type of application that satisfies the conditions of Corollary 5-1 can also be handled by unbiased minimum-variance techniques; hence, for such applications there may not be much point in using the stochastic-gradient estimation algorithm in Eq. (5.3-28).

5-3. Exactly where, in the determination of bias, does the analysis in Section 5.3 break down for Class-3 identification problems?

5-4. For the application in Example 1 of Section 5.1, show that the spacing parameter ℓ must be chosen so that $\ell \geq 1$.

5-5. Demonstrate that the identification-error system in Eq. (5.5-5) reduces to the deterministic identification-error system in Eq. (4.3-2) when no random phenomena are present.

5-6. Using the results obtained in Section 5.5, prove mean-square convergence of the unbiased stochastic-gradient algorithm for Class-2 identification problems. [*Hint:* The presence of M and $E\{\underline{x}(k)v(k)\}$ in Eq. (5.5-5) adds eleven more terms to Eq. (5.5-9), each one of which must be bounded.]

5-7. Prove Theorem 5-5.

5-8. For R(k) in Eq. (5.5-23), prove that $R(k)\Omega$ is positive definite.

5-9. Under what conditions, if any, on c, c_1, and p do the following choices for $\rho(k)$ satisfy the four conditions in Eq. (5.5-38): (a) $\rho(k) = c/(k + c_1)^p$, and (b) $\rho(k) = c/(c_1 + k^p)$?

5-10. Determine M and $\underline{\lambda}$ for the following fourth-order abstract digital system:

$$\begin{pmatrix} x_1(k+1) \\ x_2(k+1) \\ x_3(k+1) \\ x_4(k+1) \end{pmatrix} = \left(\begin{array}{c:ccc} 0 & & & \\ 0 & & I_3 & \\ 0 & & & \\ \hdashline -a_4 & -a_3 & -a_2 & -a_1 \end{array}\right) \begin{pmatrix} x_1(k) \\ x_2(k) \\ x_3(k) \\ x_4(k) \end{pmatrix} + \begin{pmatrix} b_1 \\ b_2 \\ b_3 \\ b_4 \end{pmatrix} m(k) + \begin{pmatrix} d_1 \\ d_2 \\ d_3 \\ d_4 \end{pmatrix} \omega(k)$$

where

$$y(k) = x_1(k) \quad ,$$

$$y_m(k) = y(k) + n_y(k) \quad ,$$

$$m_m(k) = m(k) + n_m(k) \quad ,$$

and $\omega(k)$, $n_y(k)$, and $n_m(k)$ are zero-mean independent discrete white processes for which $E\{\omega^2(k)\} = q$, $E\{n_y^2(k)\} = \sigma_y^2$, and $E\{n_m^2(k)\} = \sigma_m^2$.

5-11. From the expressions for M and $\underline{\lambda}$ obtained in Example 3 of Section 5.6 and Problem 5-10, obtain M and $\underline{\lambda}$ for abstract digital systems of order n, where n is arbitrary.

5-12. Determine M and $\underline{\lambda}$ associated with identifying a_1, a_2, and a_3 in the following finite-difference equations:

(a) $y(k + 3) + a_1 y(k + 2) + a_2 y(k + 1) + a_3 y(k) = \omega(k + 1) + \omega(k)$;

(b) $y(k + 3) + a_1 y(k + 2) + a_2 y(k + 1) + a_3 y(k) = \omega(k + 2) - \omega(k)$.

5-13. Another way to write the finite-difference equation in Eq. (5.1-28) is

$$y(k) + a_1 y(k - 1) + \cdots + a_n y(k - n) = \omega(k - n) \quad .$$

(a) Assuming that only noisy measured values of y(j) are available, reformulate the problem of identifying $a_1, a_2, \ldots, a_n$, as in Example 2 of Section 5.1.

(b) Attempt to compute $E\{\underline{x}(k)v(k)\}$ for this new formulation. Explain the difficulty that occurs. Show that this difficulty is not encountered when $E\{\underline{x}(k)v(k)\}$ is computed for the formulation given in Example 2 of Section 5.1.

5-14. In Problem 1-19 it was shown that $\dot{\theta}$ satisfies the following finite-difference equation:

$$\dot{\theta}(k + 2) = \phi_1\dot{\theta}(k + 1) + \phi_2\dot{\theta}(k) + \phi_3\delta(k) + \phi_4\delta(k + 1) + \omega(k + 1) - p_3\omega(k) \quad .$$

(a) Show that

$$M = \begin{pmatrix} -\sigma_\omega^2 & 0 & 0 & 0 \\ 0 & 0 & 0 & 0 \\ 0 & 0 & 0 & 0 \\ 0 & 0 & 0 & 0 \end{pmatrix}$$

and

$$\underline{\lambda} = (\sigma_\omega^2 \quad 0 \quad 0 \quad 0)'$$

where $E\{\omega^2(k)\} = \sigma_\omega^2$.

(b) Obtain the scalar algorithms for estimation of ϕ_1, ϕ_2, ϕ_3, and ϕ_4, assuming only noisy measurements of $\dot{\theta}$ and δ are available.

5-15. Determine $\underline{\lambda}^*$ and M^*, as in Example 2 of Section 5-6, for the following fourth-order system:

$$y(k + 4) + a_1 y(k + 3) + a_2 y(k + 2) + a_3 y(k + 1) + a_4 y(k) = \omega(k) \quad .$$

5-16. From the expressions for M^* and $\underline{\lambda}^*$ obtained in Example 2 of Section 5.6 and Problem 5-15, obtain M^* and $\underline{\lambda}^*$ for the finite-difference equation, of order n, in Eq. (5.6-41).

5-17. Prove that the approach to identifying the coefficients of a finite-difference equation that uses the solution to the equation as its starting point (see Section 1.3.3) leads to Class-1 identification problems.

5-18. Write out the algorithms for identifying the coefficients in the following finite-difference equations:

(a) $y(k + 3) + a_1 y(k + 2) + a_2 y(k + 1) + a_3 y(k) = m(k) + \omega(k + 1) + \omega(k)$;

(b) $y(k + 3) + a_1 y(k + 2) + a_2 y(k + 1) + a_3 y(k) = b_1^o m(k + 2) + b_3^o m(k) + \omega(k + 2) - \omega(k)$.

Use the formulation for identifying the unknown a and b^o coefficients, that is based on the solution to the finite-difference equations. In these equations m is a test signal, whereas ω is a random forcing function that cannot be measured. Only noisy measurements of y and m are available. What is ℓ?

5-19. Write the algorithms for identifying $\phi_1, \ldots, \phi_4$ in the finite-difference equation in Problem 5-14, when these parameters are identified from a formulation that is based on the solution to the finite-difference equation. Treat δ as a test signal. What is ℓ? Compare the algorithms obtained in Problem 5-14 and this problem on the basis of time required for identification, a priori information required for implementation, and computational requirements.

5-20. We wish to identify the coefficients a_1, a_2, and b_2 in the following second-order differential equation:

$$\ddot{y}(t) + a_1\dot{y}(t) + a_2 y(t) = b_2 m(t) + d_2\dot{\omega}(t) + d_3\omega(t) \quad .$$

The function m(t) is a test signal, whereas $\omega(t)$ is a random forcing function that cannot be measured. Only noisy measured values of y(t) and m(t) are available at discrete instants of time.

(a) Obtain the finite-difference equation representation for this differential equation. Obtain explicit equations for a_1, a_2, and b_2 in terms of the coefficients of the finite-difference equation.

(b) Obtain the algorithm for estimating the coefficients of the finite-difference equation from the formulation that is based on the structure of the finite-difference equation. What are M and $\underline{\lambda}$? What is ℓ?

(c) Obtain the algorithm for estimating the coefficients of the finite-difference equation from the formulation that is based on the solution to the finite-difference equation. What is ℓ?

(d) Suppose that noisy measurements of $\ddot{y}(t)$ and $\dot{y}(t)$ are available in addition to the measurements of y(t) and m(t). Explain how to identify a_1, a_2, and b_2.

5-21. Let us compare the computational requirements in the scalar measurement situation for stochastic-gradient and sequential unbiased minimum-variance algorithms. We shall make this comparison on the basis of the total number of additions, subtractions, multiplications, and divisions required for the implementation of each algorithm (see Problem 4-12 for additional related discussions on logic time).

(a) Let A(n), M(n), and D(n) denote the total number of additions, multiplications, and divisions, respectively. Compute A(n), M(n), and D(n) for the stochastic-gradient algorithm in Eqs. (5.4-6) and (5.7-4) (Class-1 identification algorithm), and for the sequential unbiased minimum-variance algorithm in Eqs. (3.7-5)-(3.7-7) [setting $\underline{z}(k) = z(k)$]. These quantities will be explicit functions of the number of unknown parameters, n.

(b) Same as (b) in Problem 4-12.

(c) Let $C_s(n)$ and $C_u(n)$ denote C(n) for the stochastic-gradient and unbiased minimum-variance algorithms. Plot $C_u(n)/C_s(n)$ for n = 1, 2, ..., 20.

5-22. Let us compare the Class-1 and Class-2 stochastic gradient algorithms in Eqs. (5.4-6) and (5.4-15) on the basis of relative computing times.

(a) Compute A(n), M(n), and D(n) (see Problem 5-21) for the Class-2 algorithm in Eq. (5.4-15), using the expression for R(k) in Eq. (5.7-4).

(b) Repeat (a) for the Class-1 algorithm in Eq. (5.4-6) (this may already have been accomplished if the reader has done Problem 5-21).

(c) Let $C_{s1}(n)$ and $C_{s2}(n)$ denote C(n) for the Class-1 and Class-2 stochastic-gradient algorithms [see (b) in Problem 4-12 for a definition of C(n)]. Plot $C_{s2}(n)/C_{s1}(n)$ for n = 1, 2, ..., 20.

5-23. We are given the following first-order system, including noisy measurements of x(t) at $t = t_i$, i = 0, 1, ..., M:

$$\dot{x}(t) = f[x(t), t]$$

$$x(0) \text{ unknown},$$

and $0 \le t \le t_f$. By quasilinearization and superposition [see Problem 1-12], $x(t)$ can be obtained sequentially, as

$$x_{\ell+1}(t) = x_{\ell+1}(0)h_{\ell+1}(t) + p_{\ell+1}(t)$$

where $h_{\ell+1}(t)$ and $p_{\ell+1}(t)$ are functions which are obtained by means of numerical integration.

(a) Suppose that numerical integration errors are so small that they can be neglected. Show that the estimation of $x_{\ell+1}(0)$ from the M + 1 noisy measurements is a Class-1 identification problem. Let σ_x^2 denote the variance of the noise associated with the measured value of x. Show that $x_{\ell+1}(0)$ can be estimated by means of the following algorithm:

$$\begin{aligned}\hat{x}_{\ell+1}(k + 1) &= [1 + R_{\ell+1}(k)\sigma_x^2]\hat{x}_{\ell+1}(k)\\ &\quad + R_{\ell+1}(k)h_{\ell+1}(k)\tilde{z}_{\ell+1}(k)\end{aligned}$$

for k = 0, 1, 2, ..., where

$$\begin{aligned}\tilde{z}_{\ell+1}(k) &= z_{\ell+1}(k) - \hat{z}_{\ell+1}(k)\\ &= [x_{\ell+1}(k) + n_{x_{\ell+1}}(k)]\\ &\quad - [h_{\ell+1}(k)\hat{x}_{\ell+1}(k) + p_{\ell+1}(k)]\end{aligned}$$

and

$$R_{\ell+1}(k) = 1/h_{\ell+1}^2(k) \qquad \text{for } k \le k'.$$

(b) Suppose that the numerical integration errors are random, with known statistics; that is to say,

$$h_{\ell+1}(t) = h_{\ell+1}^*(t) + n_{h_{\ell+1}}(t)$$

and

$$p_{\ell+1}(t) = p_{\ell+1}^*(t) + n_{p_{\ell+1}}(t)$$

where $n_{h_{\ell+1}}$ and $n_{p_{\ell+1}}$ are zero-mean random processes and

$$E\{n_{h_{\ell+1}}^2(k)\} = \sigma_n^2 \qquad E\{n_{p_{\ell+1}}^2(k)\} = \sigma_p^2 \quad .$$

Show that the estimation of $x_{\ell+1}(0)$ from the M + 1 noisy measurements is a Class-2 identification problem. What is TB? Give the stochastic-gradient algorithm for estimating $x_{\ell+1}(0)$.

5-24. Van der Pol's equation is

$$\ddot{x}(t) - \varepsilon\dot{x}(t)[1 - \frac{1}{3}\dot{x}^2(t)] + x(t) = m(t) \quad .$$

Noisy measurements of x(t) and m(t) are available at $t = t_i$, i = 0, 1, ..., M. The parameter ε is to be identified using quasilinearization [see Problem 1-26(c)] and a stochastic-gradient algorithm.

(a) Demonstrate that the estimate of ε is a Class-2 identification problem.

(b) Give all the necessary details to program a stochastic-gradient algorithm for estimation of ε.

(c) Draw a flow-graph clearly showing the *two* approximation loops in this estimation.

5-25. Throughout this chapter we have assumed that $E\{v(k)\} = 0$. Let us investigate the situation when this assumption is not true.

(a) Assume that $E\{v(k)\} = v_0$ where v_0 is known a priori. How is the Class-1 algorithm modified in this situation?

(b) Assume that $E\{v(k)\} = v_1$ where v_1 is constant but is unknown. How is the Class-1 algorithm modified so that estimates of v_1 can be obtained as well as estimates of $\underline{\theta}$?

(c) Assume that $E\{v(k)\} = v_1$ where v_1 is constant but is unknown. Now, however, we do not wish to estimate v_1. How is the estimation procedure modified so that only estimates of $\underline{\theta}$ are obtained? (*Hint:* Construct a different measurement z(k), one in which the effects of v_1 are not present.)

5-26. Demonstrate, by going through the derivations in Sections 5.3, 5.4, and 5.5 step by step, that the results in this chapter can be used when $\underline{\theta}$ is *random*.

5-27. (Research Project.) The parameter estimation-error covariance matrix is a by-product of the unbiased minimum-variance estimation technique; however, it is never computed for the stochastic-gradient technique. If it could be computed from a priori information, then it would be helpful in providing the designer with measures of confidence in the estimates obtained from the stochastic-gradient algorithms. Can a recursive expression be obtained for $E\{\tilde{\underline{\theta}}(k+\ell)\tilde{\underline{\theta}}'(k+\ell)\}$ for Class-1 and Class-2 algorithms? Can a recursive expression be obtained for $\mathrm{Tr}[E\{\tilde{\underline{\theta}}(k+\ell)\tilde{\underline{\theta}}'(k+\ell)\}]$ for these algorithms? If so, then study the sensitivity of $E\{\tilde{\underline{\theta}}(k+\ell)\tilde{\underline{\theta}}'(k+\ell)\}$ or $\mathrm{Tr}[E\{\tilde{\underline{\theta}}(k+\ell)\tilde{\underline{\theta}}'(k+\ell)\}]$ to errors in Σ_n, M, and $\underline{\lambda}$.

REFERENCES

5-1. J. M. Mendel, Sequential identification by means of gradient learning algorithms, in *Pattern Recognition and Machine Learning* (K. S. Fu, ed.), pp. 70-78, Plenum Press, New York, 1971.

5-2. A Papoulis, *Probability, Random Variables, and Stochastic Processes,* McGraw-Hill, New York, 1965.

5-3. A. M. Mood, *Introduction to the Theory of Statistics,* McGraw-Hill, New York, 1950.

5-4. M. T. Wasan, *Parametric Estimation,* McGraw-Hill, New York, 1970.

5-5. J. K. Holmes, Two stochastic approximation procedures for identifying linear systems, *IEEE Trans. Auto. Control,* AC-14, 292-295 (1969).

5-6. J. H. Venter, An extension of the Robbins-Monro procedure, *Annals Math. Stat.,* 38, 181-190 (1967).

5-7. V. N. Faddeeva, *Computational Methods of Linear Algebra,* Translated from Russian by C. D. Benster, Dover, New York, 1959.

5-8. B. Z. Vulikh, *Introduction to Functional Analysis for Scientists and Technologists,* Pergamon Press, New York, 1962.

5-9. R. Bellman, *Introduction to Matrix Analysis,* 2nd ed., McGraw-Hill, New York, 1970.

5-10. D. J. Wilde, *Optimum Seeking Methods,* Prentice-Hall, Englewood Cliffs, New Jersey, 1964.

5-11. E. Parzen, *Modern Probability Theory and Its Applications,* Wiley, New York, 1960.

5-12. H. Robbins and S. Monro, A stochastic approximation method, *Annals Math. Stat.,* 22, No. 1, 400-407 (1951).

5-13. J. Kiefer and J. Wolfowitz, Stochastic estimation of the maximum of a regression function, *Annals of Math. Stat.,* 23, 462-466 (1952).

5-14. J. R. Blum, Multidimensional stochastic approximation methods, *Annals of Math. Stat.,* 25, 737-744 (1954).

5-15. A. Dvoretzky, On stochastic approximation, *Proc. 3rd Berkeley Symp. Math. Stat. Probability,* Vol. 1, pp. 39-55, 1956.

5-16. H. J. Kushner, Hill climbing methods for the optimization of multiparameter noise disturbed systems, *Trans. ASME, J. Basic Eng.,* 85, 157-164 (1963).

5-17. R. L. Kashyap, C. C. Blaydon, and K. S. Fu, Stochastic approximation, in *Adaptive, Learning, and Pattern Recognition Systems: Theory and Applications,* (J. M. Mendel and K. S. Fu, eds.), Academic Press, New York, 1970, Chapter 9, pp. 329-355.

5-18. C. C. Blaydon, R. L. Kashyap, and K. S. Fu, Applications of the stochastic approximation methods, in *Adaptive, Learning, and Pattern Recognition Systems: Theory and Applications,* (J. M. Mendel and K. S. Fu, eds.), Academic Press, New York, 1970, Chapter 10, pp. 357-392.

5-19. G. N. Saridis and G. Stein, Stochastic Approximation algorithms for linear discrete-time system identification, *IEEE Trans. Auto. Control.,* AC-13, 515-523 (1968).

5-20. G. N. Saridis and G. Stein, A new algorithm for linear system identification, *IEEE Trans. Auto. Control,* AC-13, 592-594 (1968).

5-21. K. Kirvaitis and K. S. Fu, *Identification of Nonlinear Systems by Stochastic Approximation,* Preprints 1966 Joint Auto. Control Conf., Seattle, Washington, pp. 255-264.

5-22. D. J. Sakrison, The use of stochastic approximation to solve the system identification problem, *IEEE Trans. Auto. Control*, AC-12, 563-567 (1967).

5-23. J. M. Mendel, Identification of decomposable time-varying parameters by means of gradient algorithms, *Inform. Sci.*, (1972).

5-24. K. Knopp, *Theory and Application of Infinite Series*, Blackie, London, 1951.

5-25. A. E. Taylor, *Advanced Calculus*, Ginn, New York, 1955.

6

Estimation of Time—Varying Parameters

6.1 INTRODUCTION

In Chapters 2 through 5 our attention was directed at the identification of a vector of constant parameters, $\underline{\theta}$. In the present chapter we shall demonstrate how many of the results in these earlier chapters, as well as the approaches taken in obtaining these results, can be extended to the problem of estimating a vector of time-varying parameters, $\underline{\theta}(k)$. We suspect, intuitively, that success in applying or extending our earlier results will depend quite strongly on the rates at which the time-varying parameters $\theta_i(k)$ $(i = 1, 2, \ldots, n)$ vary. For example, it is absurd to expect to be able to estimate a parameter that varies infinitely fast.

We shall distinguish between three types of time-varying parameter information.

(i) *Type A.* $\underline{\theta} = \underline{\theta}(k)$, with no additional structural information known about $\underline{\theta}(k)$.

(ii) *Type B.* $\underline{\theta} = \underline{\theta}(k)$, where $\underline{\theta}(k)$ is *decomposable* into the product of a known information matrix $L(k)$ and a less rapidly varying vector of parameters $\underline{\beta}(k)$; that is to say,

$$\underline{\theta}(k) = L(k)\underline{\beta}(k) \quad . \tag{6.1-1}$$

(iii) *Type C.* $\underline{\theta} = \underline{\theta}(k)$, where $\underline{\theta}(k)$ is modeled by the *dynamical system*

$$\underline{\theta}(k + 1) = \Phi(k + 1, k)[\underline{\theta}(k) - \underline{\theta}_N(k)] + \underline{\theta}_N(k + 1) + \underline{\omega}(k) \tag{6.1-2}$$

where $\Phi(k + 1, k)$ is known a priori and is usually diagonal. Often $\Phi(k + 1, k)$ is a constant matrix Φ. The $\underline{\theta}_N(k)$ is a nominal time history of $\underline{\theta}(k)$ which is also assumed known a priori, and $\underline{\omega}(k)$ is a discrete vector white sequence, having the properties

$$E\{\underline{\omega}(k)\} = \underline{0} \tag{6.1-3}$$

and

$$E\{\underline{\omega}(k)\underline{\omega}'(k)\} = Q(k) \quad . \tag{6.1-4}$$

▲

Type-A information - or lack of information - about $\underline{\theta}(k)$ is sometimes all that can be assumed. On the other hand, when additional information is known about the structure of $\underline{\theta}(k)$, it is somewhat naive to assume it does not exist.

Suppose, for example, we wished to estimate the coefficients $a_1(k)$, $a_2(k)$, ..., $a_n(k)$ in the following time-varying finite-difference equation:

$$y(k + n) + a_1(k)y(k + n - 1) + \cdots + a_n(k)y(k) = \omega(k) \quad . \tag{6.1-5}$$

All we have available for the identification are noisy measurements of $y(j)$ $(j = 0, 1, \ldots)$, plus the a priori knowledge that the system is indeed time varying. For this abstract system, there is no structural information known about $a_i(k)$. Other situations where a lack of structural information occurs are those in which we are identifying a fundamental quantity such as the variable mass of a single-stage rocket. ▲

Often, a time-varying parameter is actually a product of two (or more) parameters, one of which is rapidly time varying but is known a priori, and the other of which is slowly varying but is unknown. For example, $M_\alpha(t)$ and $M_\delta(t)$ [see Section 1.1 of Chapter 1] can be written as

$$M_\alpha(t) = q(t)M_{\alpha q}(t) \tag{6.1-6}$$

and

$$M_\delta(t) = q(t)M_{\delta q}(t) \quad . \tag{6.1-7}$$

In these equations, q(t) is dynamic pressure; it varies by large amounts from one sampling instant to the next, but either may be known ahead of time from guidance information or can be measured online by means of a pressure probe.† The parameters $M_{\alpha q}(t)$ and $M_{\delta q}(t)$ vary over smaller ranges of values and usually at slower rates than do $M_\alpha(t)$ and $M_\delta(t)$. Clearly, Eqs. (6.1-6) and (6.1-7) can be combined to give

$$\begin{pmatrix} M_\alpha(t) \\ M_\delta(t) \end{pmatrix} = \begin{pmatrix} q(t) & 0 \\ 0 & q(t) \end{pmatrix} \begin{pmatrix} M_{\alpha q}(t) \\ M_{\delta q}(t) \end{pmatrix} \tag{6.1-8}$$

and this equation is exactly of the form in Eq. (6.1-1). The $M_\alpha(t)$ and $M_\delta(t)$ are decomposable by means of the known information matrix

$$L(k) = q(k)I_2 \quad . \tag{6.1-9}$$

This example illustrates that one way in which Type-B information is obtained is from the physics of a problem. Such information may often be available, just waiting to be used.

It is also possible to contrive Type-B information. To demonstrate this, let us approximate the ith component $\theta_i(t)$ of $\underline{\theta}(t)$ by an N-term expansion in the basis function $\phi_1(t), \phi_2(t), \ldots, \phi_N(t)$ (see Sections 2.4.1-2.4.3 for discussions on possible choices for these functions); that is to say,

---

† In some very high-performance vehicles it is impossible to measure or know q(t) ahead of time. In those applications, the decompositions in Eqs. (6.1-6) and (6.1-7) are not possible.

$$\theta_i(t) \cong \sum_{j=1}^{N} C_{ij}(t)\phi_j(t) \tag{6.1-10}$$

which can also be written

$$\theta_i(t) = \underline{C}_i'(t)\underline{\phi}(t) \tag{6.1-11}$$

for i = 1, 2, ..., n, where

$$\underline{C}_i(t) = [C_{i1}(t), C_{i2}(t), \ldots, C_{iN}(t)]' \tag{6.1-12}$$

and

$$\underline{\phi}(t) = [\phi_1(t), \phi_2(t), \ldots, \phi_N(t)]' \quad . \tag{6.1-13}$$

The vector $\underline{\theta}(t)$ is obtained by stacking the n equations in Eq. (6.1-11) and is given by the expression

$$\underline{\theta}(t) = L(t)\underline{C}(t) \tag{6.1-14}$$

where

$$\underline{C}(t) = \begin{pmatrix} \underline{C}_1(t) \\ \hline \underline{C}_2(t) \\ \hline \vdots \\ \hline \underline{C}_n(t) \end{pmatrix} \tag{6.1-15}$$

and[†]

$$L(t) = \left(\begin{array}{c|c|c|c} \underline{\phi}'(t) & \underline{0}_N' & \cdots & \underline{0}_N' \\ \hline \underline{0}_N' & \underline{\phi}'(t) & \cdots & \underline{0}_N' \\ \hline \vdots & \vdots & \ddots & \vdots \\ \hline \underline{0}_N' & \underline{0}_N' & \cdots & \underline{\phi}'(t) \end{array}\right) \quad . \tag{6.1-16}$$

Hopefully, $\underline{C}(t)$ will be less rapidly time varying than $\underline{\theta}(t)$ (or else, as we shall demonstrate in Section 6.2, there is not much point in using this artifice); however, we are paying a price for this decomposition. Instead of estimating the n vector $\underline{\theta}(t)$, we must

---

[†] The symbol $\underline{0}_N$ is the N × 1 zero vector.

estimate the nN vector $\underline{C}(t)$. If $\underline{\theta}(t)$ is rapidly time varying and $\underline{C}(t)$ is slowly time varying, this may be a rather small price to pay; for it may be impossible to estimate $\underline{\theta}(t)$ in any other way. ▲

Type-C information exists when we are absolutely certain that the unknown parameters are time varying and know to a certain extent the way in which they vary, but want to include a quantitative measure of our confidence in this "knowledge about their variations" into our analysis.

Letting

$$\delta\underline{\theta}(k) = \underline{\theta}(k) - \underline{\theta}_N(k) \quad , \tag{6.1-17}$$

Eq. (6.1-2) can also be written as

$$\delta\underline{\theta}(k + 1) = \Phi(k + 1, k)\delta\underline{\theta}(k) + \underline{\omega}(k) \quad . \tag{6.1-18}$$

The quantities $\underline{\theta}_N(k)$ and $\Phi(k + 1, k)$ contain our a priori knowledge about $\underline{\theta}(k)$, and $\underline{\omega}(k)$ reflects our uncertainty in this knowledge. Observe that

$$E\{\delta\underline{\theta}(k + 1)\} = \Phi(k + 1, k)E\{\delta\underline{\theta}(k)\} \quad ; \tag{6.1-19}$$

if, for example, $\Phi(k + 1, k) = \Phi$ and Φ is a stable matrix, then, for some $k > k' >> 0$,

$$E\{\delta\underline{\theta}(k)\} \to \underline{0} \quad , \tag{6.1-20}$$

which means that

$$E\{\underline{\theta}(k)\} \to \underline{\theta}_N(k) \quad . \tag{6.1-21}$$

Loosely speaking, this demonstrates that after the transients that are associated with the dynamic model for $\delta\underline{\theta}(k)$ die out, the model provides only a statistical variation of $\underline{\theta}(k)$ about $\underline{\theta}_N(k)$. The size of this variation can be controlled by increasing or decreasing $Q(k)$. We leave it as an exercise for the reader to show that

$$\mathrm{Cov}[\delta\underline{\theta}(k + 1)] = \Phi(k + 1, k)\ \mathrm{Cov}[\delta\underline{\theta}(k)]\Phi'(k + 1, k) + Q(k) \tag{6.1-22}$$

(Problem 6-1).

Modeling $\underline{\theta}(k)$ by the dynamical system in Eq. (6.1-18) causes $\underline{\theta}(k)$ to be a random process. Throughout earlier chapters in this book, we have limited our discussions to the identification of deterministic parameters. Type-C information causes us to depart somewhat from our main theme.

Let us take a closer look at the cause of $\underline{\theta}(k)$'s random behavior. We have assumed that randomness in $\underline{\theta}(k)$ is owing to modeling uncertainties; that is to say, we have purposely made a deterministic parameter appear to be random in order to obtain an additional degree of freedom in our study - Q(k). This artifice is often useful in certain error analyses. ▲

The more information that is known ahead of time about $\underline{\theta}(k)$, the more easily it can be identified. No completely rigorous theory exists for the identification of time-varying parameters. We shall borrow, adapt, and extend the estimation algorithms in Chapters 2-5 to the time-varying parameter situation. In most cases, our estimation algorithms can only be considered to be heuristic extensions of these earlier algorithms. Our confidence in them increases with increasing a priori information about $\underline{\theta}(k)$ and with decreasing time variability in $\underline{\theta}(k)$.

6.2 ESTIMATION FROM TYPE-A INFORMATION

Generalized least-squares and unbiased minimum-variance estimation cannot be used when there is no structural information about $\underline{\theta}(k)$. The reason for this can be traced back to the very origin of these estimation techniques, the concatenated measurement equation. Let us retrace the development leading up to the concatenated measurement equation in Eq. (2.2-16), but for a time-varying parameter vector.

In this case, the scalar measurement at time t_j is

$$z(t_j) = \underline{h}'(t_j)\underline{\theta}(t_j) + v(t_j) \quad . \tag{6.2-1}$$

These measurements are made at times $t_{k-L}, t_{k-L+1}, \ldots, t_k$. Collecting the L + 1 measurements, we obtain

$$\begin{aligned} z(k) &= \mathbf{h}'(k)\boldsymbol{\theta}(k) + v(k) \\ z(k-1) &= \mathbf{h}'(k-1)\boldsymbol{\theta}(k-1) + v(k-1) \\ &\ \vdots \\ z(k-L) &= \mathbf{h}'(k-L)\boldsymbol{\theta}(k-L) + v(k-L) \quad . \end{aligned} \tag{6.2-2}$$

We do not know how $\boldsymbol{\theta}(k)$, $\boldsymbol{\theta}(k-1)$, ..., $\boldsymbol{\theta}(k-L)$ are related; thus, it is not possible to express $\boldsymbol{\theta}(k)$, $\boldsymbol{\theta}(k-1)$, ..., $\boldsymbol{\theta}(k-L)$ in terms of the parameter vector at one instant of time, say t_k, which means that it is impossible to write the L + 1 equations in Eq. (6.2-2) as

$$\mathbf{Z}(k) = \mathcal{H}(k)\boldsymbol{\theta}(k) + \mathbf{V}(k) \quad . \tag{6.2-3}$$

Equation (6.2-3) is the concatenated measurement equation. ▲

Next, let us take a brief look at the difficulties that are encountered when $\boldsymbol{\theta}(k)$ is estimated by means of gradient algorithms. For convenience, our attention is directed at the deterministic situation.

The underlying formulation, statement, and solution of the estimation problem are found in Section 4.2. The important equations from that section for the time-varying parameter situation are

$$\boldsymbol{\theta}(t) = [\theta_1(t), \theta_2(t), \ldots, \theta_n(t)]' \quad , \tag{6.2-4}$$

$$y(k) = \mathbf{x}'(k)\boldsymbol{\theta}(k) \quad , \tag{6.2-5}$$

$$\hat{y}(k) = \mathbf{x}'(k)\hat{\boldsymbol{\theta}}(k) \quad , \tag{6.2-6}$$

$$\tilde{y}(k) = y(k) - \hat{y}(k) \quad , \tag{6.2-7}$$

and

$$\hat{\boldsymbol{\theta}}(k+1) = \hat{\boldsymbol{\theta}}(k) + R(k)\mathbf{x}(k)\tilde{y}(k) \tag{6.2-8}$$

for $k \geq 0$. The algorithm for estimating $\boldsymbol{\theta}(k)$ is exactly the same as the algorithm for estimating $\boldsymbol{\theta}$; however, as we demonstrate next, the identification-error system for $\tilde{\boldsymbol{\theta}}(k) = \boldsymbol{\theta}(k) - \hat{\boldsymbol{\theta}}(k)$ is not the same as the identification-error system for $\tilde{\boldsymbol{\theta}}(k) = \boldsymbol{\theta} - \hat{\boldsymbol{\theta}}(k)$.

We proceed exactly as in the development of Eq. (4.3-2). From Eqs. (6.2-8) and (6.2-5)-(6.2-7), we obtain

$$\hat{\underline{\theta}}(k+1) = \hat{\underline{\theta}}(k) + R(k)\underline{x}(k)[y(k) - \hat{y}(k)]$$

$$= \hat{\underline{\theta}}(k) + R(k)\underline{x}(k)\underline{x}'(k)[\underline{\theta}(k) - \hat{\underline{\theta}}(k)]$$

$$= \hat{\underline{\theta}}(k) + R(k)\underline{x}(k)\underline{x}'(k)\tilde{\underline{\theta}}(k) \quad . \qquad (6.2\text{-}9)$$

Adding and subtracting $\underline{\theta}(k+1)$ to the left-hand side and $\underline{\theta}(k)$ to the right-hand side of this equation, it is easily shown that

$$\tilde{\underline{\theta}}(k+1) = [I - R(k)\underline{x}(k)\underline{x}'(k)]\tilde{\underline{\theta}}(k) + \Delta\underline{\theta}(k+1, k) \qquad (6.2\text{-}10)$$

for $k \geq 0$, where

$$\Delta\underline{\theta}(k+1, k) = \underline{\theta}(k+1) - \underline{\theta}(k) \quad . \qquad (6.2\text{-}11)$$

We see that for time-varying $\underline{\theta}(k)$, the identification-error system is no longer homogeneous; it is forced by $\Delta\underline{\theta}(k+1, k)$. The notion of *bounded-input/bounded-output stability* for a forced system like Eq. (6.2-10) was introduced by means of a very simple example in Section 2.11.1. Loosely speaking, if an unforced system is stable, then that same system, when excited by a bounded input function, will have a bounded output response (*6-1* and *6-2*). We have already demonstrated uniform asymptotic stability in the large for the unforced portion of Eq. (6.2-10); hence, we can conclude that if $\Delta\underline{\theta}(k+1, k)$ is bounded then $\tilde{\underline{\theta}}(k+1)$ will be as well. Since $\tilde{\underline{\theta}}$ is the parameter estimation error between $\underline{\theta}(k)$ and $\hat{\underline{\theta}}(k)$, we would like the bound on $\tilde{\underline{\theta}}(k+1)$ to be as small as possible. Clearly, the amplitude of any bound on $\tilde{\underline{\theta}}(k+1)$ depends quite strongly on $\Delta\underline{\theta}(k+1, k)$ (Problem 6-3). For $||\tilde{\underline{\theta}}(k)||^2$ to be small, $||\Delta\underline{\theta}(k+1, k)||^2$ must also be small.

Using these arguments, we have been led to the subjective conclusion that if $\Delta\underline{\theta}(k+1, k)$ is small, then we can expect $\hat{\underline{\theta}}(k)$ to track $\underline{\theta}(k)$ well. Except for the fact that as $\Delta\underline{\theta}(k+1, k) \to \underline{0}$ Eq. (6.2-10) $\to$ Eq. (4.4-1), and that we earlier demonstrated the stability of $\tilde{\underline{\theta}}(k)$ in Eq. (4.4-1), there is nothing very rigorous in our conclusions; but to date that is where matters stand on this subject.

It is conjectured that in order to obtain more quantitative results on bounding $\Delta\underline{\theta}(k + 1, k)$, we must study finite-time stability of $\tilde{\underline{\theta}}(k)$ rather than the more conventional forms of stability described in Section 4.4.2. Clearly, we would like to be able to prespecify a value of k, say k', so that for k > k', $||\tilde{\underline{\theta}}(k)||^2 < \varepsilon$. A bound on $||\Delta\underline{\theta}||^2$ would then be a function of k' and ε as well as k.

Similar conclusions can be drawn for the stochastic situation; that is to say, we can borrow the algorithms in Eq. (5.4-6) [Class-1 identification problem] and (5.4-15) [Class-2 identification problem] to estimate $\underline{\theta}(k)$, and as long as $\Delta\underline{\theta}(k + 1, k)$ is not too large, we should be able to track $\underline{\theta}$ reasonably well.

Subjective statements such as we have made are often irritating to staunch advocates of rigor; however, if no structural information about $\underline{\theta}(k)$ is known or can be assumed [e. g., approximate $\underline{\theta}(t)$ as in Eq. (6.1-14)], there are no known rigorous results. Perhaps we are being unrealistic when we claim only Type-A information about $\underline{\theta}(k)$.

6.3 ESTIMATION FROM TYPE-B INFORMATION

Our attention is now directed to the situation, contrived or otherwise, where

$$\underline{\theta}(k) = L(k)\underline{\beta}(k) \tag{6.3-1}$$

and $\underline{\beta}(k)$ is assumed to be "slowly time varying."

We approach the estimation of $\underline{\theta}(k)$ as follows: (1) estimate $\underline{\beta}(k)$ by means of an appropriate estimation algorithm, and then (2) estimate $\underline{\theta}(k)$ from the obvious consequence of Eq. (6.3-1) that

$$\hat{\underline{\theta}}(k) = L(k)\hat{\underline{\beta}}(k) \quad . \tag{6.3-2}$$

Of course, these two steps can be combined, giving us $\hat{\underline{\theta}}(k)$ directly; but it will be conceptually helpful for us to initially consider the estimation of $\underline{\theta}(k)$ in two stages.

Unless $\underline{\beta}(k) = \underline{\beta}$, it is once again not possible to use either generalized least-squares or unbiased minimum-variance estimation algorithms to estimate $\underline{\beta}(k)$. We cannot obtain a concatenated measurement equation for time-varying parameters. If $\underline{\beta}$ is constant, which is not too likely in practical applications, we can of course use either of these algorithms. In this case,

$$\begin{aligned} \underline{Z}(k) &= H(k)\underline{\theta}(k) + \underline{V}(k) \\ &= H(k)L(k)\underline{\beta} + \underline{V}(k) \\ &= H_1(k)\underline{\beta} + \underline{V}(k) \end{aligned} \tag{6.3-3}$$

and it is clear that all we need to do to apply the algorithms of Chapters 2 and 3 is to replace $H(k)$ by $H_1(k)$ where

$$H_1(k) = H(k)L(k) \quad . \tag{6.3-4}$$

For the sequential algorithms, this means replacing $H(k + 1)$ by $H_1(k + 1)$, where

$$H_1(k + 1) = H(k + 1)L(k + 1) \quad . \tag{6.3-5}$$

In case $\underline{\beta}(k) \neq \underline{\beta}$, we must use a gradient algorithm to estimate the slowly varying $\underline{\beta}(k)$. Let us consider the deterministic situation of Chapter 4 first.

We inquire as to how the algorithm

$$\hat{\underline{\theta}}(k + 1) = \hat{\underline{\theta}}(k) + R(k)\underline{x}(k)\tilde{y}(k) \tag{6.3-6}$$

where

$$R(k) \triangleq \frac{H(k)}{\underline{x}'(k)H(k)\underline{x}(k)} \tag{6.3-7}$$

and [see Eq. (4.6-2)]

$$H(k) = \text{diag}[h_1(k), \ldots, h_n(k)] \tag{6.3-8}$$

should be modified for the estimation of $\underline{\beta}(k)$ rather than $\underline{\theta}(k)$. Observe from Eqs. (4.2-4), (4.2-5), (6.3-1), and (6.3-2) that *(6-3 - 6-6)*

$$y(k) = \underline{\theta}'(k)\underline{x}(k) = \underline{\beta}'(k\ [L'(k)\underline{x}(k)] \tag{6.3-9}$$

and

$$\hat{y}(k) = \hat{\underline{\theta}}'(k)\underline{x}(k) = \hat{\underline{\beta}}'(k)[L'(k)\underline{x}(k)] \quad . \tag{6.3-10}$$

We see that $\underline{x}(k)$ acts as the identification system input (see Fig. 4.2-1) with respect to the identification of $\underline{\theta}(k)$, and $[L'(k)\underline{x}(k)]$

acts as the system input with respect to the identification of $\underline{\beta}(k)$. Hence, replacing $\underline{x}(k)$ by $L'(k)\underline{x}(k)$ in Eqs. (6.3-6) and (6.3-7), we obtain the following algorithm for estimating $\underline{\beta}(k)$:

$$\hat{\underline{\beta}}(k + 1) = \hat{\underline{\beta}}(k) + R_{\beta P}(k)[L'(k)\underline{x}(k)]\tilde{y}(k) \tag{6.3-11}$$

where[†]

$$R_{\beta P}(k) = \frac{H(k)}{[L'(k)\underline{x}(k)]'H(k)[L'(k)\underline{x}(k)]} \tag{6.3-12}$$

▲

Next, let us consider the stochastic situation of Chapter 5. We inquire how the algorithm for Class-1 problems,

$$\hat{\underline{\theta}}(k + \ell) = [I + R(k)\Sigma_n]\hat{\underline{\theta}}(k) + R(k)\underline{r}(k)\tilde{z}(k) \tag{6.3-13}$$

where

$$R(k) = \begin{cases} \dfrac{H(k)}{\underline{r}'(k)H(k)\underline{r}(k)} & \text{for } k \leq k' \\[2ex] \dfrac{1}{k^p} \quad (\tfrac{1}{2} < p \leq 1) & \text{for } k > k' \end{cases} \tag{6.3-14}$$

and H(k) is given in Eq. (6.3-8), should be modified for the estimation of $\underline{\beta}(k)$. We leave the comparable results for Class-2 algorithms as an exercise for the reader (Problem 6-4). Observe from Eqs. (5.1-8), (5.1-13), (6.3-1), and (6.3-2) that

$$y(k) = \underline{\theta}'(k)\underline{x}(k) = \underline{\beta}'(k)[L'(k)\underline{x}(k)] \tag{6.3-15}$$

and

$$\hat{z}(k) = \hat{\underline{\theta}}'(k)\underline{r}(k) = \hat{\underline{\beta}}'(k)[L'(k)\underline{r}(k)] \quad , \tag{6.3-16}$$

which means that $\underline{x}(k)$ and $\underline{r}(k)$ act as the system and model inputs (Fig. 5.1-1) with respect to the identification of $\underline{\theta}(k)$, whereas $L'(k)\underline{x}(k)$ and $L'(k)\underline{r}(k)$ act as the system and model inputs with respect to the identification of $\underline{\beta}(k)$. The fact that $L'(k)\underline{r}(k)$ acts as the model input means

---

† The P subscript on R denotes the perfect measurement situation.

$$[L'(k)\underline{r}(k)] = [L'(k)\underline{x}(k)] + [L'(k)\underline{n}(k)] \quad ; \tag{6.3-17}$$

hence, when using Eq. (6.3-13) to estimate $\underline{\beta}(k)$, we must also modify the measurement noise covariance matrix from Σ_n to $L'(k)\Sigma_n L(k)$.

Replacing $\underline{x}(k)$ by $L'(k)\underline{x}(k)$, $\underline{r}(k)$ by $L'(k)\underline{r}(k)$, and Σ_n by $L'(k)\Sigma_n L(k)$ in Eqs. (6.3-13) and (6.3-14), we obtain the following algorithm for estimating $\underline{\beta}(k)$:

$$\hat{\underline{\beta}}(k + \ell) = [I + R_{\beta N}(k)L'(k)\Sigma_n L(k)]\hat{\underline{\beta}}(k) + R_{\beta N}(k)[L'(k)\underline{r}(k)]\tilde{z}(k) \tag{6.3-18}$$

where[†]

$$R_{\beta N}(k) = \begin{cases} \dfrac{H(k)}{[L'(k)\underline{r}(k)]'H(k)[L'(k)\underline{r}(k)]} & \text{for} \quad k \leq k' \\ \\ \dfrac{1}{k^p} \quad (\tfrac{1}{2} < p \leq 1) & \text{for} \quad k > k' \quad . \end{cases} \tag{6.3-19}$$

▲

<u>Example 1</u>. *Estimation of Time-Varying Aerodynamic Parameters from Perfect Measurements*. Here we shall consider the identification of the two time-varying aerodynamic parameters $M_\alpha(t)$ and $M_\delta(t)$ that are associated with the high-performance aerospace adaptive control system that was described in Section 1.3.5. As in Section 4.7.2, our approach is to estimate these parameters from the perfect measurement equation for $\ddot{\theta}(k)$ (see Fig. 1.3.2),

$$\ddot{\theta}(k) = M_\alpha(k)\alpha(k) + M_\delta(k)\delta(k) \quad . \tag{6.3-20}$$

Estimates of $M_\alpha(k)$ and $M_\delta(k)$ are obtained from Eqs. (6.3-2), (6.3-11), (6.3-12), (6.1-8), and (6.1-9), and are given by the following expressions (Problem 6-5):

$$\hat{M}_\alpha(k + 1) = \frac{q(k + 1)}{q(k)}\left(\hat{M}_\alpha(k) + \frac{h_1(k)\alpha(k)\tilde{y}(k)}{h_1(k)\alpha^2(k) + h_2(k)\delta^2(k)}\right) \tag{6.3-21}$$

---

† The N subscript on R denotes the <u>n</u>oisy measurement situation.

and

$$\hat{M}_\delta(k+1) = \frac{q(k+1)}{q(k)}\left(\hat{M}_\delta(k) + \frac{h_2(k)\delta(k)\tilde{y}(k)}{h_1(k)\alpha^2(k) + h_2(k)\delta^2(k)}\right) \qquad (6.3\text{-}22)$$

where

$$\tilde{y}(k) = \ddot{\theta}(k) - \hat{\ddot{\theta}}(k) \qquad (6.3\text{-}23)$$

and

$$\hat{\ddot{\theta}}(k) = \hat{M}_\alpha(k)\alpha(k) + \hat{M}_\delta(k)\delta(k) \quad . \qquad (6.3\text{-}24)$$

It is assumed that $M_\alpha(t)$, $M_\delta(t)$, $Z_\alpha(t)$, and $q(t)$ vary as shown in Fig. 6.3-1. We leave it to the reader to demonstrate that $M_{\alpha q}(t)$ and $M_{\delta q}(t)$ [see Eqs. (6.1-6) and (6.1-7)] vary over much narrower ranges of values and at slower rates than $M_\alpha(t)$ and $M_\delta(t)$ do (Problem 6-6).

All estimates were obtained for $h_2/h_1 = 1/10$ (see Section 4.7.2) and for a sequence of step commands in input normal acceleration, similar to the sequence depicted in Fig. 4.7-4. The command was incremented by 10 g every second during the 5-sec flight.

Figures 6.3-2 and 6.3-3 compare two types of estimates for $M_\alpha(t)$ and $M_\delta(t)$, respectively, each obtained for two types of control situations. The terms *a priori estimate* and *a posteriori estimate* are used here to distinguish between an estimate in which no dynamic pressure information is used and an estimate in which this information is used. A priori estimates of $M_\alpha(t)$ and $M_\delta(t)$ are obtained from Eqs. (6.3-21) and (6.3-22), respectively, by setting $q(k+1) = q(k)$. Obviously, a priori estimates are associated with Type-A information for $M_\alpha(t)$ and $M_\delta(t)$, whereas a posteriori estimates are associated with Type-B information.

The terms *preprogrammed situation* and *adaptive situation* were explained in Section 1.3.5. Briefly, in the preprogrammed situation, the control gains K_{Ni}, $K_{\dot{\alpha}}$, and K_{Na} (see Fig. 1.3-2) are preprogrammed as functions of $M_\alpha(t)$, $M_\delta(t)$, and $Z_\alpha(t)$. This can always be done in a laboratory simulation. In the adaptive situation, the three control gains are updated periodically as explicit functions of $\hat{M}_\alpha(k)$,

$\hat{M}_\delta(k)$, and $Z_\alpha(k)$ [$Z_\alpha(t)$ is assumed known]. The adaptive situation leads to a Class-3 identification problem, since the system's states are dependent upon $\hat{M}_\alpha$ and $\hat{M}_\delta$; whereas, the preprogrammed situation leads to a Class-1 identification problem.

Note how much better the a posteriori identifier is able to track both $M_\alpha(t)$ and $M_\delta(t)$; thus, in this application it is clearly advantageous to assume Type-B rather than Type-A information for $M_\alpha(t)$ and $M_\delta(t)$. Note also that the identifier works better in the adaptive situation than in the preprogrammed situation. Because no theory exists for the Class-3 identification problem, it is not apparent why this is so. ▲

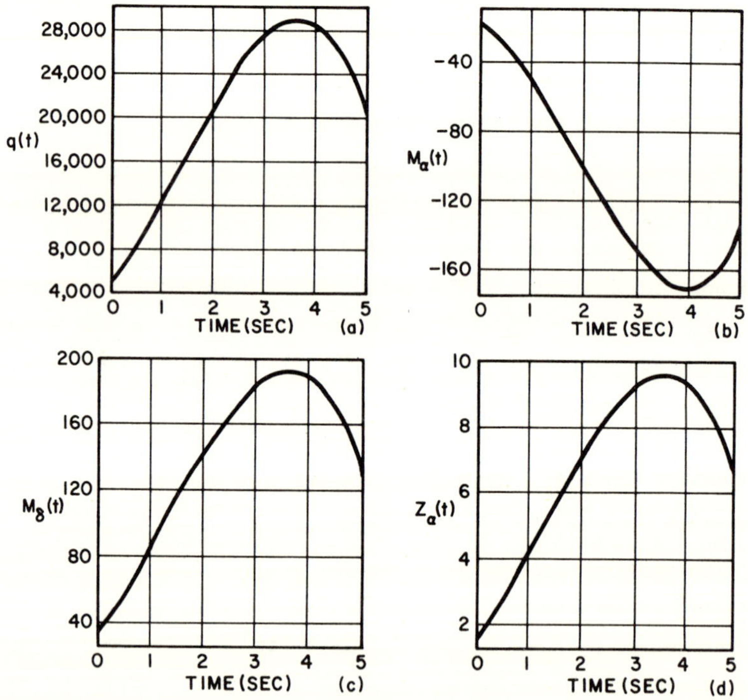

Fig. 6.3-1. Aerodynamic parameter histories.

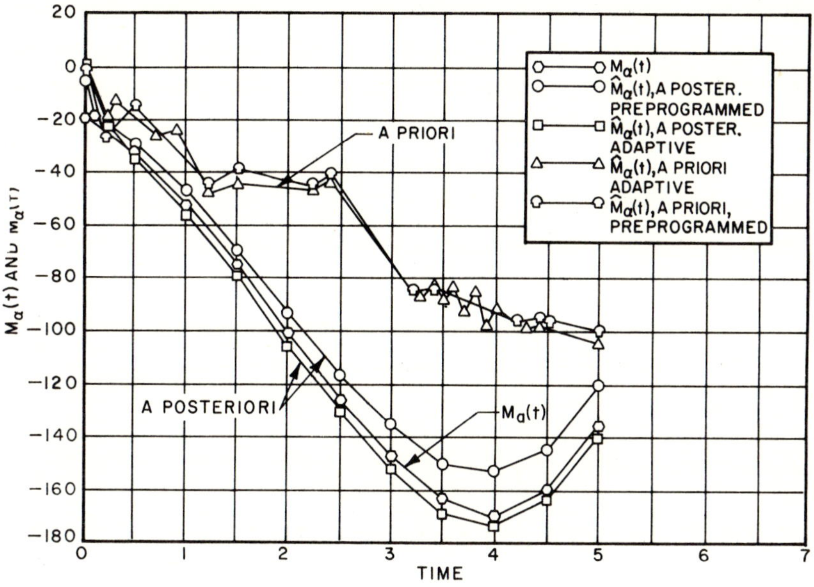

Fig. 6.3-2. Estimates of $M_\alpha(t)$.

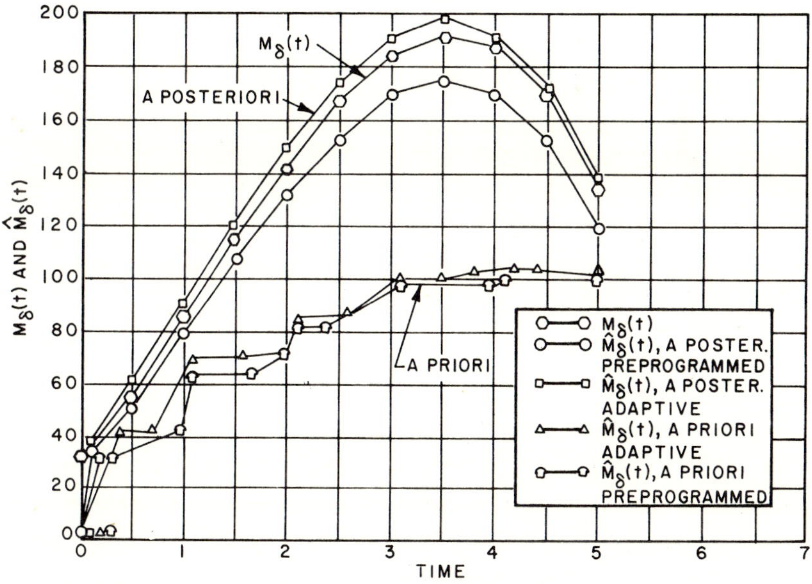

Fig. 6.3-3. Estimates of $M_\delta(t)$.

Example 2. *Estimation of Time-Varying Aerodynamic Parameters from Noisy Measurements.* Here we extend our discussions in the preceding example from the perfect measurement situation to the noisy measurement situation. The reader is referred to Example 1 in Section 5.1 for a description of this application as well as its formulation as a scalar equation error. Estimates of $M_\alpha(k)$ and $M_\delta(k)$ are obtained from Eqs. (6.3-2), (6.3-18), (6.3-19), (6.1-8), and (6.1-9), and are given by the following expressions (Problem 6-5):

$$\hat{M}_\alpha(k + 1) = \frac{q(k + 1)}{q(k)} \left[\left(1 + \frac{h_1(k)\sigma_\alpha^2}{d(k)}\right)\hat{M}_\alpha(k) + \frac{h_1(k)}{d(k)} \alpha_m(k)\tilde{z}(k) \right] \tag{6.3-25}$$

and

$$\hat{M}_\delta(k + 1) = \frac{q(k + 1)}{q(k)} \left[\left(1 + \frac{h_2(k)\sigma_\delta^2}{d(k)}\right)\hat{M}_\delta(k) + \frac{h_2(k)}{d(k)} \delta_m(k)\tilde{z}(k) \right] \tag{6.3-26}$$

where

$$\tilde{z}(k) = \ddot{\theta}_m(k) - \hat{\ddot{\theta}}(k) \quad , \tag{6.3-27}$$

$$\hat{\ddot{\theta}}(k) = \hat{M}_\alpha(k)\alpha_m(k) + \hat{M}_\delta(k)\delta_m(k) \quad , \tag{6.3-28}$$

$$d(k) = \begin{cases} h_1(k)\alpha_m^2(k) + h_2(k)\delta_m^2(k) & \text{for } k \le k' \\ k^p \quad (\frac{1}{2} < p \le 1) & \text{for } k > k' \quad , \end{cases} \tag{6.3-29}$$

and

$$\Sigma_n = \begin{pmatrix} \sigma_\alpha^2 & 0 \\ 0 & \sigma_\delta^2 \end{pmatrix} \quad . \tag{6.3-30}$$

We know that measurement-noise bias is removed from estimates of the aerodynamic parameters when they are constant (see Example 1 in Section 5.7). Is bias also removed when the identified parameters are time varying? Figure 6.3-4 demonstrates that the bias is quite small in this application; hence, we can *expect* unbiased estimates when the a posteriori identifier is used. If the three sets of

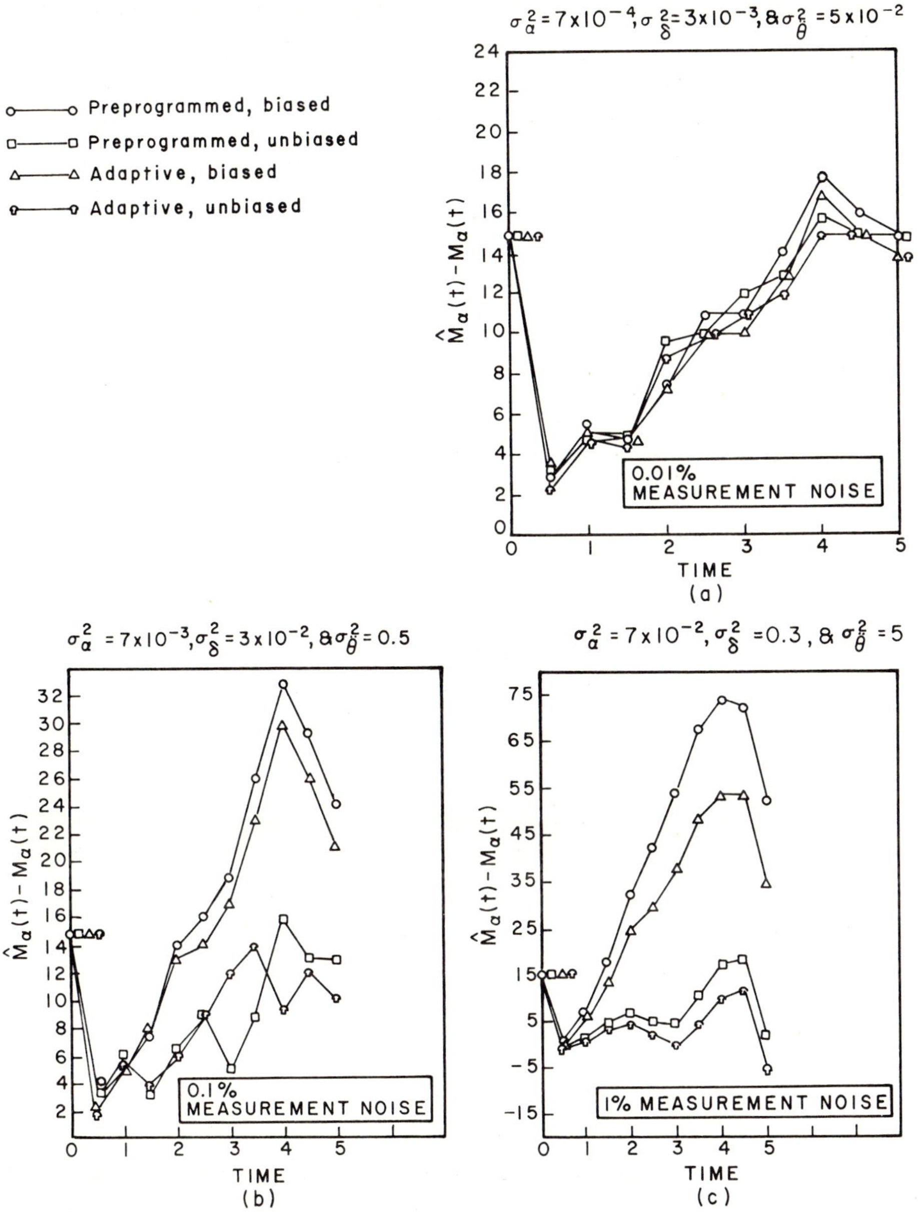

Fig. 6.3-4. Ensemble average of M_α identification error for a posteriori identification.

unbiased estimates in Figs. 6.3-4a-c are overlayed, one observes that bias is indeed small. Note also, that for very low noise levels (Fig. 6.3-4a), there is no appreciable advantage to removing the measurement-noise bias. On the other hand, for larger noise levels (Fig. 6.3-4c), it is clearly advantageous to make use of the measurement noise statistics.

Once again, we observe that in almost every case depicted in Fig. 6.3-4, identification error is smaller in the adaptive situation than for the preprogrammed situation. As noted in Example 1, no theory presently exists that explains why this is so.

In connection with our preceding discussions about measurement-noise biases, we note that when information about Σ_n is used, $\hat{M}_\alpha(k)$ and $\hat{M}_\delta(k)$ can be viewed as functions of Σ_n. A study was performed (*6-7*) to determine the sensitivities of $\hat{M}_\alpha(k)$ and $\hat{M}_\delta(k)$ to Σ_n. In all cases considered (42 in all), the sensitivities of $\hat{M}_\alpha(k)$ and $\hat{M}_\delta(k)$ to Σ_n is considerably less than unity. For example, when actual noise levels were all 10% of peak values for $\alpha(k)$, $\delta(k)$, and $\ddot{\theta}(k)$, and 5% levels were used in the identification algorithms in Eqs. (6.3-25) and (6.3-26) (which means a -50% error in Σ_n), the average percent errors in $\hat{M}_\alpha(k)$ and $\hat{M}_\delta(k)$ were less than 10%.

6.4 ESTIMATION FROM TYPE-C INFORMATION

We shall now study the important situation where, for one reason or another, it is possible to model $\underline{\theta}(k)$ by means of the equation

$$\delta\underline{\theta}(k + 1) = \Phi(k + 1, k)\delta\underline{\theta}(k) + \underline{\omega}(k) \quad . \qquad (6.4\text{-}1)$$

Let us direct our attention first to minimum-variance estimation of $\delta\underline{\theta}(k)$.

6.4.1 Minimum-Variance Estimation of $\delta\underline{\theta}(k)$

In order to make use of the a priori knowledge about $\delta\underline{\theta}(k)$ - $\Phi(k + 1, k)$ and $Q(k)$ - we will have to trace through the steps that led to the classical batch-estimation algorithm as well as the steps that led to the sequential unbiased minimum-variance estimation algorithm. Rather than doing this directly, we invoke Corollary 3-1 in Section 3.5; that is to say, we shall trace through the steps that led to both the classical batch and sequential generalized least-squares algorithms in which we set $W(k) = \mathcal{R}^{-1}(k)$.

Concatenated Measurement Equations

As in Section 2.9, we shall assume that $\underline{z}(k)$ is an $m \times 1$ vector where

$$\underline{z}(k) = \underline{y}(k) + \underline{v}(k) \quad , \tag{6.4-2}$$

and $\underline{v}(k)$ is a random vector of measurement errors having the properties

$$E\{\underline{v}(k)\} = \underline{0} \tag{6.4-3}$$

and

$$E\{\underline{v}(k)\underline{v}'(k)\} = R(k) \quad . \tag{6.4-4}$$

Here, however, each component in $\underline{y}(k)$ is related to an unknown $n \times 1$ *time-varying* parameter vector $\underline{\theta}(k)$; that is to say,

$$\underline{y}(k) = H(k)\underline{\theta}(k) \tag{6.4-5}$$

so that

$$\underline{z}(k) = H(k)\underline{\theta}(k) + \underline{v}(k) \quad . \tag{6.4-6}$$

When $\underline{\theta}(k) = \underline{\theta}_N(k)$, where $\underline{\theta}_N(k)$ is known ahead of time,

$$\underline{y}_N = H(k)\underline{\theta}_N(k) \tag{6.4-7}$$

and

$$\underline{z}_N(k) = \underline{y}_N(k) \quad . \tag{6.4-8}$$

Defining $\delta\underline{y}(k)$ and $\delta\underline{z}(k)$ as

$$\delta\underline{y}(k) = \underline{y}(k) - \underline{y}_N(k) \tag{6.4-9}$$

and

$$\delta\underline{z}(k) = \underline{z}(k) - \underline{z}_N(k) \quad , \tag{6.4-10}$$

we see that

$$\delta\underline{y}(k) = H(k)\delta\underline{\theta}(k) \tag{6.4-11}$$

and

$$\delta\underline{z}(k) = H(k)\delta\underline{\theta}(k) + \underline{v}(k) \quad . \tag{6.4-12}$$

Our objective in this section is to concatenate the L + 1 measurement-difference vectors $\delta\underline{z}(k)$, $\delta\underline{z}(k-1)$, ..., $\delta\underline{z}(k-L)$ where

$$\begin{aligned}\delta\underline{z}(k) &= H(k)\delta\underline{\theta}(k) + \underline{v}(k)\\ \delta\underline{z}(k-1) &= H(k-1)\delta\underline{\theta}(k-1) + \underline{v}(k-1)\\ &\vdots\\ \delta\underline{z}(k-L) &= H(k-L)\delta\underline{\theta}(k-L) + \underline{v}(k-L)\quad .\end{aligned} \tag{6.4-13}$$

In order to accomplish this objective, we shall need the solution to Eq. (6.4-1). We must express $\delta\underline{\theta}(k-1)$, ..., $\delta\underline{\theta}(k-L)$ as functions of $\delta\underline{\theta}(k)$. Comparing Eq. (6.4-1) with Eq. (1.3-41) in Chapter 1, whose solution is given in Eq. (1.3-44), we conclude that

$$\delta\underline{\theta}(k) = \Phi(k, j)\delta\underline{\theta}(j) + \sum_{i=j+1}^{k} \Phi(k, i)\underline{\omega}(i-1) \tag{6.4-14}$$

for all $k \geq j+1$. Since we are interested in solutions of Eq. (6.4-1) for $t = t_{k-1}, t_{k-2}, \ldots, t_{k-L}$, a more useful version of Eq. (6.4-14) is obtained by solving for $\delta\underline{\theta}(j)$, as[†]

$$\delta\underline{\theta}(j) = \Phi(j, k)\delta\underline{\theta}(k) - \sum_{i=j+1}^{k} \Phi(j, i)\underline{\omega}(i-1) \tag{6.4-15}$$

for all $j \leq k-1$. Observe that $\delta\underline{\theta}(k)$ is common to each $\delta\underline{\theta}(j)$.

From Eqs. (6.4-13) and (6.4-15), we see that

$$\begin{aligned}\delta\underline{z}(k-j) &= H(k-j)\delta\underline{\theta}(k-j) + \underline{v}(k-j)\\ &= H(k-j)\Phi(k-j, k)\delta\underline{\theta}(k)\\ &\qquad - H(k-j)\sum_{i=k-j+1}^{k} \Phi(k-j, i)\underline{\omega}(i-1) + \underline{v}(k-j)\end{aligned} \tag{6.4-16}$$

for all $j \geq 1$; thus, the L + 1 equations in Eq. (6.4-13) can now be written as (Problem 6-7)

[†] In the derivation of Eq. (6.4-15), we make use of the well-known property for the transition matrix $\Phi(k, j)$ that $\Phi^{-1}(k, j) = \Phi(j, k)$.

$$\delta\underline{\mathcal{Z}}(k) = \mathcal{H}(k)\delta\underline{\theta}(k) + \underline{\mathcal{V}}(k) - \mathcal{M}(k)\underline{\bar{\omega}}(k - 1) \tag{6.4-17}$$

where

$$\delta\underline{\mathcal{Z}}(k) = [\delta\underline{z}'(k) \mid \delta\underline{z}'(k - 1) \mid \cdots \mid \delta\underline{z}'(k - L)]' \quad , \tag{6.4-18}$$

$$\mathcal{H}(k) = \begin{pmatrix} H(k) \\ \hdashline H(k - 1)\Phi(k - 1, k) \\ \hdashline \vdots \\ \hdashline H(k - L)\Phi(k - L, k) \end{pmatrix} \quad , \tag{6.4-19}$$

$$\underline{\mathcal{V}}(k) = [\underline{v}'(k) \mid \underline{v}'(k - 1) \mid \cdots \mid \underline{v}'(k - L)]' \quad , \tag{6.4-20}$$

$$\mathcal{M}(k) = \left(\begin{array}{c:c:c:c} 0 & 0 & \cdots & 0 \\ \hdashline H(k-1)\Phi(k-1, k) & 0 & \cdots & 0 \\ \hdashline \begin{array}{l} H(k-2) \\ \times\Phi(k-2, k-1) \end{array} & \begin{array}{l} H(k-2) \\ \times\Phi(k-2, k) \end{array} & \cdots & 0 \\ \hdashline \vdots & \vdots & \ddots & \vdots \\ \hdashline \begin{array}{l} H(k-L) \\ \times\Phi(k-L, k-L+1) \end{array} & \begin{array}{l} H(k-L) \\ \times\Phi(k-L, k-L+2) \end{array} & \cdots & \begin{array}{l} H(k-L) \\ \times\Phi(k-L, k) \end{array} \end{array}\right), \tag{6.4-21}$$

and

$$\underline{\bar{\omega}}(k - 1) = [\underline{\omega}'(k - 1) \mid \underline{\omega}'(k - 2) \mid \cdots \mid \underline{\omega}'(k - L)]' \quad . \tag{6.4-22}$$

The term $\underline{\bar{\omega}}(k - 1)$ is an $Lm \times 1$ vector and $\mathcal{M}(k)$ is $(L + 1)m \times Lm$.

Equation (6.4-17) can also be written as

$$\delta\underline{\mathcal{Z}}(k) = \mathcal{H}(k)\delta\underline{\theta}(k) + \underline{\mathcal{V}}_1(k) \tag{6.4-23}$$

where

$$\underline{\mathcal{V}}_1(k) = \underline{\mathcal{V}}(k) - \mathcal{M}(k)\underline{\bar{\omega}}(k - 1) \quad . \tag{6.4-24}$$

In this form, Eq. (6.4-23) is analogous to the concatenated measurement equation in Eq. (2.9-6).

Observe that the statistics of the $(L + 1)m \times 1$ noise vector $\underline{V}_1(k)$ are

$$E\{\underline{V}_1(k)\} = \underline{0} \tag{6.4-25}$$

and

$$E\{\underline{V}_1(k)\underline{V}_1'(k)\} \triangleq R_1(k) = R(k) - M(k)\mathcal{Q}(k - 1)M'(k) \tag{6.4-26}$$

where $R(k)$ is defined in Eq. (3.2-12), and

$$\mathcal{Q}(k - 1) = \left(\begin{array}{c|c|c|c} Q(k-1) & 0 & \cdots & 0 \\ \hline 0 & Q(k-2) & \cdots & 0 \\ \hline \vdots & \vdots & \ddots & \vdots \\ \hline 0 & 0 & \cdots & Q(k-L) \end{array}\right) . \tag{6.4-27}$$

Finally, observe from Eq. (6.4-24) that $\underline{V}_1(k)$ is independent of $\underline{\omega}(k)$.

<u>New Notation</u>

Not only can Eq. (6.4-23) be used as the basis for estimating $\delta\underline{\theta}(k)$, but it can also be used as the basis for estimating $\delta\underline{\theta}(k + 1)$; for, from Eq. (6.1-18),

$$\delta\underline{\theta}(k) = \Phi(k, k + 1)\delta\underline{\theta}(k + 1) - \Phi(k, k + 1)\underline{\omega}(k) \tag{6.4-28}$$

and, therefore,

$$\begin{aligned} \delta\underline{Z}(k) &= H(k)\Phi(k, k + 1)\delta\underline{\theta}(k + 1) - H(k)\Phi(k, k + 1)\underline{\omega}(k) + \underline{V}_1(k) \\ &= H_2(k)\delta\underline{\theta}(k + 1) + \underline{V}_2(k) \end{aligned} \tag{6.4-29}$$

where

$$H_2(k) = H(k)\Phi(k, k + 1) \tag{6.4-30}$$

and

$$\underline{V}_2(k) = \underline{V}_1(k) - H_2(k)\underline{\omega}(k) \quad . \tag{6.4-31}$$

In Chapters 2 and 3, $\hat{\underline{\theta}}(k)$ denoted the estimate of $\underline{\theta}$ [or $\underline{\theta}(k)$] based on $\underline{z}(k)$, $\underline{z}(k - 1)$, ..., $\underline{z}(k - L)$, whereas $\hat{\underline{\theta}}(k + 1)$ denoted the estimate based on $\underline{z}(k + 1)$, $\underline{z}(k)$, ..., $\underline{z}(k - L)$. If $\delta\underline{\theta}(k)$ is estimated from Eq. (6.4-23), the estimate will be based on $\delta\underline{z}(k)$,

$\delta\underline{z}(k-1), \ldots, \delta\underline{z}(k-L)$ and the notation $\delta\hat{\underline{\theta}}(k)$ for this estimate is consistent with our earlier convention. If, on the other hand, $\delta\underline{\theta}(k+1)$ is estimated from Eq. (6.4-29), its estimate will also be based on $\delta\underline{z}(k), \delta\underline{z}(k-1), \ldots, \delta\underline{z}(k-L)$, and clearly the notation $\delta\hat{\underline{\theta}}(k+1)$ for this estimate is no longer consistent with our earlier convention.

We shall need some new notation in order to distinguish two types of estimates, *a priori* and *a posteriori estimates*. In the spirit of modern state estimation, we shall use the notation $\delta\hat{\underline{\theta}}(k|j)$ to denote the estimate of $\delta\underline{\theta}(k)$ at t_k based on the measurements up to and including the jth - $\underline{z}(k-L), \underline{z}(k-L+1), \ldots, \underline{z}(k-L+j)$. This conditional notation for an estimate was first introduced by Kalman (*6-8*) in his pioneering work on minimum-variance state estimation. Although at first it may appear to be somewhat cumbersome, it is actually quite concise.

Using this notation, $\delta\hat{\underline{\theta}}(k|k)$ is the estimate of $\delta\underline{\theta}(k)$ that is obtained from Eq. (6.4-23), and $\delta\hat{\underline{\theta}}(k+1|k)$ is the estimate of $\delta\underline{\theta}(k+1)$ that is obtained from Eq. (6.4-29). The vector $\delta\hat{\underline{\theta}}(k+1|k)$ is often referred to as the *single-stage predicted estimate* of $\delta\underline{\theta}(k+1)$, or the *a priori estimate* of $\delta\underline{\theta}(k+1)$; whereas $\delta\hat{\underline{\theta}}(k+1|k+1)$ is often referred to as the *filtered estimate* of $\delta\underline{\theta}(k+1)$, or, the *a posteriori estimate* of $\delta\underline{\theta}(k+1)$. Clearly, the difference between $\delta\hat{\underline{\theta}}(k+1|k+1)$ and $\delta\hat{\underline{\theta}}(k+1|k)$ is that the most recent measurement $\underline{z}(k+1)$ is used to obtain $\delta\hat{\underline{\theta}}(k+1|k+1)$, whereas it is not used in the determination of $\delta\hat{\underline{\theta}}(k+1|k)$.

Classical a Priori and a Posteriori Batch Estimation Algorithms

The classical batch estimation algorithm for $\delta\underline{\theta}(k|k)$ is obtained by comparing Eqs. (6.4-23) and (2.9-6), and by using the algorithm in Eq. (2.9-11) setting $W(k) = R_1(k)$:

$$\delta\hat{\underline{\theta}}(k|k) = [H'(k)R_1^{-1}(k)H(k)]^{-1}H'(k)R_1^{-1}(k)\delta\underline{Z}(k) \quad . \tag{6.4-32}$$

In a similar manner, the classical batch estimation algorithm for $\delta\hat{\underline{\theta}}(k+1|k)$ is

$$\delta\hat{\underline{\theta}}(k + 1|k) = [H_2'(k)R_2^{-1}(k)H_2(k)]^{-1}H_2'(k)R_2^{-1}(k)\delta\underline{Z}(k) \qquad (6.4\text{-}33)$$

where, from Eqs. (6.4-31) and (6.4-26) and the fact that $\underline{V}_1(k)$ and $\underline{\omega}(k)$ are independent,

$$R_2(k) = E\{\underline{V}_2(k)\underline{V}_2'(k)\} = R_1(k) + H_2(k)Q(k)H_2'(k) \quad . \qquad (6.4\text{-}34)$$

Letting

$$P(k|k) = [H'(k)R_1^{-1}(k)H(k)]^{-1} \qquad (6.4\text{-}35)$$

and

$$P(k + 1|k) = [H_2'(k)R_2^{-1}(k)H_2(k)]^{-1} \quad , \qquad (6.4\text{-}36)$$

these estimates can also be written more compactly as

$$\delta\hat{\underline{\theta}}(k|k) = P(k|k)H'(k)R_1^{-1}(k)\delta\underline{Z}(k) \qquad (6.4\text{-}37)$$

and

$$\delta\hat{\underline{\theta}}(k + 1|k) = P(k + 1|k)H_2'(k)R_2^{-1}(k)\delta\underline{Z}(k) \qquad (6.4\text{-}38)$$

for $k \geq L$.

We suspect that since $\delta\hat{\underline{\theta}}(k|k)$ and $\delta\hat{\underline{\theta}}(k + 1|k)$ are both functions of $\delta\underline{Z}(k)$ they are closely related. We shall demonstrate a very simple relationship between these estimates; however, first we shall prove the lemma:

Lemma 6-1. *Matrices* $P(k|k)$ *and* $P(k + 1|k)$ *are related to each other by the expression (6-9)*

$$P(k + 1|k) = \Phi(k + 1, k)P(k|k)\Phi'(k + 1, k) + Q(k) \quad . \qquad (6.4\text{-}39)$$

Proof. In this proof, we shall make use of the following matrix-inversion lemma (Problem 2-15 in Chapter 2):

$$(I_n + XY)^{-1} = I_n - X(I_m + YX)^{-1}Y \quad , \qquad (6.4\text{-}40)$$

in which X is $n \times m$ and Y is $m \times n$.

From Eqs. (6.4-36) and (6.4-34),

$$P(k + 1|k) = \{H_2'(k)[R_1(k) + H_2(k)Q(k)H_2'(k)]^{-1}H_2(k)\}^{-1} \quad ; \qquad (6.4\text{-}41)$$

hence

$$P^{-1}(k + 1|k) = H_2'(k)[I_x + R_1^{-1}(k)H_2(k)Q(k)H_2'(k)]^{-1}R_1^{-1}(k)H_2(k) \qquad (6.4\text{-}42)$$

where I_x is the $(L + 1)m \times (L + 1)m$ identity matrix. Equation (6.4-40) is now applied to the first inverse on the right-hand side of Eq. (6.4-42) by associating $R_1^{-1}(k)H_2(k)$ with X and $Q(k)H_2'(k)$ with Y; thus,

$$P^{-1}(k+1|k) = H_2'(k)\{I_x - R_1^{-1}(k)H_2(k)[I_y + Q(k)H_2'(k)R_1^{-1}(k)H_2(k)]^{-1} \times Q(k)H_2'(k)\}R_1^{-1}(k)H_2(k) \tag{6.4-43}$$

where I_y is the $n \times n$ identity matrix. The right-hand side of this equation can be greatly simplified by means of the following algebraic manipulations:

$$\begin{aligned}
P^{-1}(k+1|k) &= H_2'R_1^{-1}H_2 - H_2'R_1^{-1}H_2(I_y + QH_2'R_1^{-1}H_2)^{-1}QH_2'R_1^{-1}H_2 \\
&= H_2'R_1^{-1}H_2 - H_2'R_1^{-1}H_2\{[(H_2'R_1^{-1}H_2)^{-1} \\
&\quad + Q](H_2'R_1^{-1}H_2)\}^{-1}QH_2'R_1^{-1}H_2 \\
&= H_2'R_1^{-1}H_2 - [(H_2'R_1^{-1}H_2)^{-1} + Q]^{-1}QH_2'R_1^{-1}H_2 \\
&= [(H_2'R_1^{-1}H_2)^{-1} + Q]^{-1}\{[(H_2'R_1^{-1}H_2)^{-1} \\
&\quad + Q]H_2'R_1^{-1}H_2 - QH_2'R_1^{-1}H_2\} \\
&= [(H_2'R_1^{-1}H_2)^{-1} + Q]^{-1} \quad .
\end{aligned} \tag{6.4-44}$$

Hence

$$\begin{aligned}
P(k+1|k) &= (H_2'R_1^{-1}H_2)^{-1} + Q \\
&= [\Phi'(k, k+1)H'(k)R_1^{-1}(k)H(k)\Phi(k, k+1)]^{-1} + Q(k) \\
&= \Phi(k+1, k)[H'(k)R_1^{-1}(k)H(k)]^{-1}\Phi'(k+1, k) + Q(k) \\
&= \Phi(k+1, k)P(k|k)\Phi'(k+1, k) + Q(k) \quad ,
\end{aligned} \tag{6.4-45}$$

which follows from Eq. (6.4-35). ▲

Now that we have related $P(k|k)$ to $P(k + 1|k)$ we relate $\delta\hat{\underline{\theta}}(k|k)$ to $\delta\hat{\underline{\theta}}(k + 1|k)$, in the following theorem.

Theorem 6-1. *The estimates* $\delta\underline{\hat{\theta}}(k|k)$ *and* $\delta\underline{\hat{\theta}}(k+1|k)$*, given in Eqs. (6.4-37) and (6.4-38), respectively, are related to each other by the expression (6-9)*

$$\delta\underline{\hat{\theta}}(k+1|k) = \Phi(k+1, k)\delta\underline{\hat{\theta}}(k|k) \quad . \qquad (6.4\text{-}46)$$

Proof. From Eqs. (6.4-38) and (6.4-34),

$$\begin{aligned}\delta\underline{\hat{\theta}}(k+1|k) &= P(k+1|k)H_2'(k)R_2^{-1}(k)\delta\underline{Z}(k)\\ &= P(k+1|k)H_2'(R_1 + H_2QH_2')^{-1}\delta\underline{Z}\\ &= P(k+1|k)H_2'(I_x + R_1^{-1}H_2QH_2')^{-1}R_1^{-1}\delta\underline{Z} \quad . \qquad (6.4\text{-}47)\end{aligned}$$

As in the proof of Lemma 6-1, we apply the matrix inversion lemma in Eq. (6.4-40) to $(I_x + R_1^{-1}H_2QH_2')^{-1}$ to obtain

$$\begin{aligned}\delta\underline{\hat{\theta}}(k+1|k) &= P(k+1|k)H_2'\{I_x - R_1^{-1}H_2[I_y + QH_2'R_1^{-1}H_2]^{-1}QH_2'\}R_1^{-1}\delta\underline{Z}\\ &= P(k+1|k)H_2'R_1^{-1}\delta\underline{Z} - P(k+1|k)H_2'R_1^{-1}H_2'[I_y\\ &\quad + QH_2'R_1^{-1}H_2]^{-1}QH_2'R_1^{-1}\delta\underline{Z}\\ &= P(k+1|k)H_2'R_1^{-1}\delta\underline{Z} - P(k+1|k)[(H_2'R_1^{-1}H_2)^{-1}\\ &\quad + Q]^{-1}QH_2'R_1^{-1}\delta\underline{Z}\\ &= P(k+1|k)H_2'R_1^{-1}\delta\underline{Z} - P(k+1|k)P^{-1}(k+1|k)QH_2'R_1^{-1}\delta\underline{Z}\\ &= [P(k+1|k) - Q]H_2'R_1^{-1}\delta\underline{Z}\\ &= \Phi P(k|k)\Phi' H_2'R_1^{-1}\delta\underline{Z}\\ &= \Phi(k+1, k)P(k|k)\Phi'(k+1, k)\Phi'(k, k+1)\\ &\quad \times H'(k)R_1^{-1}(k)\delta\underline{Z}(k)\\ &= \Phi(k+1, k)P(k|k)H'(k)R_1^{-1}(k)\delta\underline{Z}(k) \quad , \qquad (6.4\text{-}48)\end{aligned}$$

where we have made use of Eq. (6.4-44) and the well-known facts for transition matrices *(6-1)* that

$$\Phi(t_2, t_1)\Phi(t_1, t_0) = \Phi(t_2, t_0) \tag{6.4-49}$$

and

$$\Phi(t, t) = I \quad . \tag{6.4-50}$$

Comparing Eqs. (6.4-48) and (6.4-37), we see that

$$\delta\hat{\underline{\theta}}(k + 1|k) = \Phi(k + 1, k)\delta\hat{\underline{\theta}}(k|k) \tag{6.4-51}$$

which was to be proved. ▲

In retrospect, the relationship between $\delta\hat{\underline{\theta}}(k|k)$ and $\delta\hat{\underline{\theta}}(k + 1|k)$ in Eq. (6.4-46) is obvious. Both estimates make use of the same collection of data, $\delta\underline{Z}(k)$; thus, the only possible way in which they can be made to differ is through application of the model for $\delta\underline{\theta}(k)$ in Eq. (6.4-1). Since $E\{\underline{\omega}(k)\} = \underline{0}$ and $E\{\delta\underline{\theta}(k + 1)\} = \Phi(k + 1, k) \times E\{\delta\underline{\theta}(k)\}$, it is clear that one way to update $\delta\hat{\underline{\theta}}(k|k)$ is to premultiply it by $\Phi(k + 1, k)$; but this product is $\delta\hat{\underline{\theta}}(k + 1|k)$, precisely.

Sequential Estimation Algorithm

For computational reasons that were discussed in Section 2.7, we would like to be able to obtain $\delta\hat{\underline{\theta}}(k + 1|k + 1)$ directly from the preceding estimate and $\delta\underline{z}(k + 1)$. *In the present situation, the preceding estimate is* $\delta\hat{\underline{\theta}}(k + 1|k)$. The sequential unbiased minimum-variance estimation algorithm for $\delta\hat{\underline{\theta}}(k + 1|k + 1)$ is obtained from Eqs. (3.7-5)- 3.7-7), as

$$\delta\hat{\underline{\theta}}(k + 1|k + 1) = \delta\hat{\underline{\theta}}(k + 1|k) + K^o(k + 1)[\delta\underline{z}(k + 1) - H(k + 1)\delta\hat{\underline{\theta}}(k + 1|k)] \quad , \tag{6.4-52}$$

$$K^o(k + 1) = P(k + 1|k)H'(k + 1)[H(k + 1)P(k + 1|k)H'(k + 1) + R(k + 1)]^{-1} \quad , \tag{6.4-53}$$

and

$$P(k + 1|k + 1) = [I - K^o(k + 1)H(k + 1)]P(k + 1|k) \tag{6.4-54}$$

for $k \geq L$ (or $k \geq 0$ when the sequential startup technique described in Section 2.8 is used). Obviously, in order to complete the calculations for $\delta\hat{\underline{\theta}}(k + 1|k + 1)$, we require the following additional equations:

$$\delta\hat{\underline{\theta}}(k + 1|k) = \Phi(k + 1, k)\delta\hat{\underline{\theta}}(k|k) \tag{6.4-55}$$

and

$$P(k + 1|k) = \Phi(k + 1, k)P(k|k)\Phi'(k + 1, k) + Q(k) \quad . \tag{6.4-56}$$

We have just derived the *discrete Kalman filter* for estimating $\underline{\theta}(k)$. The reader should compare Eqs. (6.4-52)-(6.4-56) with Eqs. (3.8-10)-(3.8-14) to see that this is so.†

Eqs. (6.4-52)-(6.4-56) should be solved in the following order: $P(k|k) \to P(k + 1|k) \to K^o(k + 1) \to \delta\hat{\underline{\theta}}(k + 1|k) \to \delta\hat{\underline{\theta}}(k + 1|k + 1) \to P(k + 1|k + 1)$. If k_0 denotes the starting value of k, then we must know $P(k_0|k_0)$ and $\hat{\underline{\theta}}(k_0|k_0)$ a priori in order to utilize the sequential estimation algorithm.

Let us consider the special case when $\underline{\theta}_N(k) = \underline{\theta}$. The vector $\delta\hat{\underline{\theta}}(k|k)$ provides us with an optimal estimate of deviations from $\underline{\theta}$. This corresponds to the situation where we are fairly certain that the unknown parameters are constant, but want to include a quantitative measure of our confidence in this "feeling" into our analysis. Unfortunately, if $\underline{\theta}$ is unknown then it is not possible to extract $\hat{\underline{\theta}}(k|k)$ from $\delta\hat{\underline{\theta}}(k|k)$. Of course, we could obtain an unbiased minimum-variance estimate of $\underline{\theta}$ using the algorithms in Chapter 3 and equate this estimate with $\underline{\theta}_N$. It would then follow that

$$\hat{\underline{\theta}}(k|k) = \hat{\underline{\theta}}(k) + \delta\hat{\underline{\theta}}(k|k) \quad . \tag{6.4-57}$$

In this equation, $\hat{\underline{\theta}}(k)$ is obtained by means of the algorithms in Chapter 3, whereas $\delta\hat{\underline{\theta}}(k|k)$ is obtained by means of the algorithm derived in this section.

---

† Many different derivations of the discrete Kalman filter can be found in the literature. Most of these assume that $\underline{\omega}(k)$ is gaussian (*6-10*, for example); our derivation is free from such an assumption.

Applicability

The algorithms we have just obtained are, of course, limited to the same types of identification problems to which generalized least-squares and minimum-variance algorithms are limited (see Sections 2.13 and 3.12). This means that they cannot be used to estimate the coefficients of finite-difference equations that are driven by random forcing functions; but they can be used to estimate the coefficients in a superposition summation or to estimate the coefficients in an approximation to a curve.

6.4.2 Gradient Estimation of $\delta\underline{\theta}(k)$

We shall demonstrate that the possibility for estimating $\delta\underline{\theta}(k)$ by means of gradient estimation algorithms depends upon interactions between certain signals which appear in the scalar equation-error identification system in Fig. 5.1-1. Our attention is directed to the stochastic situation since, even if $\underline{x}(k)$ and $y(k)$ can be measured[†] without errors, a stochastic output-errorlike noise will be present at the output of the "Identification representation" block owing to $\underline{\omega}(k)$ which appears in the a priori model of $\delta\underline{\theta}(k)$ in Eq. (6.4-1).

In the notation of Fig. 5.1-1, we see that

$$z(k) = y(k) + v(k) = \underline{x}'(k)\underline{\theta}(k) + v(k) \quad . \tag{6.4-58}$$

We also know, from Eq. (6.4-14), that

$$\delta\underline{\theta}(k) = \Phi(k, 0)\delta\underline{\theta}(0) + \sum_{i=1}^{k} \Phi(k, i)\underline{\omega}(i - 1) \quad . \tag{6.4-59}$$

Additionally, when $\underline{\theta}(k) = \underline{\theta}_N(k)$

$$z_N(k) = \underline{x}'(k)\underline{\theta}_N(k) \quad , \tag{6.4-60}$$

which means that

$$z(k) - z_N(k) \triangleq \delta z(k) = \underline{x}'(k)\delta\underline{\theta}(k) + v(k) \quad . \tag{6.4-61}$$

---

† As in Chapter 5, our attention is directed once again at the scalar measurement situation.

Combining Eqs. (6.4-61) and (6.4-59), we find

$$\delta z(k) = \underline{x}'(k)\Phi(k, 0)\delta\underline{\theta}(0) + \underline{x}'(k) \sum_{i=1}^{k} \Phi(k, i)\underline{\omega}(i - 1) + v(k) \tag{6.4-62}$$

which we shall write as

$$\delta z(k) = \underline{\chi}'(k)\delta\underline{\theta}(0) + v^*(k) \quad , \tag{6.4-63}$$

where

$$\underline{\chi}'(k) = \underline{x}'(k)\Phi(k, 0) \quad , \tag{6.4-64}$$

$$v^*(k) = v_{\omega x}(k) + v(k) \quad , \tag{6.4-65}$$

and

$$v_{\omega x}(k) = \underline{x}'(k) \sum_{i=1}^{k} \Phi(k, i)\underline{\omega}(i - 1) \quad . \tag{6.4-66}$$

[The term $v_{\omega x}(k)$ can also be written directly in terms of $\underline{\chi}(k)$, as follows:

$$v_{\omega x}(k) = \underline{x}'(k)\Phi(k, 0) \sum_{i=1}^{k} \Phi(0, k)\Phi(k, i)\underline{\omega}(i - 1)$$

$$= \underline{\chi}'(k) \sum_{i=1}^{k} \Phi(0, i)\underline{\omega}(i - 1) \quad . \tag{6.4-67}$$]

Equation (6.4-63) is the starting point for estimating $\delta\underline{\theta}(0)$ by means of unbiased stochastic-gradient algorithms. The estimate of $\delta\underline{\theta}(k)$, $\delta\hat{\underline{\theta}}(k)$, is obtained from the estimate of $\delta\underline{\theta}(0)$, $\delta\hat{\underline{\theta}}_0(k)$, and Eq. (6.4-59). Because $\underline{\omega}(j)$ is a zero mean process, we estimate $\delta\underline{\theta}(k)$ as

$$\delta\hat{\underline{\theta}}(k) = \Phi(k, 0)\delta\hat{\underline{\theta}}_0(k) \quad . \tag{6.4-68}$$

In order to obtain an unbiased estimate of $\delta\underline{\theta}(0)$, we must ascertain whether this identification problem is Class 1 or Class 2. Since $\underline{\chi}(k)$ and $v^*(k)$ are now analogous to $\underline{x}(k)$ and $v(k)$, we must take a close look at $E\{\underline{\chi}(k)v^*(k)\}$. If $E\{\underline{\chi}(k)v^*(k)\} = \underline{0}$, we will have a Class-1 identification problem; but if $E\{\underline{\chi}(k)v^*(k)\} \neq \underline{0}$, we will have a Class-2 problem, and in this case will will have to evaluate this expectation explicitly, in order to remove its deleterious effects from estimates of $\delta\underline{\theta}(0)$.

From Eqs. (6.4-65), (6.4-66), and (6.4-64) we determine that

$$E\{\underline{\chi}(k)v^*(k)\} = \Phi'(k, 0)[E\{\underline{x}(k)v(k)\} + E\{\underline{x}(k)v_{\omega x}(k)\}]$$

$$= \Phi'(k, 0)\left[E\{\underline{x}(k)v(k)\} + E\{\underline{x}(k)\underline{x}'(k) \sum_{i=1}^{k} \Phi(k, i)\underline{\omega}(i - 1)\}\right] . \qquad (6.4\text{-}69)$$

Clearly, *if* $\underline{x}(k)$ *and* $\underline{\omega}(j)$, $j = 0, 1, \ldots, k - 1$, *are independent then*

$$E\{\underline{\chi}(k)v^*(k)\} = \Phi'(k, 0)E\{\underline{x}(k)v(k)\} \qquad (6.4\text{-}70)$$

and whether or not the identification problem is Class 1 or Class 2 depends, once again, on whether or not $\underline{x}(k)$ *and* $v(k)$ *are independent.*

Two applications where $\underline{x}(k)$ and $\underline{\omega}(j)$, $j = 0, 1, \ldots, k - 1$, are independent are: identification of weights in a superposition summation, and identification of coefficients in an approximation to a curve or multivariable function.

If $\underline{x}(k)$ *and* $\underline{\omega}(j)$, $j = 0, 1, \ldots, k - 1$, *are not independent, then we have a new situation, different from either Class 1 or Class 2 problems.* In this case, the upper part of Fig. 5.2-1 must be modified to the structure depicted in Fig. 6.4-1.

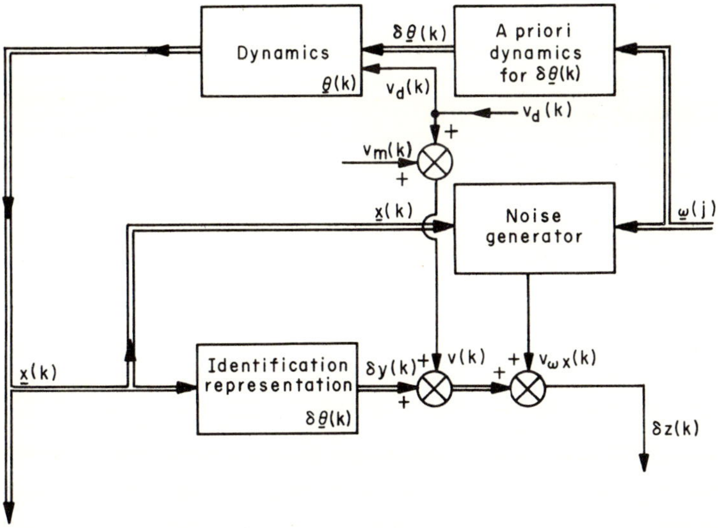

Fig. 6.4-1. Modified upper portion of Fig. 5.2-1 scalar equation-error identification system.

Coupling exists between the dynamics block that generates $\underline{x}(k)$ and the dynamics block that generates $\delta\underline{\theta}(k)$. In order to determine $\underline{x}(k)$, the a priori dynamic model for $\delta\underline{\theta}(k)$ must be augmented to the dynamic model for $\underline{x}(k)$. Even before $\delta\underline{\theta}(k)$ was modeled, we understood that $\underline{x}(k)$ is a function of $\underline{\theta}(k)$. Modeling the time variations of $\underline{\theta}(k)$ by means of a dynamic model has the unfortunate effect of making $\underline{\theta}(k)$ [or $\delta\underline{\theta}(k)$] a state, just as the elements in the "Dynamics" block are states. This means that $\underline{x}(k)$ is now a function of the *states* $\underline{\theta}(k)$; and, therefore, the *augmented dynamic system is nonlinear.*[†] The analysis of such a system is beyond the scope of this book.

When $\underline{x}(k)$ and $\underline{\omega}(j)$, $j = 0, 1, \ldots, k - 1$, are independent, the box labeled "Dynamics" in Fig. 6.4-1 is not present. It is for this reason that analysis is possible. The term $\delta\underline{\theta}(0)$ is estimated by modifying Eq. (5.4-15) to the measurement difference equation for $\delta z(k)$ in Eq. (6.4-63); that is to say,

$$\delta\hat{\underline{\theta}}_0(k + \ell) = [I + R_1(k)\Sigma_n^1(k) - R_1(k)M_1(k)]\delta\hat{\underline{\theta}}_0(k)$$
$$+ R_1(k)[\Phi'(k, 0)\underline{r}(k)]\tilde{z}(k) - R_1(k)\underline{\lambda}_1(k) \qquad (6.4\text{-}71)$$

for $k = 0, \ell, 2\ell, \ldots,$ where

$$\Sigma_n^1(k) = \Phi'(k, 0)\Sigma_n\Phi(k, 0) \quad , \qquad (6.4\text{-}72)$$

$$M_1(k) = \Phi'(k, 0)M(k) \quad , \qquad (6.4\text{-}73)$$

$$\underline{\lambda}_1(k) = \Phi'(k, 0)\underline{\lambda}(k) \quad , \qquad (6.4\text{-}74)$$

and

$$R_1(k) = R(k)\Big|_{\underline{r}(k) \text{ replaced by } \Phi'(k, 0)\underline{r}(k)} \quad . \qquad (6.4\text{-}75)$$

---

† If, for example, $\underline{x}(k + 1) = \Phi[\underline{\theta}(k)]\underline{x}(k) + \underline{\psi}m(k)$, where Φ is an explicit function of $\underline{\theta}(k)$ and $\underline{\theta}(k)$ is modeled according to Eq. (6.1-2), then $\underline{x}(k + 1) = \underline{f}[\underline{x}(k), \underline{\theta}(k)] + \underline{\psi}m(k)$ and this equation is nonlinear. The augmented dynamic system is comprised of the nonlinear equation for $\underline{x}(k + 1)$ and Eq. (6.1-2); hence, the augmented dynamic system is also nonlinear.

Here M and $\underline{\lambda}$ [see Eq. (5.4-8)] are both functions of time, since the unknown parameter vector is time varying; hence, in this case we are assuming that

$$\underline{\text{TB}} = E\{\underline{x}(k)v(k)\} = \underline{\lambda}_1(k) + M(k)\delta\underline{\theta}(0) \quad . \qquad (6.4\text{-}76)$$

For a Class-1 identification problem, both $\underline{\lambda}_1(k)$ and M(k) are zero and the algorithm in Eq. (6.4-71) simplifies to

$$\delta\hat{\underline{\theta}}_0(k + \ell) = [I + R_1(k)\Sigma_n^1(k)]\delta\hat{\underline{\theta}}_0(k)$$

$$+ R_1(k)[\Phi'(k, 0)\underline{r}(k)]\tilde{z}(k) \quad . \qquad (6.4\text{-}77)$$

To recapitulate, we have shown that gradient estimation of $\delta\underline{\theta}(k)$ is possible when $\underline{x}(k)$ and $\underline{\omega}(j)$ are statistically independent. The gradient estimation algorithms for this situation are obtained quite easily from the algorithms in Chapter 5. When $\underline{x}(k)$ and $\underline{\omega}(j)$ are not statistically independent, the parameter estimation problem is nonlinear and its solution falls outside of the scope of this book. The estimation of $\delta\underline{\theta}(k)$ in the latter situation is difficult and is currently being studied by researchers throughout the world.

PROBLEMS

6-1. For Type-C information, if $P(k) \triangleq \text{Cov}[\delta\underline{\theta}(k)]$, show that

$$P(k + 1) = \Phi(k + 1, k)P(k)\Phi'(k + 1, k) + Q(k) \quad .$$

6-2. Consider the scalar situation where $\underline{\theta} = \theta$. Using the results of Problem 6-1, obtain the steady-state value of σ_θ^2. Show how $Q(k) \triangleq q$ can be designed from a specification on σ_θ^2, that $\sigma_\theta^2 \leq \zeta$ where ζ is given a priori. How might these results be extended to the vector parameter situation?

6-3. Beginning with Eq. (6.2-10), determine conditions on $\Delta\underline{\theta}(k + 1, k)$ and R(k) so that $||\tilde{\underline{\theta}}(k + 1)||^2 \to 0$ as $k \to \infty$. (*Hint:* Use Venter's theorem.) Comment on the practical significance of these conditions.

6-4. Obtain the stochastic-gradient algorithm for estimating $\underline{\beta}(k)$ when the identification problem is Class 2. The results obtained are comparable to those obtained for Class-1 identification problems in Eqs. (6.3-18) and (6.3-19).

6-5. Derive Eqs. (6.3-21) and (6.3-22) and Eqs. (6.3-25) and (6.3-26).

6-6. Show that $M_{\alpha q}(t)$ and $M_{\delta q}(t)$ vary over smaller ranges and at slower rates than do $M_{\alpha}(t)$ and $M_{\delta}(t)$ depicted in Fig. 6.3-1.

6-7. Derive Eq. (6.4-17).

6-8. The sequential unbiased minimum-variance estimator, when Type-C information is known about $\underline{\theta}(k)$, is shown in Section 6.4 to be a discrete Kalman filter. For this filter to be "optimal," $H(k + 1)$, $R(k + 1)$, $\Phi(k + 1, k)$, and $Q(k)$ must be known exactly. Any less knowledge leads to suboptimal filters. As in Section 3.10, we can obtain recursive equations for the sensitivity of the filter to wrong (imperfect) information about any or all of these quantities.

Now, however, in addition to distinguishing between actual and modeled measurement systems, we must also distinguish between actual and modeled dynamical systems for $\underline{\theta}(k)$. The *actual dynamical system* is given by Eqs. (6.4-1), (6.1-3), and (6.1-4). The *modeled dynamical system* is described by a similar set of equations,

$$\delta\underline{\theta}_m(k + 1) = \Phi_m(k + 1, k)\delta\underline{\theta}_m(k) + \underline{\omega}(k) \quad ,$$

$$E\{\underline{\omega}(k)\} = \underline{0} \quad ,$$

and

$$E\{\underline{\omega}(k)\underline{\omega}'(k)\} = Q_m(k) \quad ,$$

in which subscripted quantities are always to be associated with the modeled dynamical system.

(a) What are the five equations for the suboptimal discrete Kalman filter?

(b) Group all uncertain elements from $H(k + 1)$, $R(k + 1)$, $\Phi(k + 1, k)$, and $Q(k)$ into the vector $\underline{\alpha}$. What is the maximum dimension of $\underline{\alpha}$?

(c) The conditional parameter estimation-error covariance matrix $P(k|\underline{\alpha})$ is

$$P(k|\underline{\alpha}) = E\{[\delta\underline{\theta}(k) - \delta\hat{\underline{\theta}}_m(k)][\delta\underline{\theta}(k) - \delta\hat{\underline{\theta}}_m(k)]'|\underline{\alpha}\} \quad .$$

Obtain sequential equations for $E\{\delta\underline{\theta}(k)\delta\underline{\theta}'(k)|\underline{\alpha}\}$, $E\{\delta\underline{\theta}(k)\delta\underline{\theta}_m'(k)|\underline{\alpha}\}$, and $E\{\delta\hat{\underline{\theta}}_m(k)\delta\hat{\underline{\theta}}_m'(k)|\underline{\alpha}\}$.

(d) Draw a flow chart showing the order in which computations must be performed to compute $P(k|\underline{\alpha})$.

(e) The effects of differences between the actual and modeled systems can be obtained by computing the sensitivity matrix $S(k+1|\underline{\alpha})$, where

$$S(k + 1|\underline{\alpha}) = P(k + 1) - P(k + 1|\underline{\alpha}) \quad .$$

Draw a flow chart showing the order in which computations must be performed to compute $S(k + 1|\underline{\alpha})$. Observe that $S(k + 1|\underline{\alpha})$ is difficult to compute, even for the simplest of systems.

6-9. Consider the scalar measurement equation

$$z(k + 1) = \theta(k + 1) + v(k + 1)$$

where

$$\theta(k + 1) = \theta(k) + \omega(k) \quad .$$

(a) Show that

$$P(k + 1|k + 1) = \frac{R[P(k|k) + Q]}{P(k|k) + Q + R}$$

where all quantities are scalars.

(b) Assume that $P(k + 1|k + 1)$ reaches a steady-state value $\overline{P}$ for large values of k. Show that

$$\overline{P} = \frac{-Q + (Q^2 + 4RQ)^{1/2}}{2} \quad .$$

(c) Let us study the sensitivity of $\overline{P}$ to small errors in Q and R by looking at $d\overline{P}$, where

$$d\overline{P} = \frac{\partial\overline{P}}{\partial Q}dQ + \frac{\partial\overline{P}}{\partial R}dR \quad .$$

Compute $\partial\overline{P}/\partial Q$ and $\partial\overline{P}/\partial R$ and discuss the behaviors of these sensitivity functions with respect to Q and R.

6-10. Prove that $P(k + 1|k) = E\{\delta\tilde{\underline{\theta}}(k + 1|k)\delta\tilde{\underline{\theta}}'(k + 1|k)\}$ and $P(k + 1|k + 1) = E\{\delta\tilde{\underline{\theta}}(k + 1|k + 1)\delta\tilde{\underline{\theta}}'(k + 1|k + 1)\}$.

6-11. Prove that $P(k + 1|k + 1) < P(k + 1|k)$.

6-12. Hill's equation is [see Problem 1-26(d)]

$$\ddot{x}(t) - ax(t) + bp(t)x(t) = m(t)$$

where $p(t)$ is a known periodic function, say $\sin \omega_0(t)$. Using quasilinearization and gradient algorithms, explain how to identify a and b from noisy measurements of $x(t)$ at $t = t_i$, $i = 0, 1, \ldots, M$. Assume that numerical integration errors are negligible.

6-13. Let us consider the identification of $a(t)$ in the first-order nonlinear differential equation

$$\dot{x}(t) = f[x(t), a(t)]$$

$$x(0) = x_0 \qquad \text{where } x_0 \text{ is known},$$

with $0 \leq t \leq t_f$, and $a(t)$ is unknown. Our approach is to approximate $a(t)$ by means of a truncated series, i. e.,

$$a(t) \approx \sum_{j=0}^{N} C_j \psi_j(t)$$

where the basis functions $\psi_0, \psi_1, \ldots, \psi_N$ are specified a priori (see Section 2.4). Let $\underline{C} = (C_0, C_1, \ldots, C_N)'$, so that $a(t)$ can be viewed as a function of $\underline{C}$; thus,

$$\dot{x}(t) = f[x(t), \underline{C}] \quad .$$

Show how to estimate $\underline{C}$ from the M + 1 measurements using quasilinearization (see Problem 1-23) and gradient algorithms, when these measurements are noisy and when numerical integration errors are negligible.

REFERENCES

6-1. L. A. Zadeh and C. A. Desoer, *Linear System Theory, the State Space Approach*, McGraw-Hill, New York, 1963.

6-2. R. J. Schwarz and B. Friedland, *Linear Systems*, McGraw-Hill, New York, 1965.

6-3. J. M. Mendel, *A Priori and a Posteriori Identification of Time-Varying Parameters*, presented at the Second Hawaii Intern. Conf. System Sci., Univ. of Hawaii, Honolulu, Hawaii, January 1969.

6-4. J. M. Mendel, Sequential identification by means of gradient learning algorithms, in *Pattern Recognition and Machine Learning* (K. S. Fu, ed.), pp. 70-78, Plenum Press, New York, 1971.

6-5. J. M. Mendel, Gradient identification for linear systems, in *Adaptive, Learning, and Pattern Recognition Systems: Theory and Applications* (J. M. Mendel and K. S. Fu, eds.), Chapter 6, pp. 209-242, Academic Press, New York, 1970.

6-6. J. M. Mendel, Identification of decomposable time-varying parameters by means of gradient algorithms, *Inform. Sci.*, 5 (1971).

6-7. J. M. Mendel, Unpublished memorandum (1969).

6-8. R. E. Kalman, A new approach to linear filtering and prediction problems, *ASME Trans., J. Basic Eng., Ser. D*, 82, 34-45 (1960).

6-9. I. A. Gura, An algebraic solution of the state estimation problem, *AIAA J.*, 7, 1242-1247 (1969).

6-10. J. S. Meditch, *Stochastic Optimal Linear Estimation and Control*, McGraw-Hill, New York, 1969.

Appendix A

Representations of Abstract Digital Systems

Theorem. *Given the following state space and finite-difference equation representations for an abstract digital system:*

$$\begin{pmatrix} x_1(k+1) \\ x_2(k+1) \\ \vdots \\ x_n(k+1) \end{pmatrix} = \left(\begin{array}{c|ccc} 0 & & & \\ 0 & & I_{n-1} & \\ \vdots & & & \\ \hline -a_n & -a_{n-1} & \cdots & -a_1 \end{array}\right) \begin{pmatrix} x_1(k) \\ x_2(k) \\ \vdots \\ x_n(k) \end{pmatrix}$$

$$+ \begin{pmatrix} b_1 \\ b_2 \\ \vdots \\ b_n \end{pmatrix} m(k) + \begin{pmatrix} d_1 \\ d_2 \\ \vdots \\ d_n \end{pmatrix} \omega(k) \qquad \text{(A-1)}$$

and

$$y(k+n) + a_1 y(k+n-1) + \cdots + a_n y(k) = b_n^o m(k+n-1)$$
$$+ b_{n-1}^o m(k+n-2) + \cdots + b_1^o m(k) + d_n^o \omega(k+n-1)$$
$$+ d_{n-1}^o \omega(k+n-2) + \cdots + d_1^o \omega(k) \quad . \qquad \text{(A-2)}$$

Symbol $m(k)$ *represents a controllable signal (e. g., test signal or feedback control signal) whereas* $\omega(k)$ *represents an uncontrollable signal (e. g., random disturbance). The* representations in Eqs. (A-1) and (A-2) are equivalent when

$$x_1(k) = y(k)$$
$$x_2(k) = y(k+1) - b_1 m(k) - d_1 \omega(k)$$
$$x_3(k) = y(k+2) - b_1 m(k+1) - b_2 m(k) - d_1 \omega(k+1) - d_2 \omega(k)$$
$$\vdots \qquad \text{(A-3)}$$
$$x_n(k) = y(k+n-1) - b_1 m(k+n-2) - b_2 m(k+n-3) - \cdots$$
$$- b_{n-1} m(k) - d_1 \omega(k+n-2) - d_2 \omega(k+n-3) - \cdots$$
$$- d_{n-1} \omega(k) \quad ,$$

$$\underline{b} = (b_1, b_2, \ldots, b_n)' = T^{-1} \underline{b}^o \qquad \text{(A-4)}$$

$$\underline{d} = (d_1, d_2, \ldots, d_n)' = T^{-1} \underline{d}^o \qquad \text{(A-5)}$$

$$T = \begin{pmatrix} 1 & 0 & 0 & \cdots & 0 & 0 \\ a_1 & 1 & 0 & \cdots & 0 & 0 \\ a_2 & a_1 & 1 & \cdots & 0 & 0 \\ \vdots & \vdots & \vdots & \ddots & \vdots & \vdots \\ a_{n-2} & a_{n-3} & a_{n-4} & \cdots & 1 & 0 \\ a_{n-1} & a_{n-2} & a_{n-3} & \cdots & a_1 & 1 \end{pmatrix} \quad . \qquad \text{(A-6)}$$

Proof. We give the proof of this result for the case when $\omega(k) = 0$, all k. The proof for the case when $\omega(k) \neq 0$ parallels our development that follows. Our approach is to derive Eq. (A-2) from Eq. (A-1).

To begin, let $k = k + 1$ in the last equation in Eq. (A-3), and solve the resulting equation for $y(k + n)$:

$$\begin{aligned} y(k + n) = {} & x_n(k + 1) + b_1 m(k + n - 1) + b_2 m(k + n - 2) \\ & + b_3 m(k + n - 3) + \cdots + b_{n-2} m(k + 2) \\ & + b_{n-1} m(k + 1) \quad . \end{aligned} \tag{A-7}$$

We also know from the last equation in Eq. (A-1) that

$$\begin{aligned} x_n(k + 1) = {} & -a_n x_1(k) - a_{n-1} x_2(k) - a_{n-2} x_3(k) - \cdots \\ & - a_2 x_{n-1}(k) - a_1 x_n(k) + b_n m(k) \quad ; \end{aligned} \tag{A-8}$$

hence, substituting this expression into Eq. (A-7) and making use of the first $n - 1$ equations in Eq. (A-3), we find that

$$\begin{aligned} y(k + n) = {} & -a_n x_1(k) - a_{n-1} x_2(k) - a_{n-2} x_3(k) - \cdots - a_2 x_{n-1}(k) \\ & - a_1 x_n(k) + b_n m(k) + b_{n-1} m(k + 1) + b_{n-2} m(k + 2) + \cdots \\ & + b_3 m(k + n - 3) + b_2 m(k + n - 2) + b_1 m(k + n - 1) \\ = {} & -a_n y(k) - a_{n-1}[y(k + 1) - b_1 m(k)] - a_{n-2}[y(k + 2) \\ & - b_1 m(k + 1) - b_2 m(k)] - \cdots - a_2[y(k + n - 2) \\ & - b_1 m(k + n - 3) - b_2 m(k + n - 4) - \cdots - b_{n-4} m(k + 2) \\ & - b_{n-3} m(k + 1) - b_{n-2} m(k)] - a_1[y(k + n - 1) \\ & - b_1 m(k + n - 2) - b_2 m(k + n - 3) - \cdots - b_{n-3} m(k + 2) \\ & - b_{n-2} m(k + 1) - b_{n-1} m(k)] + b_n m(k) + b_{n-1} m(k + 1) \\ & + b_{n-2} m(k + 2) + \cdots + b_3 m(k + n - 3) + b_2 m(k + n - 2) \\ & + b_1 m(k + n - 1) \quad . \end{aligned} \tag{A-9}$$

Collecting common terms, we find

$$y(k + n) + a_n y(k) + a_{n-1} y(k + 1) + a_{n-2} y(k + 2) + \cdots + a_2 y(k + n - 2) + a_1 y(k + n - 1)$$
$$= m(k)[a_{n-1}b_1 + a_{n-2}b_2 + \cdots + a_2 b_{n-2} + a_1 b_{n-1} + b_n]$$
$$+ m(k + 1)[a_{n-2}b_1 + \cdots + a_2 b_{n-3} + a_1 b_{n-2} + b_{n-1}]$$
$$+ m(k + 2)[\cdots + a_2 b_{n-4} + a_1 b_{n-3} + b_{n-2}] + \cdots$$
$$+ m(k + n - 2)[a_1 b_1 + b_2] + m(k + n - 1)[b_1] \quad , \tag{A-10}$$

which can also be written as

$$y(k + n) + a_1 y(k + n - 1) + a_2 y(k + n - 2) + \cdots + a_n y(k)$$
$$= b_1^o m(k + n - 1) + b_2^o m(k + n - 2) + \cdots + b_{n-1}^o m(k + 1)$$
$$+ b_n^o m(k) \tag{A-11}$$

where

$$\begin{aligned} b_1^o &= b_1 \\ b_2^o &= a_1 b_1 + b_2 \\ &\vdots \\ b_{n-1}^o &= a_{n-2}b_1 + \cdots + a_2 b_{n-3} + a_1 b_{n-2} + b_{n-1} \\ b_n^o &= a_{n-1}b_1 + a_{n-2}b_2 + \cdots + a_2 b_{n-2} + a_1 b_{n-1} + b_n \ . \end{aligned} \tag{A-12}$$

Clearly, Eq. (A-12) can also be written as

$$\underline{b}^o = T\underline{b} \tag{A-13}$$

▲

The results in this theorem are quite useful and can be applied in many ways. Suppose, for example, we are given the finite-difference equation in Eq. (A-2). We can obtain the solution of this equation, y(k), in the following manner: (1) transform the nth-order difference equation to the vector first-order state equation in

Eq. (A-1), using Eqs. (A-4)-(A-6); (2) obtain the solution to Eq. (A-1), as (see Section 1.3.3)

$$\underline{x}(k) = \Phi^k \underline{x}(0) + \sum_{i=1}^{k} \Phi^{k-i}[\underline{b}m(i - 1) + \underline{d}\omega(i - 1)] \tag{A-14}$$

where

$$\Phi = \begin{pmatrix} \begin{array}{c|c} 0 & \\ 0 & \\ \vdots & I_{n-1} \\ \vdots & \\ \hline -a_n & -a_{n-1} \quad \cdots \quad -a_1 \end{array} \end{pmatrix} , \tag{A-15}$$

$$\underline{b} = (b_1, b_2, \ldots, b_n)' , \tag{A-16}$$

$$\underline{d} = (d_1, d_2, \ldots, d_n)' , \tag{A-17}$$

and

$$\underline{x}(k) = [x_1(k), x_2(k), \ldots, x_n(k)]' ; \tag{A-18}$$

and (3) obtain y(k) from the fact that [see Eq. (A-3)]

$$y(k) = (1 \quad 0 \quad 0 \quad \cdots \quad 0)'\underline{x}(k) . \tag{A-19}$$

Appendix B

Discrete-Time Representations of Continuous-Time Systems

We begin with the following nth-order differential equation representation for a continuous-time system:

$$
\begin{aligned}
y^{(n)}(t) &+ a_1 y^{(n-1)}(t) + \cdots + a_{n-1}\dot{y}(t) + a_n y(t) \\
&= b_1 m^{(n-1)}(t) + \cdots + b_{n-1}\dot{m}(t) + b_n m(t) \\
&\quad + d_1 \omega^{(n-1)}(t) + \cdots + d_{n-1}\dot{\omega}(t) + d_n \omega(t) \quad . \qquad \text{(B-1)}
\end{aligned}
$$

Symbol $m(t)$ represents a controllable signal (e. g., test signal or feedback control signal), whereas $\omega(t)$ represents an uncontrollable signal (e. g., random disturbance). Our objectives are to obtain a *discrete-time state space representation* and a *finite-difference equation representation* for this system. To begin, we will establish a very useful continuous-time state space representation for it.

B.1 CONTINUOUS-TIME STATE SPACE REPRESENTATION

Many different continuous-time state space representations are possible for Eq. (B-1) (*B-1* and *B-2*, for example); however, not all of them are suitable for the parameter identification problem [our ultimate objective is to identify $a_1, a_2, \ldots, a_n, b_1, b_2, \ldots,$ and b_n (see Section 1.3.4)]. The state space representations that follow are quite useful for parameter identification. To begin, we consider the case when only m(t) and its n - 1 derivatives are present in Eq. (B-1).

Theorem B-1. *A state space representation for the continuous-time system*

$$y^{(n)}(t) + a_1 y^{(n-1)}(t) + \cdots + a_{n-1}\dot{y}(t) + a_n y(t)$$
$$= b_1 m^{(n-1)}(t) + \cdots + b_{n-1}\dot{m}(t) + b_n m(t) \tag{B.1-1}$$

is

$$\begin{pmatrix} \dot{x}_1 \\ \dot{x}_2 \\ \vdots \\ \dot{x}_n \end{pmatrix} = \left(\begin{array}{c|ccc} 0 & & & \\ 0 & & I_{n-1} & \\ \vdots & & & \\ \hline -a_n & -a_{n-1} & \cdots & -a_1 \end{array}\right) \begin{pmatrix} x_1 \\ x_2 \\ \vdots \\ x_n \end{pmatrix} + \begin{pmatrix} 0 \\ 0 \\ \vdots \\ 1 \end{pmatrix} m(t) \tag{B.1-2}$$

where

$$y(t) = b_n x_1(t) + b_{n-1} x_2(t) + \cdots + b_1 x_n(t) \quad . \tag{B.1-3}$$

Proof. We shall use the Laplace transform operator s to replace d/dt in Eq. (B.1-1); hence, (*B-3*),

$$\frac{Y(s)}{M(s)} = \frac{b_1 s^{n-1} + \cdots + b_{n-1}s + b_n}{s^n + a_1 s^{n-1} + \cdots + a_{n-1}s + a_n} \tag{B.1-4}$$

where Y(s) and M(s) are the Laplace transforms of y(t) and m(t), respectively.† Introduce the intermediate signal X(s) as follows:

$$\frac{Y(s)}{M(s)} = \frac{b_1 s^{n-1} + \cdots + b_{n-1}s + b_n}{s^n + a_1 s^{n-1} + \cdots + a_{n-1}s + a_n} \frac{X(s)}{X(s)} \quad . \tag{B.1-5}$$

---

† The reader unfamiliar with Laplace transforms can treat Eq. (B.1-4) as the *operational form* of Eq. (B.1-1).

Equating numerators and denominators on both sides of this equation, we obtain

$$Y(s) = (b_1 s^{n-1} + \cdots + b_{n-1} s + b_n) X(s) \qquad \text{(B.1-6)}$$

and

$$M(s) = (s^n + a_1 s^n + \cdots + a_{n-1} s + a_n) X(s) \quad , \qquad \text{(B.1-7)}$$

or, returning to the time-domain,

$$y(t) = b_1 x^{(n-1)}(t) + \cdots + b_{n-1}\dot{x}(t) + b_n x(t) \qquad \text{(B.1-8)}$$

and

$$x^{(n)}(t) + a_1 x^{(n-1)}(t) + \cdots + a_{n-1}\dot{x}(t) + a_n x(t) = m(t) \quad . \qquad \text{(B.1-9)}$$

We view Eqs. (B.1-8) and (B.1-9) as a system, where Eq. (B.1-9) is interpreted as an nth-order differential equation for x(t), and Eq. (B.1-8) is interpreted as a measurement equation.

Let

$$\begin{aligned} x_1 &= x(t) \\ x_2 &= \dot{x}(t) \\ &\vdots \\ x_n &= x^{(n-1)}(t) \quad , \end{aligned} \qquad \text{(B.1-10)}$$

then

$$\begin{aligned} \dot{x}_1 &= x_2 \\ \dot{x}_2 &= x_3 \\ &\vdots \\ \dot{x}_{n-1} &= x_n \\ \dot{x}_n &= -a_1 x_n - \cdots - a_n x_1 + m \end{aligned} \qquad \text{(B.1-11)}$$

which is Eq. (B.1-2), precisely. In addition, substituting Eq. (B.1-10) into Eq. (B.1-8), we obtain

$$y(t) = b_1 x_n + \cdots + b_{n-1} x_2 + b_n x_1 \quad , \qquad \text{(B.1-12)}$$

which is Eq. (B.1-3). ▲

The reasons why the state space representation in Theorem B-1 is well-suited for the parameter identification problem are: (1) the a and b parameters appear in separate equations in their original forms, and (2) the state equation is excited only by m(t) and not by m(t) and its derivatives.

Next, let us obtain a similar representation for the system in Eq. (B-1).

Theorem B-2. *A state space representation for the system in Eq. (B-1) is:*[†]

$$\dot{\underline{x}}_1(t) = A\underline{x}_1(t) + \underline{e}_n m(t) \qquad \text{(B.1-13)}$$

$$\dot{\underline{x}}_2(t) = A\underline{x}_2(t) + \underline{e}_n \omega(t) \qquad \text{(B.1-14)}$$

where

$$y(t) = \underline{b}'\underline{x}_1(t) + \underline{d}'\underline{x}_2(t) \quad , \qquad \text{(B.1-15)}$$

$$\underline{x}_1 = (x_{11}, x_{12}, \ldots, x_{1n})' \quad , \quad \underline{x}_2 = (x_{21}, x_{22}, \ldots, x_{2n})' \quad ,$$

$$A = \left(\begin{array}{c|ccc} 0 & & & \\ 0 & & I_{n-1} & \\ \vdots & & & \\ \hline -a_n & -a_{n-1} & \cdots & -a_1 \end{array}\right) \quad , \qquad \text{(B.1-16)}$$

$$\underline{b} = (b_n, b_{n-1}, \ldots, b_1)' \quad , \qquad \text{(B.1-17)}^{\ddagger}$$

$$\underline{d} = (d_n, d_{n-1}, \ldots, d_1)' \quad , \qquad \text{(B.1-18)}^{\ddagger}$$

and $\underline{e}_n$ *is the nth unit vector,* (0 0 ⋯ 0 1)'.

[†] Here m(t) is restricted to be a test signal. (What representation would be used during the design of a feedback control signal?)

[‡] Observe that, for the purposes of the developments in the present appendix, $\underline{b}$ and $\underline{d}$ are defined differently from the way in which they were defined in Appendix A.

Proof. Since Eq. (B-1) is linear, we can determine y(t) using the principle of superposition (*B-4*), as

$$y(t) = y_1(t) + y_2(t) \qquad (B.1\text{-}19)$$

where $y_1(t)$ is the portion of y(t) owing solely to effects of m(t), and $y_2(t)$ is the portion of y(t) owing solely to effects of $\omega(t)$; that is to say, $y_1(t)$ and $y_2(t)$ are determined from the equations

$$y_1^{(n)}(t) + a_1 y_1^{(n-1)}(t) + \cdots + a_{n-1}\dot{y}_1(t) + a_n y_1(t) = b_1 m^{(n-1)}(t) + \cdots + b_{n-1}\dot{m}(t) + b_n m(t) \qquad (B.1\text{-}20)$$

and

$$y_2^{(n)}(t) + a_1 y_2^{(n-1)}(t) + \cdots + a_{n-1}\dot{y}_2(t) + a_n y_2(t) = d_1 \omega^{(n-1)}(t) + \cdots + d_{n-1}\dot{\omega}(t) + d_n \omega(t) \quad . \qquad (B.1\text{-}21)$$

Clearly, Eqs. (B.1-13)-(B.1-15) result from applications of Theorem B-1 to Eqs. (B.1-20) and (B.1-21), and Eq. (B.1-19). .

Observe that Eqs. (B.1-13) and (B.1-14) are two *uncoupled* nth-order differential equations, whose solutions are coupled together in the observation equation, Eq. (B.1-15). We shall merely use the representation in Theorem B-2 as a means to an end, the end being an nth-order finite-difference equation representation of Eq. (B-1). It is very useful for this purpose.

B.2 DISCRETE-TIME STATE SPACE REPRESENTATIONS OF CONTINUOUS-TIME SYSTEMS

To begin, we consider the discretization of a somewhat more general continuous-time state equation than is Eqs. (B.1-2), (B.1-13), or (B.1-14):

$$\dot{\underline{x}}(t) = F(t)\underline{x}(t) + C(t)\underline{m}(t) + G(t)\underline{\omega}(t) \qquad (B.2\text{-}1)$$

where

$\underline{x} \equiv n \times 1$ state vector,

$\underline{\omega} \equiv r \times 1$ disturbance vector,

$\underline{m} \equiv p \times 1$ control and/or test signal vector,

$F(t) \equiv n \times n$ plant matrix,

$C(t) \equiv n \times p$ control distribution matrix,

$G(t) \equiv n \times r$ disturbance distribution matrix.

The signal $\underline{\omega}(t)$ is a gaussian white process[†] having the properties

$$E\{\underline{\omega}(t)\} = \underline{0} \tag{B.2-2}$$

and

$$E\{\underline{\omega}(t)\underline{\omega}'(\tau)\} = Q(t)\delta(t - \tau) \tag{B.2-3}$$

where $\delta(t - \tau)$ is the Dirac delta function.

It is well known, from the theory of differential equations (*B-1* and *B-7*), that the solution to Eq. (B.2-1) is

$$\underline{x}(t) = \Phi(t, t_0)\underline{x}(t_0) + \int_{t_0}^{t} \Phi(t, \tau)[G(\tau)\underline{\omega}(\tau) + C(\tau)\underline{m}(\tau)]d\tau \tag{B.2-4}$$

where $\Phi(t, \tau)$, the $n \times n$ *state transition matrix*, is found from the solution to

$$\begin{aligned} \dot{\Phi}(t, \tau) &= F(t)\Phi(t, \tau) \\ \Phi(t, t) &= I_n \quad . \end{aligned} \tag{B.2-5}$$

We assume that $\underline{m}(t)$ is a piecewise constant function of time; hence, setting $t_0 = t_k$ and $t = t_{k+1}$ in Eq. (B.2-4), we find

$$\underline{x}(t_{k+1}) = \Phi(t_{k+1}, t_k)\underline{x}(t_k) + \left(\int_{t_k}^{t_{k+1}} \Phi(t_{k+1}, \tau)C(\tau)d\tau\right)\underline{m}(t_k) + \int_{t_k}^{t_{k+1}} \Phi(t_{k+1}, \tau)G(\tau)\underline{\omega}(\tau)d\tau \tag{B.2-6}$$

which can be written as

$$\underline{x}(k + 1) = \Phi(k + 1, k)\underline{x}(k) + \Psi(k + 1, k)\underline{m}(k) + \underline{\omega}_d(k) \tag{B.2-7}$$

[†] A stochastic process $\{\underline{\omega}(t), t \in I\}$ is said to be a *gaussian white process* if, for any m time points $t_1, \ldots, t_m$ in I, where m is any integer, the m random vectors $\underline{\omega}(t_1), \ldots, \underline{\omega}(t_m)$ are independent gaussian random vectors (*B-5* and *B-6*).

where

$$\Phi(k + 1, k) = \Phi(t_{k+1}, t_k) \quad , \tag{B.2-8}$$

$$\Psi(k + 1, k) = \int_{t_k}^{t_{k+1}} \Phi(t_{k+1}, \tau)C(\tau)d\tau \quad , \tag{B.2-9}$$

and $\underline{\omega}_d(k)$ is a discrete gaussian white sequence that is statistically equivalent to

$$\int_{t_k}^{t_{k+1}} \Phi(t_{k+1}, \tau)G(\tau)\underline{\omega}(\tau)d\tau \quad ;$$

that is to say,

$$E\{\underline{\omega}_d(k)\} = \underline{0} \tag{B.2-10}$$

and

$$E\{\underline{\omega}_d(k)\underline{\omega}_d'(k)\} \triangleq Q_d(k) = \int_{t_k}^{t_{k+1}} \Phi(t_{k+1}, \tau)G(\tau)Q(\tau)G'(\tau)\Phi'(t_{k+1}, \tau)d\tau \ . \tag{B.2-11}$$

Great simplifications of these results occur when the system in Eq. (B.2-1) is time invariant, i. e., when $F(t) = F$, $C(t) = C$, $G(t) = G$, and $Q(t) = Q$.

To begin (*B-2* and *B-4*),

$$\Phi(t, \tau) = e^{F(t-\tau)} \quad ; \tag{B.2-12}$$

hence,

$$\Phi(k + 1, k) = e^{FT} \tag{B.2-13}$$

where we are assuming that

$$t_{k+1} - t_k = T \quad . \tag{B.2-14}$$

The matrix exponential is given by the infinite series (F must be a square matrix)

$$e^{FT} = I + FT + F^2 \frac{T^2}{2!} + F^3 \frac{T^3}{3!} + \cdots \tag{B.2-15}$$

and, for *sufficiently small values of* T,

$$e^{FT} \approx I + FT \quad . \tag{B.2-16}$$

This approximation for e^{FT} is fundamental to all succeeding developments. Comparable results can be obtained, in many instances, for higher-order truncations[†] of e^{FT}.

Substituting Eq. (B.2-16) into Eq. (B.2-9), when C(t) = C, we find

$$\begin{aligned}\Psi(k+1, k) &= \int_{t_k}^{t_{k+1}} \Phi(t_{k+1}, \tau)C\,d\tau = \int_{t_k}^{t_{k+1}} e^{F(t_{k+1}-\tau)} C\,d\tau \\ &\approx \int_{t_k}^{t_{k+1}} [I + F(t_{k+1} - \tau)]C\,d\tau \\ &\approx CT + FCt_{k+1}T - FC \int_{t_k}^{t_{k+1}} \tau\,d\tau \\ &\approx CT + FC\frac{T^2}{2} \\ &\approx CT + O(T^2) \quad . \end{aligned} \tag{B.2-17}$$

Proceeding in a similar manner for $Q_d(k)$ in Eq. (B.2-11), it is easily shown that

$$Q_d \approx (TG)\left(\frac{Q}{T}\right)(G'T) + O(T^2) \quad . \tag{B.2-18}$$

Hence, putting all of these results for the time-invariant system together, we have the following theorem.

Theorem B-3. *An approximate discrete-time state space representation of the linear, time-invariant system*

$$\dot{\underline{x}}(t) = F\underline{x}(t) + C\underline{m}(t) + G\underline{\omega}(t) \tag{B.2-19}$$

that is good to $O(T^2)$, *where* $T = t_{k+1} - t_k$ *and* $T <<<$, *is*

$$\underline{x}(k+1) = (I + FT)\underline{x}(k) + (CT)\underline{m}(k) + (GT)\underline{\omega}(k) \tag{B.2-20}$$

[†] Two-term, three-term, and five-term truncations of the series expansion of e^{FT} are equivalent to *Euler*, *Modified Euler*, and *Runge-Kutta* numerical integrations of Eq. (B.2-5).

where†

$$E\{\underline{\omega}(k)\} = \underline{0} \qquad \text{(B.2-21)}$$

and

$$E\{\underline{\omega}(k)\underline{\omega}'(k)\} = Q/T \quad . \qquad \text{(B.2-22)}$$

▲

Applying these results to Eqs. (B.1-13) and (B.1-14), we obtain the following corollary.

Corollary B-1. *An approximate discrete-time state-space representation of the 2n-system in Eqs. (B.1-13)-(B.1-15) is*

$$\underline{x}_1(k + 1) = (I + AT)\underline{x}_1(k) + T\underline{e}_n m(k) \qquad \text{(B.2-23)}$$

$$\underline{x}_2(k + 1) = (I + AT)\underline{x}_2(k) + T\underline{e}_n \omega(k) \qquad \text{(B.2-24)}$$

and

$$y(k) = \underline{b}'\underline{x}_1(k) + \underline{d}'\underline{x}_2(k) \quad . \qquad \text{(B.2-25)}$$

▲

Observe, also, that if $\omega(t)$ is a gaussian white process, with $E\{\omega(t)\} = 0$ and $E\{\omega(t)\omega(\tau)\} = q\delta(t - \tau)$, then $\omega(k)$ is a gaussian white sequence, with $E\{\omega(k)\} = 0$ and $E\{\omega^2(k)\} = q/T$.

B.3 FINITE-DIFFERENCE EQUATION REPRESENTATIONS OF CONTINUOUS-TIME SYSTEMS

Here we obtain finite-difference equation representations for the continuous-time systems in Eqs. (B.1-1) and (B-1). We present details for the simpler of these systems, in Eq. (B.1-1), and indicate how the results obtained may be easily generalized to the system in Eq. (B-1).

---

† We are adopting the frequently used convention of letting

$$\underline{\omega}_d(k) = (GT)\underline{\omega}(k) \quad .$$

Observe that $E\{\underline{\omega}_d(k)\underline{\omega}_d'(k)\} = GTE\{\underline{\omega}(k)\underline{\omega}'(k)\}TG' = GT(Q/T)TG'$ which is in agreement with Q_d in Eq. (B.2-18), as it must be.

Our approach is to: (1) obtain a suitable continuous-time state-space representation for Eq. (B.1-1); (2) obtain a discrete-time representation of the continuous-time state equation; and (3) obtain the finite-difference equation from the discrete-time state equation. We have already accomplished tasks (1) and (2); results are summarized in Theorem B-1 and Corollary B-1 [Eq. (B.2-23) and Eq. (B.2-25) in which $\underline{d} \triangleq \underline{0}$]. Results for task (3) are summarized in the following theorem.

Theorem B-4. *Given the discrete-time system*

$$\underline{x}(k+1) = (I + AT)\underline{x}(k) + T\underline{e}_n m(k) \tag{B.3-1}$$

and

$$y(k) = \underline{b}'\underline{x}(k) \tag{B.3-2}$$

where

$$A = \left(\begin{array}{c|ccc} 0 & & & \\ 0 & & I_{n-1} & \\ \vdots & & & \\ \hline -a_n & -a_{n-1} & \cdots & -a_1 \end{array}\right), \tag{B.3-3}$$

$$\underline{e}_n = (0, 0, \ldots, 0, 1)', \tag{B.3-4}$$

and

$$\underline{b} = (b_n, b_{n-1}, \ldots, b_1)'. \tag{B.3-5}$$

The finite-difference equation for the output signal y(k) *is*

$$y(k+n) + \alpha_1 y(k+n-1) + \cdots + \alpha_n y(k) = \beta_1 m(k+n-1) + \cdots + \beta_n m(k) \tag{B.3-6}$$

where

$$\alpha_{n-i} = \sum_{j=i}^{n-1} (-1)^{j-i} \binom{j}{j-i} a_{n-j} T^{n-j} + (-1)^{n-i} \binom{n}{n-i} \tag{B.3-7}$$

and

$$\beta_{n-i} = \sum_{j=i}^{n-1} (-1)^{j-i} \binom{j}{j-i} b_{n-j} T^{n-j} \tag{B.3-8}$$

for $i = 0, 1, \ldots, n-1$. ▲

By means of this theorem, we can first direct our attention to identification of the 2n parameters $\alpha_1, \ldots, \alpha_n, \beta_1, \ldots, \beta_n$, and then we can solve for the 2n parameters $a_1, \ldots, a_n, b_1, \ldots, b_n$ using the 2n linear algebraic equations, Eqs. (B.3-7) and (B.3-8).†

Proof. The finite-difference equation for y(k) can be obtained in a number of ways. Our approach is to use z transforms (*B-8* and *B-9*) and algebra. The z transforms of Eqs. (B.3-2) and (B.3-1) are‡:

$$Y(z) = b_n X_1(z) + b_{n-1}X_2(z) + \cdots + b_1 X_n(z) \qquad \text{(B.3-9)}$$

and

$$z\underline{X}(z) = (I + AT)\underline{X}(z) + T\underline{e}_n M(z) \quad ; \qquad \text{(B.3-10)}$$

hence,

$$\underline{X}(z) = [(z - 1)I + AT]^{-1} T\underline{e}_n M(z) \qquad \text{(B.3-11)}$$

and

$$Y(z) = (b_n \underline{e}_1' + b_{n-1}\underline{e}_2' + \cdots + b_1 \underline{e}_n')\underline{X}(z) \qquad \text{(B.3-12)}$$

where $\underline{e}_j$ is the jth unit vector. Combining Eqs. (B.3-11) and (B.3-12), we obtain

$$Y(z) = (b_n \underline{e}_1' + \cdots + b_1 \underline{e}_n')[(z - 1)I + AT]^{-1} T\underline{e}_n M(z) \qquad \text{(B.3-13)}$$

which we claim is Eq. (B.3-6), but in operator notation. Equation (B.3-13) is the *formal solution*, from which everything else follows by straightforward, though very tedious algebra.

The finite-difference equation in Eq. (B.3-6) can be obtained in exactly n + 2 steps. This is clear when one rewrites Eq. (B.3-9) as

---

† We leave it to the reader to demonstrate that the two systems of equations in Eqs. (B.3-7) and (B.3-8) are triangular; hence, their solutions are very easy to compute.

‡ The reader unfamiliar with z transforms can, for the purposes of the present analysis, treat z as a forward shift operator; i. e., $z(x) = x(k + 1)$.

$$\frac{Y(z)}{M(z)} = b_n \frac{X_1(z)}{M(z)} + b_{n-1} \frac{X_2(z)}{M(z)} + \cdots + b_1 \frac{X_n(z)}{M(z)}$$

$$= b_n \frac{N_{x_1}(z)}{D(z)} + b_{n-1} \frac{N_{x_2}(z)}{D(z)} + \cdots + b_1 \frac{N_{x_n}(z)}{D(z)} \quad \text{(B.3-14)}$$

where we have used the well-known fact, from linear systems theory, that each transfer function $X_i(z)/M(z)$ is a ratio of two polynomials in z, a numerator polynomial $N_{x_i}(z)$ that is distinctly different for each $X_i(z)$, and a denominator polynomial D(z) that is the same for each $X_i(z)$. The polynomial D(z) is the characteristic equation for the system [see Eq. (B.3-10)]†

$$[(z - 1)I + AT]\underline{X}(z) = T\underline{e}_n M(z) \quad . \quad \text{(B.3-15)}$$

The n + 2 steps required to obtain the finite-difference equation are *(1)* compute D(z); *(2) - (n + 1)* compute $N_{x_1}(z)$, ..., and $N_{x_n}(z)$; and *(n + 2)* compute Y(z) in Eq. (B.3-14).

We leave the algebraic details to the reader, since they provide no additional *insight* into the solution, other than the solution itself. ▲

The results in Theorem B-4 are generalized to the 2n system of equations, Eqs. (B.2-23) and (B.2-24), in Theorem B-5.

Theorem B-5. *Given the 2n discrete-time system*

$$\underline{x}_1(k + 1) = (I + AT)\underline{x}_1(k) + T\underline{e}_n m(k) \quad \text{(B.3-16)}$$

$$\underline{x}_2(k + 1) = (I + AT)\underline{x}_2(k) + T\underline{e}_n \omega(k) \quad \text{(B.3-17)}$$

and

$$y(k) = \underline{b}'\underline{x}_1(k) + \underline{d}'\underline{x}_2(k) \quad \text{(B.3-18)}$$

where

---

† The characteristic equation is obtained by setting the determinant of [(z - 1)I + AT] equal to zero.

$$A = \left(\begin{array}{c|ccc} 0 & & & \\ 0 & & I_{n-1} & \\ \vdots & & & \\ \hline -a_n & -a_{n-1} & \cdots & -a_1 \end{array}\right) , \tag{B.3-19}$$

$$\underline{e}_n = (0, 0, \ldots, 0, 1)' , \tag{B.3-20}$$

$$\underline{b} = (b_n, b_{n-1}, \ldots, b_1)' , \tag{B.3-21}$$

and

$$\underline{d} = (d_n, d_{n-1}, \ldots, d_1)' . \tag{B.3-22}$$

The finite-difference equation for the output signal $y(k)$ *is*

$$y(k + n) + \alpha_1 y(k + n - 1) + \cdots + \alpha_n y(k) = \beta_1 m(k + n - 1) + \cdots$$
$$+ \beta_n m(k) + \gamma_1 \omega(k + n - 1) + \cdots + \gamma_n \omega(k) \tag{B.3-23}$$

where α_{n-i} *and* β_{n-i} *are given by Eqs. (B.3-7) and (B.3-8), respectively, for* $i = 0, 1, \ldots, n - 1$, *and*

$$\gamma_{n-i} = \sum_{j=i}^{n-1} (-1)^{j-i} \binom{j}{j - i} d_{n-j} T^{n-j} \tag{B.3-24}$$

for $i = 0, 1, \ldots, n - 1$.

Proof. The proof of this theorem uses the results in Theorem B-4 and the principle of superposition, which is known to hold for linear systems (*B-2*). Assuming that $\Omega(z)$ exists,† the z transform of Eq. (B.3-18) plus superposition give

$$Y(z) = Y_1(z) + Y_2(z) \tag{B.3-25}$$

where

$$\frac{Y_1(z)}{M(z)} = b_n \frac{X_{11}(z)}{M(z)} + \cdots + b_1 \frac{X_{1n}(z)}{M(z)}$$
$$= b_n \frac{N_{x_{11}}(z)}{D(z)} + \cdots + b_1 \frac{N_{x_{1n}}(z)}{D(z)} \tag{B.3-26}$$

† $\Omega(z)$ is the z transform of $\omega(t)$.

and

$$\frac{Y_2(z)}{\Omega(z)} = d_n \frac{X_{21}(z)}{\Omega(z)} + \cdots + d_1 \frac{X_{2n}(z)}{\Omega(z)}$$
$$= d_n \frac{N_{x_{21}}(z)}{D(z)} + \cdots + d_1 \frac{N_{x_{2n}}(z)}{D(z)} \quad . \qquad \text{(B.3-27)}$$

Clearly, $Y(z)$ is obtained in exactly $2n + 4$ operations: *(1)* computation of $D(z)$; *(2) - (n + 1)* computation of $N_{x_{11}}(z)$ through $N_{x_{1n}}(z)$; *(n + 2)-(2n + 1)* computation of $N_{x_{21}}(z)$ through $N_{x_{2n}}(z)$; *(2n + 2)* computation of $Y_1(z)$; *(2n + 3)* computation of $Y_2(z)$; and *(2n + 4)* computation of $Y(z)$.

That $D(z)$ is common to both the $Y_1(z)$ and $Y_2(z)$ equations is clear from the original difference equations, in which both have the same transition matrix $I + AT$.

The details of computing the γ_{n-i} $(i = 0, 1, \ldots, n - 1)$ coefficients are identical to those for computing the β_{n-i} coefficients in Theorem B-4, and are left to the reader. ▲

Observe that even though the discrete state-space representation is of dimension 2n, the finite-difference equation representation is nth order. This must be so, since the original differential equation, Eq. (B-1), is nth order.

REFERENCES

B-1. K. Ogata, *State Space Analysis of Control Systems*, Prentice-Hall, Englewood Cliffs, New Jersey, 1967.

B-2. L. A. Zadeh and C. A. Desoer, *Linear System Theory, The State Space Approach*, McGraw-Hill, New York, 1963.

B-3. J. A. Aseltine, *Transform Method in Linear System Analysis*, McGraw-Hill, New York, 1958.

B-4. R. J. Schwarz and B. Friedland, *Linear Systems*, McGraw-Hill, New York, 1965.

B-5. A. H. Jazwinski, *Stochastic Processes and Filtering Theory*, Academic Press, New York, 1970.

B-6. J. S. Meditch, *Stochastic Optimal Linear Estimation and Control*, McGraw-Hill, New York, 1969.

B-7. L. S. Pontryagin, *Ordinary Differential Equations*, Addison-Wesley, Reading, Mass., 1962.

B-8. B. C. Kuo, *Discrete-Data Control Systems*, Prentice-Hall, Englewood Cliffs, New Jersey, 1970.

B-9. J. R. Ragazzini and G. F. Franklin, *Sampled-Data Control Systems*, McGraw-Hill, New York, 1958.

Appendix C

Parameter Transformation Matrices

We shall determine explicit expressions for the $2n \times 2n$ parameter transformation matrix $\Lambda(\hat{\underline{\theta}})$ [see Section 1.3.3 in Chapter 1], where

$$\Lambda(\hat{\underline{\theta}}) = \begin{pmatrix} \underline{\lambda}_1'(\hat{\underline{\theta}}) \\ \hline \underline{\lambda}_2'(\hat{\underline{\theta}}) \\ \hline \vdots \\ \hline \underline{\lambda}_{2n}'(\hat{\underline{\theta}}) \end{pmatrix}, \tag{C-1}$$

$$\underline{\lambda}_j'(\hat{\underline{\theta}}) = [\lambda_{j1}(\hat{\underline{\theta}}),\ \lambda_{j2}(\hat{\underline{\theta}}),\ \ldots,\ \lambda_{j,2n}(\hat{\underline{\theta}})] \tag{C-2}$$

for $j = 1, 2, \ldots, 2n$, and the $\underline{\lambda}_j(\hat{\underline{\theta}})$ must be found from the following $2n$ equations [Eqs. (1.3-76) and (1.3-83)]:

$$\begin{aligned} \hat{\theta}_1 &= \underline{h}'\underline{\psi} &&= \underline{\lambda}_1'\underline{p} \\ \hat{\theta}_2 &= \underline{h}'\Phi\underline{\psi} &&= \underline{\lambda}_2'\underline{p} \\ \hat{\theta}_3 &= \underline{h}'\Phi^2\underline{\psi} &&= \underline{\lambda}_3'\underline{p} \\ &\ \vdots && \ \vdots \\ \hat{\theta}_{2n} &= \underline{h}'\Phi^{2n-1}\underline{\psi} &&= \underline{\lambda}_{2n}'\underline{p} \end{aligned} \tag{C-3}$$

in which $\underline{h}$, Φ, and $\underline{\psi}$ must be specified a priori, and $\underline{p}$ is a 2n parameter vector. The matrix $\Lambda(\hat{\underline{\theta}})$ will be obtained for an abstract digital system and for a discrete-time representation of a continuous-time system.

C.1 THE MATRIX $\Lambda(\hat{\underline{\theta}})$ FOR AN ABSTRACT DIGITAL SYSTEM

Equivalent state space and finite-difference equation representations of an abstract digital system are derived in Appendix A. Our interest is in identifying the coefficients $a_1, a_2, \ldots, a_n, b_1, b_2, \ldots, b_n$ from measurements of an output signal y(j). Recall that

$$y(j) = \underline{h}'\underline{x}(j) \tag{C.1-1}$$

where

$$\underline{h} = (1, 0, 0, \ldots, 0)' \quad , \tag{C.1-2}$$

and

$$\underline{x}(k + 1) = \Phi\underline{x}(k) + \underline{\psi}m(k) + \underline{\gamma}\omega(k) \tag{C.1-3}$$

in which

$$\Phi = \left(\begin{array}{c|ccc} 0 & & & \\ 0 & & I_{n-1} & \\ \vdots & & & \\ \hline -a_n & -a_{n-1} & \cdots & -a_1 \end{array}\right) \quad , \tag{C.1-4}$$

$$\underline{\psi} = (b_1, b_2, \ldots, b_n)' \quad , \tag{C.1-5}$$

and

$$\underline{\gamma} = (d_1, d_2, \ldots, d_n)' \quad . \tag{C.1-6}$$

Clearly, from the definitions of $\underline{h}$, Φ, and $\underline{\psi}$,

$$\begin{aligned} \hat{\theta}_1 &= b_1 \\ \hat{\theta}_2 &= b_2 \\ &\vdots \\ \hat{\theta}_n &= b_n \end{aligned} \tag{C.1-7}$$

and

$$\hat{\theta}_{n+1} = \underline{h}'\Phi^n\underline{\psi} = -(a_n, a_{n-1}, \ldots, a_1)\underline{\psi}$$

$$= -a_n b_1 - a_{n-1}b_2 - \cdots - a_1 b_n \qquad \text{(C.1-8)}$$

which, with the help of Eq. (C.1-7), can also be written as

$$\hat{\theta}_{n+1} = -a_n\hat{\theta}_1 - a_{n-1}\hat{\theta}_2 - \cdots - a_1\hat{\theta}_n \quad . \qquad \text{(C.1-9)}$$

In order to obtain comparable expressions for $\hat{\theta}_{n+2}, \ldots, \hat{\theta}_{2n}$, let us prove the following lemma.

Lemma C-1. *For* $\underline{h}$ *and* Φ *defined in Eqs. (C.1-2) and (C.1-4), respectively,*

$$\underline{h}'\Phi^{n+j} = -\sum_{i=1}^{j} a_i\underline{h}'\Phi^{n+j-i} - (\underbrace{0, 0, \ldots, 0}_{j \text{ terms}}, \underbrace{a_n, a_{n-1}, \ldots, a_{j+1}}_{n - j \text{ terms}}) \qquad \text{(C.1-10)}$$

where $j = 0, 1, \ldots, n - 1$ *and* $a_0 \triangleq 0$.

Proof. Our proof is by the *principle of mathematical induction*. For $j = 0$, Eq. (C.1-10) reduces to

$$\underline{h}'\Phi^n = -(a_n, a_{n-1}, \ldots, a_1) \qquad \text{(C.1-11)}$$

which, from the third expression on the right-hand side of Eq. (C.1-8), is clearly the correct value for $\underline{h}'\Phi^n$. Assuming the truth of Eq. (C.1-10) for $j = \ell$, let us demonstrate its truth for $j = \ell + 1$:

$$\underline{h}'\Phi^{n+\ell+1} = \underline{h}'\Phi^{n+\ell}\Phi$$

$$= -\sum_{i=1}^{\ell} a_i\underline{h}'\Phi^{n+\ell+1-i} - (\underbrace{0, \ldots, 0}_{\ell}, \underbrace{a_n, \ldots, a_{\ell+1}}_{n-\ell})\Phi$$

$$= -\sum_{i=1}^{\ell} a_i\underline{h}'\Phi^{n+\ell+1-i} + a_{\ell+1}(a_n, a_{n-1}, \ldots, a_1)$$

$$- (\underbrace{0, \ldots, 0}_{\ell+1}, \underbrace{a_n, \ldots, a_{\ell+2}}_{n-\ell-1})$$

$$= -\sum_{i=1}^{\ell} a_i\underline{h}'\Phi^{n+\ell+1-i} - a_{\ell+1}\underline{h}'\Phi^n - (\underbrace{0, \ldots, 0}_{\ell+1}, \underbrace{a_n, \ldots, a_{\ell+2}}_{n-\ell-1})$$

$$= -\sum_{i=1}^{\ell+1} a_i\underline{h}'\Phi^{n+\ell+1-i} - (\underbrace{0, \ldots, 0}_{\ell+1}, \underbrace{a_n, \ldots, a_{\ell+2}}_{n-\ell-1}) \quad . \qquad \text{(C.1-12)}$$

In obtaining this expression we have made use of the explicit structure of Φ given by Eq. (C.1-4). The right-hand side of Eq. (C.1-12) is exactly what is obtained from Eq. (C.1-10) by setting $j = \ell + 1$; thus, we have proven the correctness of Eq. (C.1-10). ▲

The term $\underline{h}'\Phi^{n+j}\underline{\psi}$ is obtained from Eq. (C.1-10), as

$$\underline{h}'\Phi^{n+j}\underline{\psi} = -\sum_{i=1}^{j} a_i\underline{h}'\Phi^{n+j-i}\underline{\psi} - (\underbrace{0, 0, \ldots, 0}_{j}, \underbrace{a_n, \ldots, a_{j+1}}_{n-j})\underline{\psi} \quad ; \tag{C.1-13}$$

however, we also know from Eq. (C-3), that

$$\underline{h}'\Phi^{n+j}\underline{\psi} = \hat{\theta}_{n+j+1} \quad . \tag{C.1-14}$$

Using this relationship on both sides of Eq. (C.1-13), as well as the explicit expression for $\underline{\psi}$, Eq. (C.1-5), we find that

$$\hat{\theta}_{n+j+1} = -\sum_{i=1}^{j} a_i\hat{\theta}_{n+j+1-i} - \sum_{i=j+1}^{n} a_i b_{n+j+1-i} \quad . \tag{C.1-15}$$

Finally, using Eq. (C.1-7), we obtain

$$\hat{\theta}_{n+j+1} = -\sum_{i=1}^{n} a_i\hat{\theta}_{n+j+1-i} \tag{C.1-16}$$

for $j = 0, 1, \ldots, n - 1$.

Equations (C.1-7) and (C.1-16) constitute a 2n system from which we can solve for $a_1, a_2, \ldots, a_n, b_1, b_2, \ldots, b_n$. Results are summarized in the following theorem.

Theorem C-1. *For the abstract digital system in Eqs. (C.1-1)-(C.1-6), if the composite parameter vector* $\underline{p}$ *is defined as*

$$\underline{p} = (b_1, b_2, \ldots, b_n, a_1, a_2, \ldots, a_n)' \tag{C.1-17}$$

then the transformation matrix, $\Lambda(\hat{\underline{\theta}})$, *is*

$$\Lambda(\underline{\theta}) = \left(\begin{array}{c|cccc} I_n & & & 0_n & \\ \hline & \hat{\theta}_1 & \hat{\theta}_2 & \cdots & \hat{\theta}_n \\ & \hat{\theta}_2 & \hat{\theta}_3 & \cdots & \hat{\theta}_{n+1} \\ 0_n & \vdots & \vdots & \ddots & \vdots \\ & \hat{\theta}_n & \hat{\theta}_{n+1} & \cdots & \hat{\theta}_{2n} \end{array}\right) \tag{C.1-18}$$

where I_n *is the* $n \times n$ *identity matrix and* 0_n *is the* $n \times n$ *null matrix.*

▲

C.2 THE MATRIX $\Lambda(\hat{\underline{\theta}})$ FOR A DISCRETE-TIME REPRESENTATION OF A CONTINUOUS-TIME SYSTEM

Equivalent state space and finite-difference equation representations of a continuous-time system are derived in Appendix B (see Theorems B-4 and B-5 for the main results). Our interest is in identifying the coefficients $a_1, a_2, \ldots, a_n, b_1, b_2, \ldots, b_n$ in the representation (Theorem B-5)

$$\begin{pmatrix} \underline{x}_1(k+1) \\ \hline \underline{x}_2(k+1) \end{pmatrix} = \left(\begin{array}{c|c} \Phi & 0 \\ \hline 0 & \Phi \end{array}\right) \begin{pmatrix} \underline{x}_1(k) \\ \hline \underline{x}_2(k) \end{pmatrix} + \begin{pmatrix} \underline{\psi} \\ \hline \underline{0} \end{pmatrix} m(k) + \begin{pmatrix} \underline{0} \\ \hline \underline{\gamma} \end{pmatrix} \omega(k) \tag{C.2-1}$$

$$y(k) = (\underline{b}' \mid \underline{d}') \begin{pmatrix} \underline{x}_1(k) \\ \hline \underline{x}_2(k) \end{pmatrix} \tag{C.2-2}$$

where

$$\Phi = I + AT \quad , \tag{C.2-3}$$

$$\underline{\psi} = \underline{\gamma} = T\underline{e}_n \quad , \tag{C.2-4}$$

$$A = \left(\begin{array}{c|c} 0 & \\ 0 & I_{n-1} \\ \vdots & \\ \hline -a_n & -a_{n-1} \quad \cdots \quad -a_1 \end{array}\right) , \tag{C.2-5}$$

$$\underline{b} = (b_n, b_{n-1}, \ldots, b_1)' , \tag{C.2-6}$$

and

$$\underline{d} = (d_n, d_{n-1}, \ldots, d_1)' . \tag{C.2-7}$$

Equations (C.2-1) and (C.2-2) can also be written more compactly as

$$\underline{x}(k + 1) = \Phi_1\underline{x}(k) + \underline{\psi}_1 m(k) + \underline{\gamma}_1 \omega(k) \tag{C.2-8}$$

and

$$y(k) = \underline{b}_1'\underline{x}(k) , \tag{C.2-9}$$

where the relationships between the symbols in Eqs. (C.2-8) and (C.2-1), and (C.2-9) and (C.2-2) are obvious.

From Eqs. (1.3-61), (C.2-8), and (C.2-9), we determine that

$$y(k) = \sum_{i=0}^{\infty} \underline{b}_1'\Phi_1^i[\underline{\psi}_1 m(k - i - 1) + \underline{\gamma}_1 \omega(k - i - 1)] . \tag{C.2-10}$$

We leave it to the reader to show that y(k) can also be written as

$$y(k) = \sum_{i=0}^{\infty} \underline{b}'\Phi^i[\underline{\psi} m(k - i - 1) + \underline{\gamma}\omega(k - i - 1)] ; \tag{C.2-11}$$

hence, the development presented in Section 1.3.3 for an abstract digital system applies as well to our discrete representation of a continuous-time system. Observe that for this representation

$$\underline{h} = \underline{b} . \tag{C.2-12}$$

▲

It is considerably more difficult to obtain $\Lambda(\hat{\underline{\theta}})$ for this system than it was for the abstract digital system. Our approach is to state the results for n = 2, 3, and 4, and to then generalize them to arbitrary n. We give details only for n = 2.

For n = 2, Φ, $\underline{\psi}$, and $\underline{b}$ are

$$\Phi = \begin{pmatrix} 1 & T \\ -a_2T & 1 - a_1T \end{pmatrix} \triangleq \begin{pmatrix} 1 & T \\ -a_2T & a_1^* \end{pmatrix} \quad , \tag{C.2-13}$$

$$\underline{\psi} = (0,\ T)' \quad , \tag{C.2-14}$$

and

$$\underline{b} = (b_2,\ b_1)' \quad . \tag{C.2-15}$$

Computing $\underline{b}'\underline{\psi} = \hat{\theta}_1$ [see Eq. (C-3)], we find

$$\hat{\theta}_1 = b_1T \quad . \tag{C.2-16}$$

Computing $\underline{b}'\Phi$ and then $\underline{b}'\Phi\underline{\psi} = \hat{\theta}_2$, we find

$$\underline{b}'\Phi = (b_2 - b_1a_2T \quad , \quad b_2T + b_1a_1^*) \tag{C.2-17}$$

and

$$\hat{\theta}_2 = T(b_2T + b_1a_1^*) \quad . \tag{C.2-18}$$

Using Eq. (C.2-16) to eliminate b_1 from this last expression, we obtain

$$\hat{\theta}_2 = T(b_2T + \frac{1}{T}\,\hat{\theta}_1a_1^*) \quad . \tag{C.2-19}$$

Comparing Eqs. (C.2-18) and (C.2-17), we see that the latter equation can also be written, as

$$\underline{b}'\Phi = (b_2 - a_2\hat{\theta}_1 \quad , \quad \frac{1}{T}\,\hat{\theta}_2) \quad . \tag{C.2-20}$$

We proceed to the computations of $\underline{b}'\Phi^2$ and $\underline{b}'\Phi^2\underline{\psi} = \hat{\theta}_3$:

$$\underline{b}'\Phi^2 = [b_2 - a_2\hat{\theta}_1 - a_2\hat{\theta}_2 \quad , \quad T(b_2 - a_2\hat{\theta}_1) + \frac{1}{T}\,a_1^*\hat{\theta}_2] \tag{C.2-21}$$

and

$$\hat{\theta}_3 = T[T(b_2 - a_2\hat{\theta}_1) + \frac{1}{T}\,a_1^*\hat{\theta}_2] \quad . \tag{C.2-22}$$

Observe, also, that

$$\underline{b}'\Phi^2 = (b_2 - a_2\hat{\theta}_1 - a_2\hat{\theta}_2 \quad , \quad \frac{1}{T}\,\hat{\theta}_3) \quad . \tag{C.2-23}$$

Finally, we compute $\underline{b}'\Phi^3$ and $\underline{b}'\Phi^3\underline{\psi} = \hat{\theta}_4$:

$$\underline{b}'\Phi^3 = [b_2 - a_2\hat{\theta}_1 - a_2\hat{\theta}_2 - a_2\hat{\theta}_3 \quad , \quad T(h_2 - a_2\hat{\theta}_1 - a_2\hat{\theta}_2) + \frac{1}{T}\, a_1^*\hat{\theta}_3] \tag{C.2-24}$$

and

$$\hat{\theta}_4 = T[T(b_2 - a_2\hat{\theta}_1 - a_2\hat{\theta}_2) + \frac{1}{T}\, a_1^*\hat{\theta}_3] \quad . \tag{C.2-25}$$

Collecting Eqs. (C.2-16), (C.2-19), (C.2-22), and (C.2-25) together, we find that

$$\begin{pmatrix} \hat{\theta}_1 \\ \hat{\theta}_2 \\ \hat{\theta}_3 \\ \hat{\theta}_4 \end{pmatrix} = \begin{pmatrix} 1 & 0 & 0 & 0 \\ 0 & 1 & \hat{\theta}_1 & 0 \\ 0 & 1 & \hat{\theta}_2 & -\hat{\theta}_1 \\ 0 & 1 & \hat{\theta}_3 & -\sum_{i=1}^{2}\hat{\theta}_i \end{pmatrix} \begin{pmatrix} b_1T \\ b_2T^2 \\ a_1^* \\ a_2T^2 \end{pmatrix} \quad . \tag{C.2-26}$$

Comparable results for n = 3 and n = 4, are: *for* n = 3,

$$\begin{pmatrix} \hat{\theta}_1 \\ \hat{\theta}_2 \\ \hat{\theta}_3 \\ \hat{\theta}_4 \\ \hat{\theta}_5 \\ \hat{\theta}_6 \end{pmatrix} = \begin{pmatrix} 1 & 0 & 0 & 0 & 0 & 0 \\ 0 & 1 & 0 & \hat{\theta}_1 & 0 & 0 \\ 0 & 1 & 1 & \hat{\theta}_2 & -\hat{\theta}_1 & 0 \\ 0 & 1 & 2 & \hat{\theta}_3 & -\sum_{i=1}^{2}\hat{\theta}_i & -\hat{\theta}_1 \\ 0 & 1 & 3 & \hat{\theta}_4 & -\sum_{i=1}^{3}\hat{\theta}_i & -2\hat{\theta}_1 - \hat{\theta}_3 \\ 0 & 1 & 4 & \hat{\theta}_5 & -\sum_{i=1}^{4}\hat{\theta}_i & -3\hat{\theta}_1 - \hat{\theta}_2 - 2\hat{\theta}_3 \end{pmatrix} \begin{pmatrix} b_1T \\ b_2T^2 \\ b_3T^3 \\ a_1^* \\ a_2T^2 \\ a_3T^3 \end{pmatrix} \tag{C.2-27}$$

and, *for* n = 4,

$$\begin{pmatrix} \hat{\theta}_1 \\ \hat{\theta}_2 \\ \hat{\theta}_3 \\ \hat{\theta}_4 \\ \hat{\theta}_5 \\ \hat{\theta}_6 \\ \hat{\theta}_7 \\ \hat{\theta}_8 \end{pmatrix} = \begin{pmatrix} 1 & 0 & 0 & 0 & 0 & 0 & 0 & 0 \\ 0 & 1 & 0 & 0 & \hat{\theta}_1 & 0 & 0 & 0 \\ 0 & 1 & 1 & 0 & \hat{\theta}_2 & -\hat{\theta}_1 & 0 & 0 \\ 0 & 1 & 2 & 1 & \hat{\theta}_3 & -\sum_{i=1}^{2}\hat{\theta}_i & -\hat{\theta}_1 & 0 \\ 0 & 1 & 3 & 3 & \hat{\theta}_4 & -\sum_{i=1}^{3}\hat{\theta}_i & -2\hat{\theta}_1 - \hat{\theta}_2 & -\hat{\theta}_1 \\ 0 & 1 & 4 & 6 & \hat{\theta}_5 & -\sum_{i=1}^{4}\hat{\theta}_i & -3\hat{\theta}_1 - 2\hat{\theta}_2 - \hat{\theta}_3 & -3\hat{\theta}_1 - \hat{\theta}_2 \\ 0 & 1 & 5 & 10 & \hat{\theta}_6 & -\sum_{i=1}^{5}\hat{\theta}_i & -4\hat{\theta}_1 - 3\hat{\theta}_2 - 2\hat{\theta}_3 - \hat{\theta}_4 & -6\hat{\theta}_1 - 3\hat{\theta}_2 - \hat{\theta}_3 \\ 0 & 1 & 6 & 15 & \hat{\theta}_7 & -\sum_{i=1}^{6}\hat{\theta}_i & -5\hat{\theta}_1 - 4\hat{\theta}_2 - 3\hat{\theta}_3 - 2\hat{\theta}_4 - \hat{\theta}_5 & -10\hat{\theta}_1 - 6\hat{\theta}_2 - 3\hat{\theta}_3 - \hat{\theta}_4 \end{pmatrix} \begin{pmatrix} b_1T \\ b_2T^2 \\ b_3T^3 \\ b_4T^4 \\ a_1^* \\ a_2T^2 \\ a_3T^3 \\ a_4T^4 \end{pmatrix} \tag{C.2-28}$$

Observe from Eqs. (C.2-26)-(C.2-28) that the composite parameter vector is, in each case:

$$\text{for } n = 2, \quad \underline{p} = (b_1T,\ b_2T,\ a_1^*,\ a_2T^2)' \quad , \tag{C.2-29}$$

$$\text{for } n = 3, \quad \underline{p} = (b_1T,\ b_2T^2,\ b_3T^3,\ a_1^*,\ a_2T^2,\ a_3T^3)' \quad , \tag{C.2-30}$$

$$\text{and for } n = 4, \quad \underline{p} = (b_1T,\ b_2T^2,\ b_3T^3,\ b_4T^4,\ a_1^*,\ a_2T^2,\ a_3T^3,\ a_4T^4)' \quad . \tag{C.2-31}$$

It is clear what $\Lambda(\hat{\underline{\theta}})$ is for n = 2-4 from Eqs. (C.2-26)-(C.2-28), respectively.

By observing the formation of the $\Lambda(\hat{\underline{\theta}})$ matrices for n = 2, 3, and 4 we obtain the following algorithm.

Algorithm for Constructing $\Lambda(\hat{\underline{\theta}})$

The elements of $\Lambda(\hat{\underline{\theta}})$ denoted λ_{ij} are obtained from the expression (see Fig. C-1, as well);

$$\lambda_{11} = 1 \quad , \tag{C.2-32}$$

$$\lambda_{1j} = 0 \quad j = 2, 3, \ldots, 2n \quad , \tag{C.2-33}$$

$$\lambda_{i1} = 0 \quad i = 2, 3, .. , 2n \quad , \tag{C.2-34}$$

$$\lambda_{i2} = 1 \quad i = 2, 3, \ldots, 2n \quad , \tag{C.2-35}$$

$$\lambda_{ij} = \lambda_{i-1,j} + \lambda_{i-1,j-1} \quad \begin{cases} i = 2, 3, \ldots, 2n \\ j = 3, 4, \ldots, n \end{cases} \tag{C.2-36}$$

$$\lambda_{i,n+1} = \hat{\theta}_{i-1} \quad i = 2, 3, \ldots, 2n \quad , \tag{C.2-37}$$

$$\lambda_{i,n+2} = \lambda_{i-1,n+2} - \lambda_{i-1,n+1} \quad i = 2, 3, \ldots, 2n \tag{C.2-38}$$

and

$$\lambda_{i,j} = \lambda_{i-1,j} + \lambda_{i-1,j-1} \quad \begin{cases} i = 2, 3, \ldots, 2n \\ j = n + 3, n + 4, \ldots, 2n \end{cases} \tag{C.2-39}$$

▲

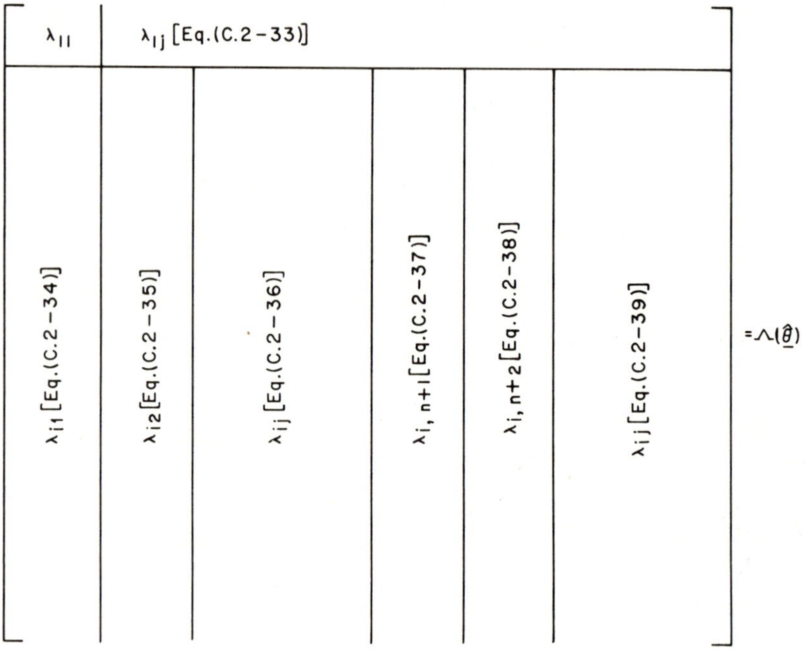

Fig. C-1. Construction of $\Lambda(\hat{\underline{\theta}})$.

We summarize our results in Theorem C-2.

Theorem C-2. *For the discrete-time representation of the continuous-time system in Eqs. (C.2-1)-(C.2-7), if the composite parameter vector* $\underline{p}$ *is defined as*

$$\underline{p} = (b_1T,\ b_2T^2,\ \ldots,\ b_nT^n,\ a_1^*,\ a_2T^2,\ \ldots,\ a_nT^n)' \tag{C.2-40}$$

where

$$a_1^* = 1 - a_1T \quad , \tag{C.2-41}$$

then the transformation matrix $\Lambda(\hat{\underline{\theta}})$ *is given by Eqs. (C.2-32)-(C.2-39).*

Appendix D

Lyapunov's Main Stability Theorem: Statement and Proof

__Theorem.__ *Consider the discrete-time, free dynamic system*

$$\underline{x}(t_{k+1}) = \underline{f}[\underline{x}(t_k), t_k] \tag{D-1}$$

where

$$\underline{f}(\underline{0}, t_k) = \underline{0} \tag{D-2}$$

for all t_k. Suppose *there exists a scalar function* $V(\underline{x}, t_k)$ *such that*

$$V(\underline{0}, t_k) = 0 \tag{D-3}$$

for all t_k, *and the following also prevails.*

(i) The $V(\underline{x}, t_k)$ *is positive definite; i. e., there exists a continuous nondecreasing scalar function* α *such that* $\alpha(0) = 0$ *and, for all* t_k *and all* $\underline{x} \neq \underline{0}$,

$$V(\underline{x}, t_k) \geq \alpha(||\underline{x}||) > 0 \quad . \tag{D-4}$$

(ii) There exists a continuous scalar function γ *such that* $\gamma(0) = 0$ *and, for all* t_k *and* $\underline{x} \neq \underline{0}$,

---

† This appendix follows closely Refs. *D-1* and *D-2*.

$$\Delta V(\underline{x}, t_k) = \text{rate of increase of } V \text{ along motion starting at } \underline{x}(t_k)$$

$$= V[\underline{\phi}(t_{k+1}; \underline{x}, t_k), t_{k+1}] - V(\underline{x}, t_k)$$

$$\leq -\gamma(||\underline{x}||) < 0 \quad . \tag{D-5}$$

(iii) There exists a continuous, nondecreasing scalar function β *such that* $\beta(0) = 0$ *and, for all* t_k *and* $\underline{x} \neq \underline{0}$,

$$V(\underline{x}, t_k) \leq \beta(||\underline{x}||) \quad . \tag{D-6}$$

(iv) $\alpha(||\underline{x}||) \to \infty$ *when* $||\underline{x}|| \to \infty$.

Then *the equilibrium state* $\underline{x}_e = \underline{0}$ *is uniformly asymptotically stable in the large and* $V(\underline{x}, t_k)$ *is a Lyapunov function of the system in Eq. (D-1).*

Proof. Using (ii) we see that V is decreasing along any motion; hence,[†]

$$V[\underline{\phi}(t_j; \underline{x}_0, t_0), t_j] - V(\underline{x}_0, t_0) = \sum_{\tau=t_0}^{t_j} \Delta V[\underline{\phi}(\tau; \underline{x}, t_0), \tau] < 0 \tag{D-7}$$

for $t_j > t_0$.

(a) *To prove uniform stability,* consider any $\varepsilon > 0$. Take $\delta(\varepsilon) > 0$ such that $\beta(\delta) < \alpha(\varepsilon)$ [see Fig. D-1]. This is possible because β is continuous and $\beta(0) = 0$ [condition (iii)]. Then if $||\underline{x}_0|| \leq \delta$, t_0 being arbitrary, we have - using (ii) and Eq. (D-7) [see Fig. D-1] - for all $t_j \geq t_0$

$$\alpha(\varepsilon) > \beta(\delta) \geq V(\underline{x}_0, t_0) \geq V[\underline{\phi}(t_j; \underline{x}_0, t_0), t_j]$$

$$\geq \alpha[||\underline{\phi}(t_j; \underline{x}_0, t_0)||] \quad . \tag{D-8}$$

But since α is nondecreasing and positive, this implies

$$||\underline{\phi}(t_j; \underline{x}_0, t_0)|| \leq \varepsilon \quad \text{for} \quad t \geq t_0, \quad ||\underline{x}_0|| \leq \delta(\varepsilon) \quad ; \tag{D-9}$$

[†] We are assuming $t_{k+1} - t_k = 1$. See footnote ‡ page 198.

hence, we have shown that for each real number $\varepsilon > 0$, there is a real number $\delta(\varepsilon) > 0$ such that $||\underline{x}_0|| \leq \delta(\varepsilon)$ implies $||\underline{\phi}(t_j; \underline{x}_0, t_0)|| \leq \varepsilon$ for all $t \geq t_0$. Thus we have proved uniform stability (see stability definition 2, in Section 4.4.2). Note that (iv) is not needed and it is sufficient to have $\Delta V(\underline{x}, t_k) \leq 0$.

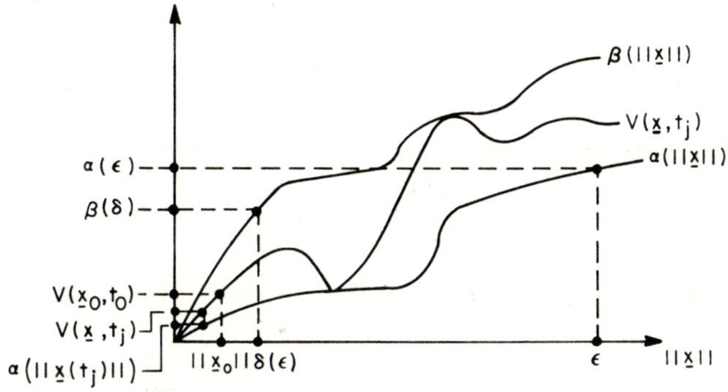

Fig. D-1. Curves of $\alpha(||\underline{x}||)$, $\beta(||\underline{x}||)$, and $V(\underline{x}, t_j)$ for (a) of proof.

(b) Now we prove that $||\underline{\phi}(t_j; \underline{x}_0, t_0)|| \to 0$ with $t_j \to \infty$ uniformly in t_0 and $||\underline{x}_0|| \leq r$.

Take any positive constant c_1 and find an $r > 0$ satisfying $\beta(r) < \alpha(c_1)$ [Fig. D-2a].[†] Take any initial state $||\underline{x}_0|| \leq r$. By part (a) of the proof, $||\underline{\phi}(t_j; \underline{x}_0, t_0)|| \leq c_1$ for all $t_j \geq t_0$, t_0 being arbitrary.

Now take any $0 < \mu \leq ||\underline{x}_0||$ [Fig. D-2a]. Find a $\nu(\mu) > 0$ such that $\beta(\nu) < \alpha(\mu)$. Denote by $c_2(\mu, r) > 0$ the minimum of the continuous function $\gamma(||\underline{x}||)$ on the compact set $\nu(\mu) \leq ||\underline{x}|| \leq c_1(r)$ [Fig. D-2b]. We arbitrarily define the time $T(\mu, r)$ as

$$T(\mu, r) \triangleq \beta(r)/c_2(\mu, r) > 0 \quad . \tag{D-10}$$

We shall use the *method of contradiction* to prove that $||\underline{\phi}(t_j; \underline{x}_0, t_0)|| \to 0$ with $t_j \to \infty$. Suppose that $||\underline{\phi}(t_j; \underline{x}_0, t_0)|| > \nu$

[†] The quantity r plays the role of δ in this part of the proof.

over the interval $t_0 \leq t_j \leq t_\ell \triangleq t_0 + T(\mu, r)$. Again, by (ii) and Eq. (D-7), we have

$$V[\underline{\phi}(t_\ell; \underline{x}_0, t_0), t_\ell] - V(\underline{x}_0, t_0) = \sum_{\tau=t_0}^{t_\ell} \Delta V[\underline{\phi}(\tau; \underline{x}, t_0), \tau]$$

$$\leq -\gamma[||\underline{x}(\tau)||](t_\ell - t_0)$$

$$\leq -c_2(\mu, r)(t_\ell - t_0) \quad . \tag{D-11}$$

From Fig. D-2a, we see that

$$V(\underline{x}_0, t_0) \leq \beta(r) \quad ; \tag{D-12}$$

hence,

$$V[\underline{\phi}(t_\ell; \underline{x}_0, t_0), t_\ell] \leq \beta(r) - c_2(\mu, r)(t_\ell - t_0)$$

$$= \beta(r) - c_2(\mu, r)T(\mu, r)$$

$$= 0 \tag{D-13}$$

which contradicts condition (i).

It must be, therefore, that for some t_j in the interval (t_0, t_ℓ), say t_m, we have $||\underline{x}_m|| = ||\underline{\phi}(t_m; \underline{x}_0, t_0)|| = \nu$. Proceeding as we did in part (a) of this proof when we obtained Eq. (D-8), we find (see also Fig. D-2a)

$$\alpha(\mu) > \beta(\nu) \geq V(\underline{x}_m, t_m) \geq V[\underline{\phi}(t_j; \underline{x}_m, t_m), t_j]$$

$$\geq \alpha[||\underline{\phi}(t_j; \underline{x}_m, t_m)||] > 0 \tag{D-14}$$

for all $t \geq t_m$, which means that

$$||\underline{\phi}(t_j; \underline{x}_0, t_0)|| < \mu \qquad \text{for all} \quad t_j \geq t_0 + T(\mu, r) \geq t_m \quad , \tag{D-15}$$

which proves uniform asymptotic stability.

(c) To prove uniform asymptotic stability in the large, i. e., that the constant r can be chosen to be arbitrarily large, observe that by (iv) there exists for any r a constant $c_1(r)$ such that $\beta(r) < \alpha(c_1)$. Moreover, uniform boundedness is automatic, since $c_1(r)$

does not depend upon t_0. We have thus *proved uniform asymptotic stability in the large* for the equilibrium state $\underline{x}_e = \underline{0}$ of the system in Eq. (D-1).

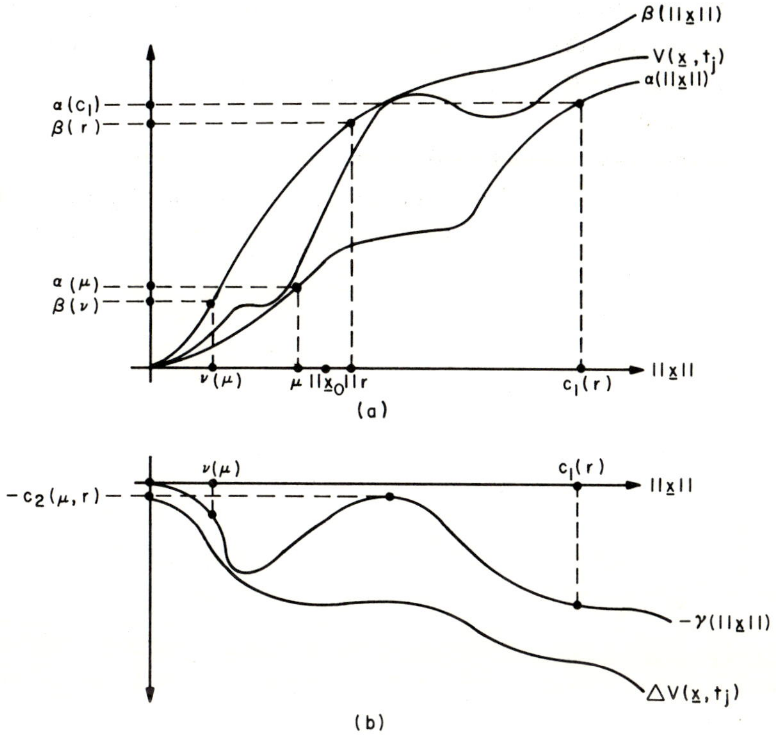

Fig. D-2. (a) Curves of $\alpha(||\underline{x}||)$, $\beta(||\underline{x}||)$, and $V(\underline{x}, t_j)$ for (b) of proof; (b) curves of $-\gamma(||\underline{x}||)$ and $\Delta V(\underline{x}, t_j)$ for (b) of proof.

▲

The conditions of this theorem are sufficient but not necessary for system stability; thus, if it is not possible to find a $V(\underline{x}, t)$ that satisfies all of the theorem's conditions, we cannot conclude anything about the system's stability. For additional discussions on Lyapunov's second method the reader is referred to Refs. *D-1* - *D-4*.

REFERENCES

D-1. R. E. Kalman and J. E. Bertram, Control system analysis and design via the second method of Liapunov: I. Continuous-time systems, *ASME J. Basic. Eng.*, Ser. D, 82, 371-393 (1960).

D-2. R. E. Kalman and J. E. Bertram, Control system analysis and design via the second method of Liapunov: II. Discrete-time systems, *ASME J. Basic Eng.*, Ser. D, 82, 392-400 (1960).

D-3. W. Hahn, *Theory and Applications of Liapunov's Direct Method*, Prentice-Hall, Englewood Cliffs, New Jersey, 1963.

D-4. J. P. La Salle and S. Lefschetz, *Stability by Liapunov's Direct Method with Applications*, Academic Press, New York, 1961.

Appendix E

Some Facts About Conditional Expectations

We assume that the reader is familiar with probability notions for sequences of vector random variables and random processes (*E-1* - *E-3*, for example).

Let $\underline{x}$ be a random n vector and $\underline{y}$ be a random m vector, so that $f(\underline{x}, \underline{y})$, $f(\underline{x})$, and $f(\underline{y})$ denote the *joint probability density functions of* $\underline{x}$ *and* $\underline{y}$, the *marginal probability density function of* $\underline{x}$, and the *marginal probability density function of* $\underline{y}$, respectively, where, for example,

$$f(\underline{y}) = \int_{-\infty}^{\infty} \cdots \int_{-\infty}^{\infty} f(\underline{x}, \underline{y})dx_1 \; dx_2 \; \cdots \; dx_n \quad . \tag{E-1}$$

The term $f(\underline{x}|\underline{y})$ denotes the *conditional probability density function* of $\underline{x}$ given $\underline{y}$, and

$$f(\underline{x}|\underline{y}) = \frac{f(\underline{x}, \underline{y})}{f(\underline{y})} \quad . \tag{E-2}$$

The *expected value* of a scalar or vector-valued function $g(\cdot)$ of the random n vector $\underline{x}$ is

$$E\{g(\underline{x})\} = \int_{-\infty}^{\infty} \cdots \int_{-\infty}^{\infty} g(\underline{x})f(\underline{x}) \; dx_1 \; dx_2 \; \ldots \; dx_n \quad . \tag{E-3}$$

Similarly,

$$E\{g(\underline{x}, \underline{y})\} = \int_{-\infty}^{\infty} \cdots \int_{-\infty}^{\infty} g(\underline{x}, \underline{y}) f(\underline{x}, \underline{y}) \, dx_1 \cdots dx_n \, dy_1 \cdots dy_m . \tag{E-4}$$

The conditional expected value of a scalar or vector-valued function $g(\cdot)$ of a random n vector $\underline{x}$ with respect to another random m vector $\underline{y}$ is

$$E\{g(\underline{x}) | \underline{y}\} = \int_{-\infty}^{\infty} \cdots \int_{-\infty}^{\infty} g(\underline{x}) f(\underline{x}|\underline{y}) \, dx_1 \cdots dx_n . \tag{E-5}$$

Observe that $f(\underline{x}|\underline{y})$ and subsequently $E\{g(\underline{x})|\underline{y}\}$ depend on the realizations of $\underline{y}$ and, as a result, are themselves random variables. They are functions of the random vector $\underline{y}$.

If $\underline{x}$ and $\underline{y}$ are *statistically independent*, then

$$f(\underline{x}, \underline{y}) = f(\underline{x}) f(\underline{y}) \tag{E-6}$$

and, therefore,

$$f(\underline{x}|\underline{y}) = f(\underline{x}) . \tag{E-7}$$

We summarize seven useful properties of conditional expectation in the following theorem.

Theorem. *Let* $\underline{x}$, $\underline{y}$, *and* $\underline{z}$ *be jointly distributed random vectors;* c *and* h *fixed constants; and* $g(\cdot)$ *a scalar-valued function. Assume* $E\{\underline{x}\}$, $E\{\underline{z}\}$, *and* $E\{g(\underline{y})\underline{x}\}$ *exist. Then (E-4 and E-5)*

$$E\{\underline{x}|\underline{y}\} = E\{\underline{x}\} \quad \textit{if } \underline{x} \textit{ and } \underline{y} \textit{ are independent,} \tag{E-8}$$

$$E\{g(\underline{y})\underline{x}|\underline{y}\} = g(\underline{y}) E\{\underline{x}|\underline{y}\} , \tag{E-9}$$

$$E\{c|\underline{y}\} = c , \tag{E-10}$$

$$E\{g(\underline{y})|\underline{y}\} = g(\underline{y}) , \tag{E-11}$$

$$E\{c\underline{x} + h\underline{z}|\underline{y}\} = cE\{\underline{x}|\underline{y}\} + hE\{\underline{z}|\underline{y}\} , \tag{E-12}$$

$$E\{\underline{x}\} = E\{E\{\underline{x}|\underline{y}\}\} \quad \textit{where the outer expectation is with respect to } \underline{y}, \text{ and} \tag{E-13}$$

$$E\{g(\underline{y})\underline{x}\} = E\{g(\underline{y})E\{\underline{x}|\underline{y}\}\} \quad \textit{where the outer expectation is with respect to } \underline{y}. \tag{E-14}$$

Proof. We shall prove Eqs. (E-8), (E-12), and (E-14), and leave the other proofs to the reader.

(a) From Eq. (E-5), with $g(\underline{x}) = \underline{x}$, and Eq. (E-7),

$$E\{\underline{x}|\underline{y}\} = \int_{-\infty}^{\infty} \cdots \int_{-\infty}^{\infty} \underline{x}f(\underline{x}|\underline{y}) \; dx_1 \cdots dx_n$$

$$= \int_{-\infty}^{\infty} \cdots \int_{-\infty}^{\infty} \underline{x}f(\underline{x}) \; dx_1 \cdots dx_n$$

$$= E\{\underline{x}\} \quad . \tag{E-15}$$

(b) From Eqs. (E-4) and (E-5), and Eq. (E-1),

$$E\{c\underline{x} + h\underline{z}|\underline{y}\} = \int_{-\infty}^{\infty} \cdots \int_{-\infty}^{\infty} (c\underline{x} + h\underline{z})f(\underline{x}, \underline{z}|\underline{y}) \times dx_1 \cdots dx_n \; dz_1 \cdots dz_\ell$$

$$= \int_{-\infty}^{\infty} \cdots \int_{-\infty}^{\infty} c\underline{x}f(\underline{x}, \underline{z}|\underline{y})d\underline{x}d\underline{z} + \int_{-\infty}^{\infty} \cdots \int_{-\infty}^{\infty} h\underline{z}f(\underline{x}, \underline{z}|\underline{y})d\underline{x}d\underline{z}$$

$$= \int_{-\infty}^{\infty} \cdots \int_{-\infty}^{\infty} c\underline{x}f(\underline{x}|\underline{y})d\underline{x} + \int_{-\infty}^{\infty} \cdots \int_{-\infty}^{\infty} h\underline{z}f(\underline{z}|\underline{y})d\underline{z}$$

$$= cE\{\underline{x}|\underline{y}\} + hE\{\underline{z}|\underline{y}\} \quad . \tag{E-16}$$

(c) From Eqs. (E-4) and (E-2),

$$E\{g(\underline{y})\underline{x}\} = \int_{-\infty}^{\infty} \cdots \int_{-\infty}^{\infty} g(\underline{y})\underline{x}f(\underline{x}, \underline{y})d\underline{x}d\underline{y}$$

$$= \int_{-\infty}^{\infty} \cdots \int_{-\infty}^{\infty} g(\underline{y})\underline{x}f(\underline{x}|\underline{y})f(\underline{y})d\underline{x}d\underline{y}$$

$$E\{g(\underline{y})\underline{x}\} = \int_{-\infty}^{\infty} \cdots \int_{-\infty}^{\infty} g(\underline{y})\left[\int_{-\infty}^{\infty} \cdots \int_{-\infty}^{\infty} \underline{x} f(\underline{x}|\underline{y}) d\underline{x}\right] f(\underline{y}) d\underline{y}$$

$$= \int_{-\infty}^{\infty} \cdots \int_{-\infty}^{\infty} g(\underline{y}) E\{\underline{x}|\underline{y}\} f(\underline{y}) d\underline{y}$$

$$= E\{g(\underline{y}) E\{\underline{x}|\underline{y}\}\} \quad . \tag{E-17}$$

REFERENCES

E-1. A. Papoulis, *Probability, Random Variables, and Stochastic Processes*, McGraw-Hill, New York, 1965.

E-2. E. Parzen, *Stochastic Processes*, Holden-Day, San Francisco, 1962.

E-3. F. A. Graybill, *An Introduction to Linear Statistical Models*, Vol. I, McGraw-Hill, New York, 1961.

E-4. A. H. Jazwinski, *Stochastic Processes and Filtering Theory*, Academic Press, New York, 1970.

E-5. J. S. Meditch, *Stochastic Optimal Linear Estimation and Control*, McGraw-Hill, New York, 1969.

Appendix F

Venter's Theorem

In this appendix we shall prove the following theorem.

Theorem [Venter's Theorem (*F-1*)]. If

$$\beta(k + 1) \leq [1 - \alpha(k)]\beta(k) + w(k) \tag{F-1}$$

and

(i) $\sum_{k=0}^{\infty} \alpha(k) \to \infty$,

(ii) $\alpha(k) \to 0$ *as* $k \to \infty$,

(iii) $\alpha(k)$, $w(k)$, *and* $\beta(k) \geq 0$,

(iv) $\sum_{k=0}^{\infty} w(k) < \infty$,

then

$$\beta(k) \to 0 \quad as \quad k \to \infty \quad . \tag{F-2}$$

▲

The proof of this theorem includes a study into the limiting behavior of the product $\Pi_{i=1}^{k} [1 - \alpha(i)]$ as $k \to \infty$. The following lemma provides us with conditions on $\alpha(i)$ such that $\Pi_{i=1}^{k}$ $[1 - \alpha(i)]$ converges to zero as $k \to \infty$:

<u>Lemma</u>. If $0 < \alpha(i) < 1$ *for* $i > I$, *and* $\sum_{i=I+1}^{\infty} \alpha(i) \to \infty$, then

$$\lim_{k\to\infty} \prod_{i=1}^{k} [1 - \alpha(i)] \to 0 \quad . \tag{F-3}$$

Proof.

$$\prod_{i=1}^{k} [1 - \alpha(i)] = \prod_{i=1}^{I} [1 - \alpha(i)] \prod_{i=I+1}^{k} [1 - \alpha(i)]$$

$$= g(I) \prod_{i=I+1}^{k} [1 - \alpha(i)] \quad . \tag{F-4}$$

Now

$$\prod_{i=I+1}^{k} [1 - \alpha(i)] = \exp\left\{\sum_{i=I+1}^{k} \ln[1 - \alpha(i)]\right\} \quad , \tag{F-5}$$

which follows from the facts that

$$\exp\{\ln[1 - \alpha(I + 1)] + \cdots + \ln[1 - \alpha(k)]\}$$

$$= \exp\{\ln[1 - \alpha(I + 1)]\} \cdots \exp\{\ln[1 - \alpha(k)]\}$$

$$= [1 - \alpha(I + 1)] \cdots [1 - \alpha(k)]$$

$$= \prod_{i=I+1}^{k} [1 - \alpha(i)] \quad . \tag{F-6}$$

Now, for $0 < \alpha(i) < 1$,

$$\ln[1 - \alpha(i)] = -\alpha(i) + \phi(i)\alpha^2(i) \tag{F-7}$$

where $\phi(i)$ is a bounded sequence (*F-2*); hence,

$$\prod_{i=I+1}^{k} [1 - \alpha(i)] = \exp\left\{\sum_{i=I+1}^{k} [\phi(i)\alpha^2(i) - \alpha(i)]\right\}$$

$$= \exp\left\{- \sum_{i=I+1}^{k} \alpha(i)\right\} \exp\left\{\sum_{i=I+1}^{k} \phi(i)\alpha^2(i)\right\} \quad . \tag{F-8}$$

Substituting this expression into Eq. (F-4) and passing to the limit, we obtain the result in Eq. (F-3), since

$$\lim_{k\to\infty} \left\{ \exp[-\sum_{i=I+1}^{k} \alpha(i)] \right\} \to 0 \tag{F-9}$$

by virtue of our assumption that $\sum_{i=I+1}^{\infty}\alpha(i)$ diverges. ▲

We now turn to the proof of Venter's theorem.†

Proof. The proof proceeds in three major steps. First, two solution-type inequalities are derived from Eq. (F-1). Second, the existence of an upper bound on $\beta(k + 1)$ is established in the limit as $k \to \infty$. Finally, it is shown that this limiting upper bound must approach zero.

A. Derivation of Two Solution-Type Inequalities

Infinite Product Solution

For k = 1 and k = 2, Eq. (F-1) becomes

$$\beta(2) \leq [1 - \alpha(1)]\beta(1) + w(1) \quad , \tag{F-10}$$

$$\begin{aligned} \beta(3) &\leq [1 - \alpha(2)]\beta(2) + w(2) \\ &\leq [1 - \alpha(2)][1 - \alpha(1)]\beta(1) \\ &\quad + [1 - \alpha(2)]w(1) + w(2) \end{aligned} \tag{F-11}$$

where we have used the fact, from condition (iii) in the statement of the theorem, that $\beta(2) \geq 0$. Generalizing, we obtain

$$\begin{aligned} \beta(k + 1) &\leq \prod_{i=1}^{k} [1 - \alpha(i)]\beta(1) \\ &\quad + \sum_{i=1}^{k-1} w(i) \prod_{j=i+1}^{k} [1 - \alpha(j)] + w(k) \end{aligned} \tag{F-12}$$

which is an infinite product (as $k \to \infty$) solution to our original inequality.

---

† This proof was communicated to the author by Dr. Jack Holmes.

Infinite Sum Solution

Equation (F-1) can also be written as

$$\beta(k + 1) \leq \beta(k) - \alpha(k)\beta(k) + w(k) \quad . \tag{F-13}$$

For k = 1 and 2, we have

$$\beta(2) \leq \beta(1) - \alpha(1)\beta(1) + w(1) \tag{F-14}$$

$$\begin{aligned}\beta(3) &\leq \beta(2) - \alpha(2)\beta(2) + w(2) \\ &\leq \beta(1) - \alpha(1)\beta(1) + w(1) \\ &\quad - \alpha(2)\beta(2) + w(2) \quad .\end{aligned} \tag{F-15}$$

Generalizing, we obtain

$$\beta(k + 1) \leq \beta(1) - \sum_{i=1}^{k} \alpha(i)\beta(i) + \sum_{i=1}^{k} w(i) \tag{F-16}$$

which is an infinite sum (as $k \to \infty$) solution to Eq. (F-1).

B. Existence of an Upper Bound

Passing to the limit in Eq. (F-12), we find

$$\begin{aligned}\lim_{k\to\infty} \beta(k + 1) &\leq \beta(1) \lim_{k\to\infty} \prod_{i=1}^{k} [1 - \alpha(i)] \\ &\quad + \lim_{k\to\infty} \left\{ \sum_{i=1}^{k-1} w(i) \sum_{j=i+1}^{k} [1 - \alpha(j)] + w(k) \right\} \quad .\end{aligned} \tag{F-17}$$

By condition (ii) it must be that for some value of i, say i = I, we have $0 < \alpha(i) < 1$; hence,

$$\begin{aligned}\lim_{k\to\infty} \beta(k + 1) &\leq \beta(1) \prod_{i=1}^{I} [1 - \alpha(i)] \lim_{k\to\infty} \prod_{i=I+1}^{k} [1 - \alpha(i)] \\ &\quad + \lim_{k\to\infty} \left\{ \sum_{i=1}^{k-1} w(i) \prod_{j=i+1}^{k} [1 - \alpha(j)] + w(k) \right\} \quad .\end{aligned} \tag{F-18}$$

The first term in this expression converges to zero (Lemma), and since $w(k) \geq 0$ and $\sum_{i=0}^{\infty} w(i) < \infty$ [conditions (iii) and (iv)], the second term is bounded from above. Denoting the value of this bound as B, where the exact value of B is immaterial, we have demonstrated that

$$\lim_{k\to\infty} \beta(k+1) \le B < \infty \quad , \tag{F-19}$$

which establishes the existence of a limiting upper bound on $\beta(k+1)$.

Zero Upper Bound

We shall prove that

$$\lim_{k\to\infty} \beta(k+1) \to 0 \tag{F-20}$$

by the *method of contradiction*. Assume that

$$\lim_{k\to\infty} \beta(k+1) = \beta^* > 0 \tag{F-21}$$

and let

$$0 < \gamma < \beta^* \quad . \tag{F-22}$$

Clearly, for some value of k, say k_0,

$$\beta(k+1) \ge \gamma \qquad \text{for all} \quad k \ge k_0 \quad . \tag{F-23}$$

We now use the following version of Eq. (F-16):

$$\beta(k+1) \le \beta(k_0) - \sum_{i=k_0}^{k} \alpha(i)\beta(i) + \sum_{i=k_0}^{k} w(i) \tag{F-24}$$

which becomes, upon application of Eq. (F-23),

$$\beta(k+1) \le \beta(k_0) - \gamma \sum_{i=k_0}^{k} \alpha(i) + \sum_{i=k_0}^{k} w(i) \quad ; \tag{F-25}$$

hence,

$$\lim_{k\to\infty} \beta(k+1) \le B^* - \gamma \lim_{k\to\infty} \sum_{i=k_0}^{k} \alpha(i) \tag{F-26}$$

where

$$B^* = \beta(k_0) + \lim_{k\to\infty} \sum_{i=k_0}^{k} w(i) \le \beta(k_0) + w^* \ge 0 \tag{F-27}$$

since, by conditions (iv) and (iii), $\sum_{i=0}^{\infty} w(i) < \infty$, $\beta(k_0) \ge 0$, and $w(i) \ge 0$.

For some value of k, the right-hand side of Eq. (F-26) must become negative, since $\sum_{i=k_0}^{\infty} \alpha(i) \to \infty$; but this means that $\lim_{k\to\infty} \beta(k + 1) < 0$, which contradicts our assumption in Eq. (F-21). From condition (iii), we also know that $\lim_{k\to\infty} \beta(k + 1)$ cannot be negative; thus, having ruled out all positive and negative values for $\lim_{k\to\infty} \beta(k + 1)$, but having established the existence of the limit, it can only be that

$$\lim_{k\to\infty} \beta(k + 1) \to 0 \quad . \tag{F-28}$$

REFERENCES

F-1. J. H. Venter, An extension of the Robbins-Monro procedure, *Annals Math. Stat.*, 38, 181-190 (1967).

F-2. K. Knopp, *Theory and Application of Infinite Series*, Blackie, London, 1951.

Appendix G

Upper Bound on $\overline{\|\tilde{\underline{\theta}}(k+\ell)\|^2}$

Equation (5.5-20) gives the following upper bound for $\overline{\|\tilde{\underline{\theta}}(k+\ell)\|^2}$:

$$\begin{aligned}\overline{\|\tilde{\underline{\theta}}(k+\ell)\|^2} \le \overline{\|\tilde{\underline{\theta}}(k)\|^2} &+ \text{U.B.}(\overline{T2}) + \text{U.B.}(\overline{T3}) \\ &+ \text{U.B.}(\overline{T4}) - 2\ \text{L.B.}(|\overline{T5} - \overline{T6}|) \\ &+ 2\ \text{U.B.}(\overline{T7}) + 2\ \text{U.B.}(\overline{T8}) \\ &+ 2\ \text{U.B.}(\overline{T9}) + 2\ \text{U.B.}(\overline{T10}) \qquad \text{(G-1)}\end{aligned}$$

where T2, T3, ..., T10 are defined in Eqs. (5.5-10)-(5.5-19). We shall compute the upper and lower bounds on the right-hand side of Eq. (G-1) in this appendix. It is assumed that the reader is acquainted with the following well-known inequalities (*G-1*):

(i) *Schwarz inequality*

$$|\underline{b}_1'\underline{b}_2|^2 \le \|\underline{b}_1\|^2\|\underline{b}_2\|^2 \qquad \text{(G-2)}$$

(ii) *Triangle inequality*

$$\|\underline{b}_1 + \underline{b}_2\|^2 \le \|\underline{b}_1\|^2 + \|\underline{b}_2\|^2 \qquad \text{(G-3)}$$

(iii) *Cauchy inequality*

$$||A\underline{b}||^2 \leq ||A||^2||\underline{b}||^2 \tag{G-4}$$

(iv) *Schwarz inequality for expectations*

$$|E\{\underline{b}_1'\underline{b}_2\}|^2 \leq E\{||\underline{b}_1||^2\}E\{||\underline{b}_2||^2\} \tag{G-5}$$

(v) *Quadratic form inequalities*

$$\underline{b}'A\underline{b} \leq \lambda_M||\underline{b}||^2 \quad \text{and} \tag{G-6}$$

$$\underline{b}'A\underline{b} \geq \lambda_m||\underline{b}||^2 \tag{G-7}$$

where λ_M is the maximum eigenvalue of A, and λ_m is the minimum eigenvalue of A. ▲

G-1. U.B.($\overline{T2}$)

$$T2 = ||R(k)\Sigma_n\tilde{\underline{\theta}}(k)||^2 \quad ; \tag{G-8}$$

hence, from two applications of the Cauchy inequality to this expression, we find

$$T2 \leq ||R(k)||^2||\Sigma_n||^2||\tilde{\underline{\theta}}(k)||^2 \quad . \tag{G-9}$$

Taking the expectation of both sides, we have

$$\overline{T2} \leq ||R(k)||^2||\Sigma_n||^2\overline{||\tilde{\underline{\theta}}(k)||^2} \quad . \tag{G-10}$$

▲

G-2. U.B.($\overline{T3}$)

$$T3 = ||R(k)\underline{r}(k)\underline{r}'(k)\tilde{\underline{\theta}}(k)||^2 \quad ; \tag{G-11}$$

hence,

$$T3 \leq ||R(k)||^2||\underline{r}(k)\underline{r}'(k)\tilde{\underline{\theta}}(k)||^2 \tag{G-12}$$

so that

$$\overline{T3} \leq ||R(k)||^2\overline{T3'} \tag{G-13}$$

where

$$\overline{T3'} = \overline{||\underline{r}(k)\underline{r}'(k)\tilde{\underline{\theta}}(k)||^2}$$

$$= E\{\tilde{\underline{\theta}}'(k)||\underline{r}(k)||^2\underline{r}(k)\underline{r}'(k)\tilde{\underline{\theta}}(k)\}$$

$$= E\{\tilde{\underline{\theta}}'(k)E\{||\underline{r}(k)||^2\underline{r}(k)\underline{r}'(k)|\tilde{\underline{\theta}}(k)\}\tilde{\underline{\theta}}(k)\} \qquad \text{(G-14)}$$

in which the outer expectation is with respect to $\tilde{\underline{\theta}}(k)$, or $\hat{\underline{\theta}}(k)$, since $\tilde{\underline{\theta}}(k) = \underline{\theta} - \hat{\underline{\theta}}(k)$ and $\underline{\theta}$ is deterministic [see Eq. (E-14) in Appendix E]. Let

$$E\{||\underline{r}(k)||^2\underline{r}(k)\underline{r}'(k)|\hat{\underline{\theta}}(k)\} = B \quad , \qquad \text{(G-15)}$$

then

$$\overline{T3'} = E\{\tilde{\underline{\theta}}'(k)B\tilde{\underline{\theta}}(k)\} \quad . \qquad \text{(G-16)}$$

Consider

$$T3' = \tilde{\underline{\theta}}'(k)B\tilde{\underline{\theta}}(k) \quad . \qquad \text{(G-17)}$$

From Eq. (G-6), we find

$$T3' \leq \lambda_M||\tilde{\underline{\theta}}(k)||^2 \qquad \text{(G-18)}$$

where λ_M is the maximum eigenvalue of B; hence,

$$\overline{T3'} \leq \lambda_M\overline{||\tilde{\underline{\theta}}(k)||^2} \qquad \text{(G-19)}$$

and

$$\overline{T3} \leq \lambda_M||R(k)||^2\overline{||\tilde{\underline{\theta}}(k)||^2} \quad . \qquad \text{(G-20)}$$

▲

G-3. U.B.($\overline{T4}$)

$$T4 = ||R(k)\underline{f}(k)||^2 \qquad \text{(G-21)}$$

where

$$\underline{f}(k) = [\underline{r}(k)\underline{r}'(k) - \Sigma_n - \underline{r}(k)\underline{x}'(k)]\underline{\theta} - \underline{r}(k)v(k) \quad . \qquad \text{(G-22)}$$

From the Cauchy and triangle inequalities, we find

$$T4 \leq ||R(k)||^2||\underline{f}(k)||^2$$

$$\leq ||R(k)||^2[||\underline{r}(k)\underline{r}'(k)\underline{\theta}||^2 + ||\Sigma_n\underline{\theta}||^2$$

$$+ ||\underline{r}(k)\underline{x}'(k)\underline{\theta}||^2 + ||\underline{r}(k)v(k)||^2] \quad . \qquad \text{(G-23)}$$

We leave it to the reader to show that if, in addition to the four major assumptions listed in Section 5.3.2,

$$E\{v^2(k)\} < \infty \quad , \tag{G-24}$$

$$E\{||\underline{r}(k)||^2\underline{r}(k)\underline{r}'(k)|\hat{\underline{\theta}}(k)\} < \infty \quad , \tag{G-25}$$

and

$$E\{||\underline{r}(k)||^2 v^2(k)|\hat{\underline{\theta}}(k)\} < \infty \quad , \tag{G-26}$$

then

$$\overline{T4} \leq c_1||R(k)||^2 \tag{G-27}$$

where $c_1 < \infty$. It is not necessary to know the exact value of c_1. ▲

G-4. L.B.($|\overline{T5} - \overline{T6}|$)

$$T5 = \tilde{\underline{\theta}}'(k)R(k)\Sigma_n\tilde{\underline{\theta}}(k) \tag{G-28}$$

and

$$T6 = \tilde{\underline{\theta}}'(k)R(k)\underline{r}(k)\underline{r}'(k)\tilde{\underline{\theta}}(k) \quad . \tag{G-29}$$

Consider $\overline{T6}$ first:

$$\begin{aligned}\overline{T6} &= E\{\tilde{\underline{\theta}}'(k)R(k)\underline{r}(k)\underline{r}'(k)\tilde{\underline{\theta}}(k)\} \\ &= E\{\tilde{\underline{\theta}}'(k)R(k)E\{\underline{r}(k)\underline{r}'(k)|\hat{\underline{\theta}}(k)\}\tilde{\underline{\theta}}(k)\}\end{aligned} \tag{G-30}$$

where the outer expectation is with respect to $\tilde{\underline{\theta}}(k)$, or $\hat{\underline{\theta}}(k)$. The inner expectation was computed in Section 5.3.2, and is given in Eq. (5.3-20); hence,

$$\overline{T6} = E\{\tilde{\underline{\theta}}'(k)R(k)(\Omega + \Sigma_n)\tilde{\underline{\theta}}(k)\} \quad . \tag{G-31}$$

Clearly, then

$$\overline{T5} - \overline{T6} = -E\{\tilde{\underline{\theta}}'(k)R(k)\Omega\tilde{\underline{\theta}}(k)\} \quad . \tag{G-32}$$

Let us assume that

$$R(k)\Omega \text{ is positive definite.} \tag{G-33}$$

This assumption is examined in great detail in Section 5.5.4. By this assumption and Eq. (G-7), we conclude that

$$|\overline{T5} - \overline{T6}| \geq \lambda_2\overline{||\tilde{\underline{\theta}}(k)||^2} \tag{G-34}$$

where λ_2 is the minimum eigenvalue of $R(k)\Omega$. ▲

G-5. U.B.$(\overline{T7})$

$$T7 = \tilde{\underline{\theta}}'(k)R(k)\underline{f}(k) \quad ; \tag{G-35}$$

hence,

$$\begin{aligned}\overline{T7} &= E\{\tilde{\underline{\theta}}'(k)R(k)\underline{f}(k)\} \\ &= E\{\tilde{\underline{\theta}}'(k)R(k)E\{\underline{f}(k)\,|\,\hat{\underline{\theta}}(k)\}\}\end{aligned} \tag{G-36}$$

where the outer expectation is with respect to $\tilde{\underline{\theta}}(k)$, or $\hat{\underline{\theta}}(k)$. We leave it to the reader to show from the definition of $\underline{f}(k)$ and the major assumptions listed in Section 5.3.2 that

$$E\{\underline{f}(k)\,|\,\hat{\underline{\theta}}(k)\} = \underline{0} \quad ; \tag{G-37}$$

hence,

$$\overline{T7} = 0 \quad . \tag{G-38}$$

▲

G-6. U.B.$(\overline{T8})$

$$T8 = \tilde{\underline{\theta}}'(k)\Sigma_n R^2(k)\underline{r}(k)\underline{r}'(k)\tilde{\underline{\theta}}(k) \quad . \tag{G-39}$$

Using Schwarz's inequality for expectations, we find

$$\begin{aligned}\overline{T8} &\le \left[E\{||\tilde{\underline{\theta}}(k)\Sigma_n R(k)||^2\}E\{||R(k)\underline{r}(k)\underline{r}'(k)\tilde{\underline{\theta}}(k)||^2\}\right]^{1/2} \\ &\le \left[(\overline{T2})(\overline{T3})\right]^{1/2} \quad .\end{aligned} \tag{G-40}$$

Substituting the upper bounds in Eqs. (G-10) and (G-20) for $\overline{T2}$ and $\overline{T3}$, into this last expression, we find

$$\overline{T8} \le \lambda_M^{1/2}||\Sigma_n||\;||R(k)||^2\overline{||\tilde{\underline{\theta}}(k)||^2} \quad . \tag{G-41}$$

▲

G-7. U.B.$(\overline{T9})$

$$T9 = \tilde{\underline{\theta}}'(k)\Sigma_n R^2(k)\underline{f}(k) \quad ; \tag{G-42}$$

hence,

$$\begin{aligned}\overline{T9} &= E\{\tilde{\underline{\theta}}'(k)\Sigma_n R^2(k)\underline{f}(k)\} \\ &= E\{\tilde{\underline{\theta}}'(k)\Sigma_n R^2 E\{\underline{f}(k)\,|\,\hat{\underline{\theta}}(k)\}\} \\ &= 0\end{aligned} \tag{G-43}$$

where we have used Eq. (G-37). ▲

G-8. U.B.($\overline{T10}$)

$$T10 = \tilde{\underline{\theta}}'(k)\underline{r}(k)\underline{r}'(k)R^2(k)\underline{f}(k) \quad . \tag{G-44}$$

Using Schwarz's inequality for expectations, we find

$$\begin{aligned}\overline{T10} &\leq \left[E\{||R(k)\underline{r}(k)\underline{r}'(k)\tilde{\underline{\theta}}(k)||^2\}E\{||R(k)\underline{f}(k)||^2\}\right]^{1/2} \\ &\leq \left[(\overline{T3})(\overline{T4})\right]^{1/2} \quad .\end{aligned} \tag{G-45}$$

Substituting the upper bounds in Eqs. (G-20) and (G-27) for $\overline{T3}$ and $\overline{T4}$, into this last expression, we find

$$\overline{T10} \leq c_2||R(k)||^2\left[\overline{||\tilde{\underline{\theta}}(k)||^2}\right]^{1/2} \tag{G-46}$$

where

$$c_2 = (\lambda_M c_1)^{1/2} \quad . \tag{G-47}$$

A more useful upper bound for $\overline{T10}$, is†

$$\overline{T10} \leq c_2||R(k)||^2\left[1 + \overline{||\tilde{\underline{\theta}}(k)||^2}\right] \quad . \tag{G-48}$$

▲

Having computed all of the bounds required for Eq. (G-1), that equation becomes

$$\begin{aligned}\overline{||\tilde{\underline{\theta}}(k + \ell)||^2} &\leq \overline{||\tilde{\underline{\theta}}(k)||^2} + ||R(k)||^2||\Sigma_n||^2\overline{||\tilde{\underline{\theta}}(k)||^2} \\ &\quad + \lambda_M||R(k)||^2\overline{||\tilde{\underline{\theta}}(k)||^2} + c_1||R(k)||^2 \\ &\quad - 2\lambda_2\overline{||\tilde{\underline{\theta}}(k)||^2} \\ &\quad + 2\lambda_M^{1/2}||\Sigma_n||\ ||R(k)||^2\overline{||\tilde{\underline{\theta}}(k)||^2} \\ &\quad + 2c_2||R(k)||^2\left[1 + \overline{||\tilde{\underline{\theta}}(k)||^2}\right] \\ &\leq \{1 + ||R(k)||^2\left[||\Sigma_n||^2 + \lambda_M\right. \\ &\quad \left. + 2\lambda_M^{1/2}||\Sigma_n|| + 2c_2\right] - 2\lambda_2\}\overline{||\tilde{\underline{\theta}}(k)||^2} \\ &\quad + (c_1 + 2c_2)||R(k)||^2\end{aligned}$$

† If $0 < a < 1$, then $\sqrt{a} \leq 1 + a$; the inequality also can be used if $a \geq 1$.

$$\overline{||\tilde{\underline{\theta}}(k + \ell)||^2} \le \left[1 + c_3||R(k)||^2 - 2\lambda_2\right]\overline{||\tilde{\underline{\theta}}(k)||^2}$$
$$+ c_1^*||R(k)||^2 \qquad \text{(G-49)}$$

where

$$c_1^* = c_1 + 2c_2 \qquad \text{(G-50)}$$

and

$$c_3 = ||\Sigma_n||^2 + 2\lambda_M^{1/2}||\Sigma_n|| + \lambda_M \quad . \qquad \text{(G-51)}$$

Equation (G-49) is the desired upper bound on $\overline{||\tilde{\underline{\theta}}(k + \ell)||^2}$.

REFERENCE

G-1. M. Abromowitz and I. A. Stegun, *Handbook of Mathematical Functions,* National Bureau of Standards, Applied Math. Ser., 55, p. 11 (1964).

Author Index

Subject Index